T0401888

Heat Stress and Animal Productivity

Anjali Aggarwal • Ramesh Upadhyay

Heat Stress and Animal Productivity

 Springer

Anjali Aggarwal
Dairy Cattle Physiology Division
Principal Scientist, National Dairy
    Research Institute
Karnal, Haryana, India

Ramesh Upadhyay
Dairy Cattle Physiology Division
Principal Scientist, National Dairy
    Research Institute
Karnal, Haryana, India

ISBN 978-81-322-0878-5   ISBN 978-81-322-0879-2 (eBook)
DOI 10.1007/978-81-322-0879-2
Springer New Delhi Heidelberg New York Dordrecht London

Library of Congress Control Number: 2012956024

Printed on acid-free paper

Springer is part of Springer Science+Business Media (www.springer.com)

Introduction

Stress is a broad term and is described as the cumulative detrimental effect of a variety of factors on the health and performance of animals. Yousef (1985) defined stress as the magnitude of forces external to the body which tend to displace its systems from their resting or ground state. In view of this, heat stress for the dairy cow can be understood to indicate all high temperature-related forces that induce adjustments occurring from the subcellular to the whole-animal level to help the cow avoid physiological dysfunction and for it to better fit its environment. Environmental factors such as ambient temperature, solar radiation and humidity have direct and indirect effects on animals (Collier et al. 1982). Heat stress occurs in animals when there is an imbalance between heat production within the body and its dissipation. Thermoregulation is the means by which an animal maintains its body temperature. It involves a balance between heat gain and heat loss. Under heat stress, a number of physiological and behavioural responses vary in intensity and duration in relation to the animal genetic make-up and environmental factors. Climatic, environmental, nutritional, physical, social or physiological stressors are likely to reduce welfare and performance of animals (Freeman 1987). Heat stress is one of the most important stressors especially in hot regions of the world. The endeavour by homeotherms to stabilise body temperature within fairly narrow limits is essential to control biochemical reactions and physiological processes associated with normal metabolism (Shearer and Beede 1990). The general homeostatic responses to thermal stress in mammals include reduction in faecal and urinary water losses, reduction in feed intake and production and increased sweating, respiratory rates and heart rates. In response to stress, mammals set physical, biochemical and physiological processes into play to try and counteract the negative effects of heat stress and maintain thermal equilibrium. Adaptation to heat stress requires the physiological integration of many organs and systems, namely, endocrine, cardio-respiratory and immune system (Altan et al. 2003). Direct effects involve heat exchanges between the animal and the surrounding environment that are related to radiation, temperature, humidity and wind speed. Under present climate conditions, the lack of ability of animals to dissipate the environmental heat determines that, in many areas in the world, animals suffer heat stress during, at least, part of the year. Heat stress has a variety of detrimental effects on livestock (Fuquay 1981), with significant effects on milk production and reproduction in dairy cows (Johnson 1987; Valtorta and Maciel 1998). Dairy cattle show signs of heat stress when THI is higher than 72 (Armstrong 1994).

The comfort limit depends on level of production. Animals presenting higher level of production are more sensitive to heat stress (Johnson 1987). Heat stress also lowers natural immunity, making animals more vulnerable to disease in the following days and weeks. The decrease in fertility is caused by elevated body temperature that influences ovarian function, oestrous expression, oocyte health and embryonic development. The biological mechanism by which heat stress impacts production and reproduction is partly explained by reduced feed intake but also includes altered endocrine status, reduction in rumination and nutrient absorption and increased maintenance requirements resulting in a net decrease in nutrient/energy availability for production. Since climate change could result in an increase of heat stress, all methods to help animals cope with or, at least, alleviate the impacts of heat stress could be useful to mitigate the impacts of global change on animal responses and performance. Three basic management schemes for reducing the effect of thermal stress have been suggested: (a) physical modification of the environment, (b) genetic development of less sensitive breeds and (c) improved nutritional management schemes.

Shades are the most simple method to reduce the impact of high solar radiation. Shades can be either natural or artificial. Tree shades have proved to be more efficient. Air moving is an important factor in the relief of heat stress, since it affects convective and, according to air humidity, evaporative heat losses. Where possible, natural ventilation should be maximised by constructing open-sided constructions. Improved systems capable of either cooling the animal directly or cooling the surrounding environment are necessary to better control the animal's body temperature and maintain production in hot and hot-humid climates. With cooling devices, the temperature in the animal sheds may be kept low to cool animal, and THI can be kept around 72. Ration modification can help minimise the drop in milk production that hot weather causes. Decreasing the forage to concentrate ratio (feeding more concentrate) can result in more digestible rations that may be consumed in greater amounts. Among the genetic adaptations that have developed in zebu cattle during its evolution has been the acquisition of genes for thermotolerance and disease resistance. Thus, an alternative scheme to crossbreeding for utilising the zebu genotype for livestock production in hot climates is to incorporate those zebu genes that confer thermotolerance into European breeds while avoiding undesirable genes.

References

Altan O, Pabuccuoglu A, Alton A, Konyalioglu S, Bayraktar H (2003) Effect of heat stress on oxidative stress, lipid peroxidation and some stress parameters in broilers. Br Poult Sci 4:545–550

Armstrong DV (1994) Heat stress interaction with shade and cooling. J Dairy Sci 77:2044–2050

Collier RJ, Beede DK, Thatcher WW, Israel LA, Wilcox CJ (1982) Influences of environment and its modification on dairy animal health and production. Journal of Dairy Science 65:2213–2227

Freeman BM (1987) The stress syndrome. World's Poult Sci J 43:15–19

Fuquay JW (1981) Heat stress as it affects animal production. J Anim Sci 52:164–174

Johnson HD (1987) Bioclimate effects on growth, reproduction and milk production. In: Bioclimatology and the adaptation of livestock. Elsevier, Amsterdam (Part II, Chapter 3)

Shearer JK, Beede DK (1990) Thermoregulation and physiological responses of dairy cattle in hot weather. Agri-Pract 11:5–17

Valtorta SE, Maciel M (1998) Respuesta reproductiva. In: Producción de leche en verano. Centro de publicaciones de la Secretaría de Extensión de la UNLitoral, Santa Fe, Argentina, pp 64–76

Valtorta SE, Gallardo MR, Castro HC, Castelli MC (1996) Artificial shade and supplementation effects on grazing dairy cows in Argentina. Trans ASAE 39:233–236

Yousef MK (1985) Basic principles. Stress physiology in livestock, vol 1. CRC Press, Boca Raton

Foreword

Climate change is the most serious long-term challenge to be faced by farmers and livestock owners around the world, as it is likely to impact livestock production and health. The IPCC predicts that by 2100, the increase in global surface temperature may be between 1.8 and 4.0°C. Even if global temperature increases by 1.5–2.5°C, approximately 20–30% of livestock and animal species are expected to be at risk of extinction. Increased number of thermal stress days with higher temperatures and humidity particularly due to global warming are also likely to favor growth of vectors and pests, challenging the health of livestock. It has been found that the livestock of tropics are more resilient to environmental and climatic stress due to their genotype and capacity to interact with the environment. Buffaloes and high-producing cows suffer most at high ambient temperatures, and severe heat stress often leads to loss in their reproductive and productive performance.

Heat stress negatively impacts livestock performance in most areas of the world, as it reduces milk production in animals with high genetic merit. Strategies to alleviate metabolic and environmental heat loads in high-milk-producing animals particularly during early lactation need to be elucidated and developed. The identification of heat-stressed livestock and understanding the biological mechanism(s) by which thermal stress reduces milk production and reproductive functions is critical for developing novel approaches to maintain production or to minimize the reduction in productivity during hot summer months. Adaptation to heat stress requires physiological integration of many organs and systems, namely, endocrine, cardiorespiratory, urinary, and immune system. Heat stress also lowers natural immunity, making livestock more vulnerable to diseases. Further, the decrease in fertility due to elevated body temperature through its effect on ovarian function, estrous expression, oocyte health, and embryonic development has been established. It is known that during heat stress, activation of the hypothalamic–pituitary–adrenal axis and the consequent increase in plasma glucocorticoid concentrations are the most important responses of the animals to heat stress. As such, a greater understanding is required on mechanisms associated with immune suppression and hormonal changes in high-producing animals.

The supplementation of antioxidants in feed (micronutrients and vitamin E) to high-producing animals, especially during periparturient period, may help in improving the productivity by reducing the stress and risk of mastitis. Knowledge on feeding and eating rhythm and postprandial intake patterns will enable predicting diurnal patterns in rumen, post-rumen, and peripheral nutrient assimilation. These will suggest optimal, suboptimal and unfavorable times of nutrient supply to mammary cells and milk synthesis.

All attempts have to be made to alleviate the heat stress on high-producing animals. A number of animal cooling options are being used as per requirements. Air fans, wetting, evaporation to cool the air, and shade to minimize transfer of solar radiation are used to enhance heat dissipation from animals. Animals in ponds lose heat very fast primarily due to conduction and coefficient of heat transfer to water from skin.

This publication very nicely covers the effect of heat stress on animal production and also suggests various methods for alleviation of heat stress. The improved understanding of the impact of heat stress on livestock will help in developing management techniques to alleviate heat stress on dairy animals. It will also serve as a useful reading material for researchers, teachers, dairy executives, and managers.

I wish them all the best.

Director, National Dairy Research Institute,   A.K. Srivastava
Karnal-132001, India

Preface

Thermal stress is a major limiting factor in livestock production under tropical climate and also during summer season in temperate climates. Heat stress occurs when the ambient temperature lies above thermoneutral zone. It was traditionally thought that milk synthesis begins to decrease when the THI exceeds 72, but with increasing milk production, it has been observed that high-yielding dairy cows reduce milk yield at a THI of approximately 68. The animal comfort limit depends on level of production and breed of animal. Animals in higher level of production are more sensitive to heat stress. Different livestock species have different sensitivities to ambient temperature and humidity. The capacity to tolerate heat stress is much higher in zebu breeds of cattle particularly higher temperatures at low relative humidity than crossbreds of taurine breeds. This is mainly due the fact that zebu cattle can dissipate excessive heat more effectively by sweating, whereas crossbreds have relatively low ability to sweat. During hot-humid weather, the thermo-regulatory capability of cattle to dissipate heat by sweating and panting is compromised, and heat stress occurs in cattle. The water vapour content of the air plays an important role in determining the capacity to lose heat from skin and lungs. The increasing concern with the thermal comfort of dairy cows is justifiable not only for countries occupying tropical zones but also for nations in temperate zones in which high ambient temperatures are challenging livestock production system. Climate change poses formidable challenges to the development of livestock sector all over the world as it is likely to aggravate the heat stress on livestock, adversely affecting their productive and reproductive performance. Since climate change could result in an increase of heat stress, all methods to help animals cope with or, at least, help alleviate the impacts of heat stress could be useful to mitigate the impacts of global change on animal responses and performance.

The authors wish to thank Dr. A.K. Srivastava, Director of the Institute, for his inspiration and development of the publication. The authors are grateful for contribution and support from their colleagues and friends. The authorities at Indian Council of Agricultural Research, Dr. S. Ayyappan, Secretary (DARE) and Director General (ICAR), and Dr. K.M.L. Pathak, Deputy Director General (Animal Science), are duly acknowledged.

Anjali Aggarwal and Ramesh Upadhyay

About the Book

This book is on livestock with specific reference to heat stress and its alleviation. The topic is very pertinent in light of the impacts of climate change on livestock production and health. Work related to effect of heat stress on animal productivity, immunity, and hormonal levels is discussed in detail. Heat stress occurs in animals when there is an imbalance between heat production within the body and its dissipation. Thermoregulation is the means by which an animal maintains its body temperature. Under heat stress, a number of physiological and behavioral responses vary in intensity and duration in relation to the animal genetic make-up and environmental factors. In response to stress, mammals set physical, biochemical, and physiological processes into play to try and counteract the negative effects of heat stress and maintain thermal equilibrium.

Adaptation to heat stress requires the physiological integration of many organs and systems, viz. endocrine, cardiorespiratory, and immune system. Heat stress also lowers natural immunity making animals more vulnerable to disease in the following days and weeks. The decrease in fertility is caused by elevated body temperature that influences ovarian function, estrous expression, oocyte health, and embryonic development.

The increasing concern with the thermal comfort of dairy cows is justifiable not only for countries occupying tropical zones, but also for nations in temperate zones in which high ambient temperatures are becoming an issue. Improving milk production is, therefore, an important tool for improving the quality of life particularly for rural people in developing countries. The environmental conditions necessitate reduction of heat stress due to solar radiation and heat.

This book discusses all these aspects in detail. Recent works related to effect of heat stress on animal productivity, immunity, and hormonal levels are also discussed in the book. Information on biological rhythm is also included. The book also discusses the methods for alleviation of heat stress in livestock, especially cows and buffaloes. This book would be a ready reckoner for students, researchers and academia and would pave way for further research.

Contents

Thermoregulation

Contents

Abstract

Body temperature in homeotherms is maintained by the thermoregulatory system within 1°C of its normal temperature under ambient conditions that do not impose heat stress. A rise in the core body temperature also increases heat loss by panting and sweating. These responses are physiological strategies to transfer heat from the cow's body to the environment. In order to maintain homeothermy, an animal must be in thermal equilibrium with its environment, which includes radiation, air temperature, air movement and humidity. The range of temperature within which the animal uses no additional energy to maintain its body temperature is called the thermoneutral zone (TNZ), within which the physiological costs are minimal and productivity is maximum. Temperature–humidity index (THI) is an index for assessment of the potential of an environment to induce heat stress in humans and farm animals. Heat loss via skin is more in cows and heat loss by respiration is higher in buffaloes. This is due to less number of sweat glands in buffaloes. Exposure of the animal to high environmental temperature stimulates the peripheral and core receptors to transmit nerve impulses to the specific centres in the hypothalamus, to help in preventing the rise in body temperature. The specific centres in the hypothalamus are the defensive evaporative and non-evaporative cooling systems, appetite centre and the adaptive mechanisms that cause such reactions. The suppressive impulses

A. Aggarwal and R. Upadhyay, *Heat Stress and Animal Productivity*,
DOI 10.1007/978-81-322-0879-2_1, © Springer India 2013

due to positive heat gain transmitted to the appetite centre cause a decrease in feed intake; therefore, less substrates become available for enzymatic activities, hormone synthesis and heat production, which help in cooling the body. Among the genetic adaptations that have developed in zebu cattle during its evolution have been the acquisition of genes for thermotolerance. Zebu breeds are better able to regulate body temperature in response to heat stress than are cattle from a variety of *Bos taurus* breeds of European origin.

1   Introduction

All livestock species maintain their body temperature within a close defined thermoregulatory limit regardless of the external environment variations. The most comfortable environmental temperature range for temperate dairy cattle is between 5 and 25°C, which is believed to be the thermal comfort zone (McDowell 1972) with minimal physiological cost and maximum productivity (Folk 1974). Body temperature in all homeotherms is maintained by the thermoregulatory mechanisms within ±1°C of its normal set temperature under ambient conditions and that do not impose heat stress (Bligh and Harthoorn 1965). Above this temperature (25°C), dairy cow increases respiratory frequency and rectal temperature that cause a decline in milk yield and reproductive performance (Bitman et al. 1984). A rise in the core temperature also increases heat loss by panting and sweating (Spain and Spiers 1998). These responses are considered normal thermoregulatory physiological mechanisms to transfer heat from the body to the surrounding environment. An increase in respiratory frequency with or without change in tidal volume or panting expends energy that increases the daily maintenance requirement of an animal by 7–25% (NRC 1981).

The stress is defined as the magnitude of forces external to the body which tend to displace its systems from their resting or ground state (Yousef 1985). In this light, heat stress for the dairy cows indicate all high temperature-related forces that induce adjustments occurring from the

subcellular to the whole-animal level to help the cows avoid physiological dysfunction and to suit their environment. The endeavour by thermoregulatory processes to stabilise body core temperature within a set narrow limit is essential to control biochemical reactions and physiological processes associated with normal metabolism (Shearer and Beede 1990).

2   The Thermoneutral Zone

The range of environmental temperature within which the animals use no additional energy to maintain their body temperature is called the thermoneutral zone (Fig. 1). Within the thermoneutral zone (TNZ), the physiological costs are minimal, and productivity is maximum (Du Prezz et al. 1990). Lactating dairy cows have a high metabolic rate and heat increment; therefore, they require an effective thermoregulatory mechanism to maintain their body temperature. The metabolic heat production rises due to the metabolism of nutrients, and because of this reason, the heat load increases (Kadzere et al. 2001). Both above and below TNZ, the changes in behaviour and diurnal pattern of animals are observed. This range is bounded by the lower critical temperature (LCT) and upper critical temperature (UCT) given in Fig. 1 for different species. 'Effective ambient temperature (EAT)' is the actual temperature felt by the animal and may be very different from the air temperature. The EAT of a calf housed in a clean, dry hutch bedded with straw may be 8–10°C warmer than the air temperature, and the EAT of a heifer exposed to wind and rain may be considerably lower than the ambient temperature. The upper critical temperature of dairy cattle is lower than other livestock species (Wathes et al. 1983). Generally, the TNZ range (from lower critical temperature (LCT) to upper critical temperature (UCT)) is influenced by animal age, species, breed, feed intake, diet composition, previous state of temperature acclimation or acclimatisation, production, specific housing and pen conditions, tissue insulation (fat, skin), external insulation (coat) and animal behaviour (Yousef 1985).

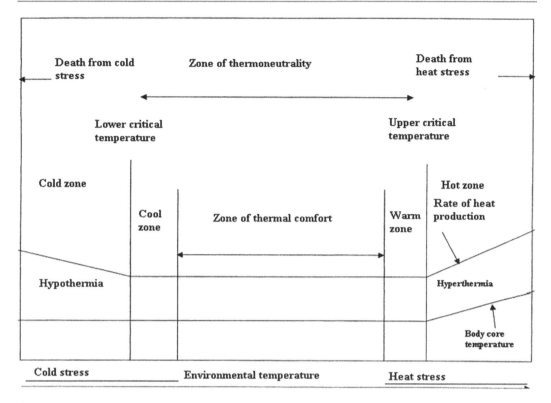

Fig. 1 Thermoneutral zone (Kadzere et al. 2001)

Igono et al. (1992) measured the highest milk production in Holstein cows under a desert environment maintained at ambient temperatures below 21°C throughout the day. McArthur and Clark (1988) indicated that the TNZ was related to the heat and water balance of animal. The farther an animal moves away from its preferred body temperature, the productive processes are likely to be affected more. McDowell et al. (1976) also reported that even small increase in core temperature has profound effects on tissue and endocrine functions that, in turn, can negatively affect animal fertility, growth, lactation and the ability to work.

2.1   Lower Critical Temperature (LCT)

The ambient temperature below which the rate of heat production of an animal under resting state increases to maintain body heat balance is the lower critical temperature (LCT). This implies that the rate of heat production is dependent upon ambient thermal demand below the LCT (Yousef 1985). Below the LCT, animal metabolism must increase to generate heat to maintain core temperature. This can be accomplished through increased energy intake. The general rule of thumb is that energy intake must increase by 1% for each degree of cold below the LCT. Non-evaporative heat loss declines as ambient temperatures rise above the LCT making the animals more dependent on peripheral vasodilatation and water evaporation to enhance heat loss and to prevent a rise in body temperature.

2.2   Upper Critical Temperature (UCT)

The UCT is the air temperature at which the animal increases heat production as a consequence of a rise in core temperature mainly due to an inadequate evaporative heat loss (Yousef 1985). Estimates of UCT are based on studies on dairy cows exposed for short periods to constant temperatures in climatic chambers (Kibler 1964).

Fig. 2 Lower and upper critical temperatures for dairy animals

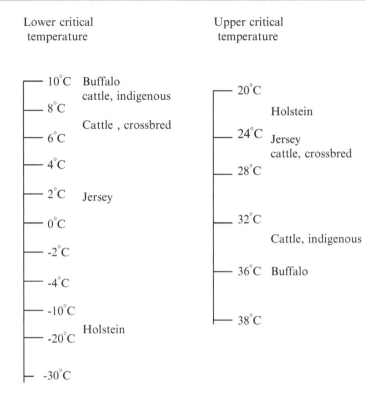

Lower critical temperature

Upper critical temperature

As the thermal load exceeds the evaporative heat loss capacity of an animal, core temperature rises and may lead to hyperthermia. Continuous rise and unabated heat may cause animal death due to hyperthermia.

At high temperatures, the potential for non-evaporative heat loss is reduced, and animals rely on the evaporative water loss to dissipate any excess heat generated by metabolism (McArthur and Clark 1988). The UCT is 25–26°C for dairy cows and most likely remain unaltered irrespective of previous acclimatisation or of their milk production (Berman et al. 1985). This is in contradiction with Yousef (1985) indicating that the TNZ varies with physiological state and other environmental conditions.

The UCT can be decided on the basis of thermoregulatory functions like increased sweating and respiratory water loss and increased body temperature (Berman et al. 1985). Evaporative water loss from the skin increases at air temperatures above 20°C (Berman 1968). The upper and lower critical temperatures for various dairy breeds have been given in Fig. 2.

2.3   Factors Affecting TNZ

- *Wind velocity* – The movement of air around an animal affects its ability to maintain body temperature. Air movement increases evaporation and heat convection. In hot temperatures, wind can increase evaporation, thereby cooling the animal. Calves and heifers less than 1 year of age can benefit from some type of protection or shelter from wind. Newborn calves are more susceptible to effects of cold than hot temperature. In very cold temperatures, wind chill is used to calculate the effective temperature.
- *Air humidity* – The level of ambient humidity can affect an animal's ability to dissipate body heat or thermoregulation, particularly in hot–humid climates. High humidity levels reduce the rate of evaporative cooling and efficiency of cooling mechanisms.
- *Precipitation* – Rain and snow wet the animal hair coat that reduces thermal insulation.
- *Radiation* – Direct solar exposure increases net heat gain by an animal. The effective

temperature increases 3–5°C in animals exposed to direct sunlight. During cold, the solar exposure may be beneficial to animal and provide fast body warm up, but during summer, direct radiation causes heat stroke at high temperatures.

- *Body hair coat* – Hair coat provides insulation to body from the outside environment and affect its ability to maintain heat balance. The hair coat helps trap air to make a layer of 'dead air' for insulation. The body insulation is hindered by soiling with manure, water, urine, mud, etc., and insulative value of the hair coat is markedly reduced. Animal with low insulation or soiled with manure and mud in cold temperatures is likely to have a higher metabolism to cope with the extra heat loss.
- *Animal shelter and bedding* – Animal shelter reduces the exposure to direct solar radiations and effects of precipitation, wind and other elements on an animal's ability to maintain thermal balance. Bedding in animal house reduces the conduction of heat from the animal into the floor or ground particularly during cold/winter.

3 Heat Stress Indicators

Temperature–humidity index (THI) is an index for assessment of the potential of an environment to induce heat stress in humans and farm animals. THI offers a method of combining two of the more important and easily measured climatic elements into a possible measure to compare temperature and humidity data and animal response at different locations. The THI may be calculated by several equations as given below:

(a) By calculating the dry and wet bulb temperatures using the following equation:

$$THI = 0.72(W + D) + 40.6 \quad \text{(McDowell et al. 1976)}$$

where W is wet bulb and
D is dry bulb temperature in °C.

    Temperature–humidity index values of 70 or less are considered comfortable for humans and animals, 75–78 stressful and values greater than 78 cause extreme distress to lactating cows which are unable to maintain their normal body temperatures.

It was observed that rectal temperature increased at THI >80, the respiration rate started to increase above a THI of 73 and increased steeply at THI >80 (Lemerle and Goddard 1986). This suggests that homeostatic mechanisms in animals, including increased respiration, can prevent a rise in rectal temperature until the THI reaches 80. This is similar to the critical THI level of 78 given by McDowell et al. (1976).

(b) By calculating the air temperature (T) and relative humidity (RH) by using the following equation:

$$THI = (1.8 \times T + 32) - (0.55 - 0.0055 \times RH) \times (1.8 \times T - 26)$$

where T is the air temperature (°C) and RH the relative humidity (%).

(Tucker et al. 2008)

The critical values for THI in dairy cows are determined as 64 for minimum, 72 as the mean value and 76 as the maximum. At a THI of 75, 50% of the human population feel uncomfortable, and this value even dairy cattle also appears to be uncomfortable (Johnson et al. 1989). A THI value of 72 equates to 25°C and 50% relative humidity. If the THI value exceeds 72, the cow suffers from heat stress and the milk production declines (West 2003). When the value is between 78 and 82, the cow is severely affected and cooling by artificial ways is necessary for animal and sustains production. If THI is above 82, the heat stress vulnerable cows may cease to produce or decline productivity, or even death may occur from heat stress (Du Prezz et al. 1990).

A significant depression in milk production and in reproduction is noticed at an average daily THI more than 75. However, some depressions may also be observed above 70 and particularly above 75 in dairy cattle producing at high levels or acclimated to a lower ambient temperature and THI. The average daily THI should be derived on basis of the average of hourly THI for accurate analysis of environmental discomfort. An average THI is normally calculated on the basis of maximum and minimum ambient temperature and is an estimate of the average THI of the day with 5–7% variations but may be suitable for determining animal production functions and impact of environment.

(c) By calculating dry bulb temperature and relative humidity by using the following equation:

$$\text{THI} = \text{db}°\text{C} - \left\{(0.31 - 0.31\,\text{RH})(\text{db}°\text{C} - 14.4)\right\},$$

(Marai et al. 2001)

where db°C = dry bulb temperature in °C and RH = relative humidity percentage (RH%)/100.

The values obtained indicate the following:

<22.2 = absence of heat stress
22.2 to <23.3 = moderate heat stress
23.3 to <25.6 = severe heat stress
25.6 and more = extreme severe heat stress

Both equations are applicable in cattle, sheep, goats and buffaloes.

(d) By observing body temperature and respiratory rate to assess heat stress level in cattle: (Perez 2000).

Body temperature and respiration rate have been recommended to be used as parameters to determine heat stress in cattle; altogether with THI values to determine and evaluate heat stress in cattle, rectal temperature (RT) and respiration rate (RR) are the most sensitive indices of assessing heat stress among the physiological reactions studied. The increase in ambient temperature from 29 to 31°C resulted in an increase in RT from 37.8 to 38.0°C and RR from 20.5 to 22.4 breaths/min in buffalo heifers and from 37.9 to 39.7°C and from 23.4 to 41.0 breaths/min, respectively, in lactating buffalo cows (Kamal and Ibrahim 1969a, b; Kamal et al. 1978). In Murrah male buffalo, 7–9-month-old calves were not able to maintain RT within normal range during summer when ambient temperature and solar radiation were maximum (44.1°C). Their RR increased 5–6

times the normal, their tongue protruded and salivary activity increased. In the heifers, the comparison showed that averages of RT, RR and pulse rate (PR) were lower in buffaloes than those of the zebu cattle, under cool, comfortable conditions. The ambient temperature was 18.5°C, and vapour pressure was 8.5 mmHg, under the cool, comfortable conditions. When exposed for 6 h to hot–humid conditions, the RT and RR of the buffaloes exceeded those of the zebu heifers towards the end of the exposure period. The ambient temperature was 40.5°C, and vapour pressure was 39.5 mmHg under the hot–humid conditions. On exposure to hot arid conditions, the buffalo heifers reacted more sharply than zebu heifers as evidenced by increase in RT, RR and respiratory volume. However, the physiological reactions were rather mild in case of the zebu heifers and more severe in the buffalo heifers as compared to the cool comfortable conditions (Das et al. 1999).

Acute heat exposure to 33–43°C and 40–60% RH of young (aged 6 months) and old Egyptian buffalo calves (aged 12 months) induced more significant increases in RT (3.4 and 3.2%) and RR (495 and 335%, respectively) than in the control. The chronic heat exposure of 6- and 12-month-old buffalo calves was accompanied by increases in RT (4.1 and 3.0%), RR (528 and 318%) and evaporative water loss (69.4 and 51.2%, respectively; Nessim 2004). In another study, it was reported that exposure of Egyptian buffaloes to direct solar radiation for 3 h continuously from 12:00 to 15:00 h caused an increase in RT (38.5–40.9°C), RR (26.1–124.8 breaths/min) and PR (64.5–82.2 pulses/min; Ashour 1990). Shafie (1993) observed the effect of heat, accompanied by direct solar radiation for 3 h from 12:00 to 15:00 h, during August in Egypt on the physiological response of buffaloes resulted in an increase in absolute values of RT (1.7°C), skin surface temperature (2.3°C), RR (70.0 breaths/min), PR (18.0 pulses/min) and arterial pressure (23.0 mmHg). Exposure of Indian buffaloes for 6 h to sun direct radiation in hot and dry season increased RT and PR by 1.4°C and 11 pulses/

min, respectively, when air temperature was 42.9°C, vapour pressure was 14.3 mmHg and wind speed was 8.6 km/h. A rise in humidity and air temperature resulted in an increase in RT and PR of the same animals, whereas higher wind speed helped the animals to dissipate heat. The average sweat gland density in buffalo was 86.94/cm². Heritability of sweat gland characteristics (length, diameter, shape, volume and density) ranged between 0.38 and 0.68 (Nagarcenkar and Sethi 1981). It was observed that exposure of swamp non-pregnant female buffaloes for 2–3 h to solar radiation at air temperature of 42.1°C increased RT from 40.1 to 41.1°C, RR from 76.7 to 128.0 breaths/min and PR from 67.0 to 84.7 pulses/min (Zhengkang et al. 1994).

The buffalo behaviour changes during direct heat exposure. Asker and Ragab (1952) observed that the Egyptian buffaloes tied in the sun during May and June from 10:00 to 12:00 h became restless and stopped rumination. Other signs of discomfort observed are kicking, nervous switching of the tail and stretching of the head, after half an hour of exposure. Other signs observed after continued exposure to solar radiations are panting (64.54–82.16 breaths/min) dribbling of saliva from their mouth, mucus from the nostrils and tears from the eyes.

4   Adaptation to High Temperature

Morphologically, buffaloes have a good coat of soft hair like that of cattle at birth and during early calfhood. The hair on the body becomes sparser and almost devoid of hair as the animal grows. The amount of hair coat retained varies considerably, depending on the breed, season, housing practices and because of its exposure to water and mud. The colour of the hair may be black, dun, creamy yellow, dark, light grey or white. Series of B (brown), C (albino), D (dilute) and E (extension) are present in buffaloes, while only A (agouti), B, C, D, E and P (pink eye) are found in cattle (Searle 1968).

Anatomically, buffalo skin is covered with a thick epidermis, containing many melanin parti-

cles that give the skin surface its characteristic black colour (Shafie 1985). The melanin particles trap the ultraviolet rays preventing them from penetrating through the dermis of the skin to the lower tissue. These rays are abundant in solar radiation in the tropics and subtropics, and excessive exposure of animal tissue could be detrimental, even resulting in skin tumours. This beneficial characteristic is reinforced by well-developed sebaceous glands, with greater secretion activity than in cattle (Shafie and Abou El-Khair 1970). These glands secrete sebum, a fatty substance which emerges on the skin's surface and covers it with a lubricant, making it slippery for water and mud. This greasy sebum, along with the thick hornified top layer of skin, prevents water and the solutes in it from being absorbed into the skin. In this way, the animal is protected from the harmful effects of any deleterious chemical compounds in the water. Moreover, the sebum layer melts during hot weather and becomes glossier to reflect many of the heat rays, thus relieving the animal from the excessive external heat load.

Particularly, buffaloes suffer in the sun, since they exhibit signs of great distress when exposed to direct solar radiation or when working in the sun during a hot weather due to the fact that their bodies absorb a great deal of solar radiation because of their dark skin and sparse coat or hair, and in addition to that, they possess a less efficient evaporative cooling system due to their rather poor sweating ability, although this ability is more efficient than in cattle. Nagarcenkar and Sethi (1981) indicated that buffaloes possessing high sweat gland density and sweating volume coefficient were more heat tolerant, the performance was superior among the more heat-tolerant buffaloes and milk production was higher in more heat-tolerant buffaloes.

Buffaloes also have different physiological adaptation to extremes of heat and cold than the various breeds of cattle. Body temperature of buffaloes is slightly lower than those of cattle, but buffalo skin is usually black and heat absorbent and has only sparsely protection by hair. Buffalo skin has one-sixth of the density of sweat glands that cattle skin has, so buffaloes dissipate heat limitedly by sweating. Excessive work or exposure causes a rise in body temperature, pulse rate and respiration rate, and general discomfort increases more quickly in buffaloes than those of cattle. During a trial in Egypt, 2-h exposure to the sun caused the temperature of the buffalo to rise by 1.3°C, whereas temperatures of cattle rose by only 0.2–0.3°C. In the shade, however, controlled field studies on Egyptian buffaloes showed that the thermoregulatory mechanism functions more efficiently in buffaloes than in cattle, when the speed of recovery from the effect of stress is taken as a measure of efficiency (Mullick 1960). Pandy and Roy (1969) confirmed these results in their studies on the changes which are orderly manifestations of various physiological adjustments necessary for adaptation to higher environmental temperature, that is, the seasonal changes in body temperature, cardiorespiratory and haematological characteristics, body water content and electrolytic status of buffaloes under conventional farm management.

Regarding the behavioural characteristics, buffaloes prefer to wallow and cool off their body in a pond rather than seek shade. They may wallow for whole day and remain immersed in water or mud and chewing with half closed eyes. When temperature and humidity are high, buffaloes wallow or roll in mud during hot or even cool periods, cold seasons. Artificial cooling also provides comfort to buffalo and alleviates body heat. Ragab et al. (1953) reported that sprinkling adult females for 2 h showed an average fall of 0.9°C in body temperature. Cockrell (1974) added that the body temperature of buffaloes in the hot sun could only be kept normal by wallowing or by quasi-continuous application of water, preferably with an air draught or wind to dry it off. Titto et al. (1996) reported that sprinkling young female buffaloes for 15 min caused a quick decrease in physiological parameters expressed as high reduction in rectal temperature and respiratory rate, and Ablas et al. (2007) concluded that water for immersion or shade is an essential benefit to buffaloes' production in warm climates. This type of behaviour is more efficient than keeping in low temperature housing, although an artificial wallow becomes fouled by excreta

unless the water is continually flowing (Cockrell 1974). Experimental evidence has indicated that for maintaining proper homeothermy, the buffalo has to be provided with wallows or showers in the summer months and be protected from cold during the winter months preferably by housing them (Aggarwal and Singh 2008). The wallowing was found to be more beneficial than showers for alleviation of heat stress in lactating Murrah buffaloes. In wallow or in shade, buffaloes cool off quickly, perhaps because the black skin, which is rich in blood vessels, conducts and radiates heat efficiently, but buffaloes cool off more quickly than cattle in the shade. In Australia, Trinidad, Florida, Malaysia and elsewhere, buffaloes grow normally without wallowing as long as adequate shade is available. In Egypt, buffaloes were raised successfully without wallowing in a desert new reclaimed land (Marai et al. 2009). Particularly, buffaloes maintained only on shaded pastures of silvopastoral systems have achieved gain of weight of 0.911 kg/day in Brazil (Castro et al. 2008). Such practices are beneficial due to the fact that the use of sprinklers is expensive and water consuming and become of utmost importance in the areas which suffer from scarcity of water.

5 Thermoregulatory Mechanisms

Similar to other homeotherms, cattle regulate their internal body temperature by dissipating the heat produced due to metabolic activity in body and maintain balance heat flow from the animal to the surrounding environment. Heat flow occurs through processes dependent on surrounding temperature (sensible heat loss, i.e. conduction, convection, radiation) and humidity (latent heat loss, evaporation through sweating and panting). The magnitude of sensible heat loss via conduction and convection is dependent on the surface area per unit body weight, the magnitude of the temperature gradient between the animal and the air and the conductance of heat from the body core to the skin and from the skin to the surrounding air. Heat exchange by radiation depends upon surface area as well as the reflective properties of the hair coat. Light-coloured hair coats and sleek and shiny hair coats reflect a greater proportion of incident solar radiation than hair coats that are dark in colour or more dense and woolly (Stewart 1953; Hutchinson and Brown 1969; Finch 1986; Hansen 1990). One physiological response to heat stress is a reduction in heat production (Kibler and Brody 1952; Seif et al. 1979), which in turn is caused in large part by a reduction in feed intake (Kibler and Brody 1952; Johnston et al. 1958; Seif et al. 1979; Lough et al. 1990), milk yield (Johnson 1965; Lough et al. 1990; Elvinger et al. 1992; Aggarwal 2004) and thyroid hormone secretion (Magdub et al. 1982; Al-Haidary et al. 2001; Aggarwal 2004). Heat stress also leads to activation of heat loss mechanisms. Blood flow to the periphery increases so that heat loss via conduction and convection is enhanced (Choshniak et al. 1982). Cattle change posture and orientation to the sun to reduce gain of heat from solar radiation. Moreover, chronic exposure to elevated environmental temperatures results in a lightening of the hair coat (Stewart and Brody 1954). Heat stress also leads to activation of evaporative heat loss mechanisms involving an increase in sweating rate and respiratory minute volume (Kibler and Brody 1952; Choshniak et al. 1982; Gaughan et al. 1999; Al-Haidary et al. 2001). About 70–85% of maximal heat loss via evaporation is due to sweating with the remainder by respiration (Kibler and Brody 1952; Finch 1986). Heat loss via skin is more in cows, and heat loss by respiration is higher in buffaloes. This is due to less number of sweat glands in buffaloes (Aggarwal and Upadhyay 1997, 1998). As air temperatures approach those of skin temperature, evaporation becomes the major route for heat exchange with the environment.

Reactions of homeotherms to moderate climatic changes, generally, are compensatory and are directed at maintaining or restoring thermal balance (West 1999). When the environmental temperature reaches near the cow's body temperature, high ambient relative humidity percentage (RH%) reduces evaporation and affects the cow's cooling capability, and core temperature increases. This occurs due to the negative effects of high humidity on dissipation of body heat because of the decline in effectiveness of radiation, conduction and convection and the efficiency of evaporative cooling (West 1993). Vaporisation from the respiratory tract and the outer body surface is negatively affected by levels of the temperature and percentage of relative humidity of the air. It has been observed that buffaloes can acclimate more to high than low temperatures (Zicarelli et al. 2005). Through natural selection, buffaloes have acquired several morphological features that allow them to adapt to hot–humid areas. For instance, melanin-pigmented skin is useful for protection against ultraviolet rays, and low hair density facilitates heat dissipation by convection and radiation. In hot–dry climates, low humidity determines intense evaporative heat loss, which in buffaloes is limited by the low number of sweat glands. In addition, respiratory evaporation is less effective than in cattle due to induced alkalosis as a consequence of a rapid increase of blood pH (Koga 1991). In hot–humid climates, evaporative heat loss is not as effective in body heat dissipation. Thus, buffaloes rely on wallowing for efficient thermoregulation, and high secretion of sebum protects the skin in the water or mud (Hafez et al. 1955). In particular, buffaloes in hot conditions increase blood volume and flow to the skin surface to maintain a high skin temperature and facilitate heat dissipation while in the mud or in the water (Koga 1999).

6   Response to Heat Stress

The change in the environmental conditions, as is the case during heat stress, affects the normal metabolic balance and produces a positive feedback when the temperature is above the thermal comfort or upper range of the tolerance. In dairy cattle, as milk production increases, metabolic heat production rises with the metabolism of large amounts of nutrients, which makes the high-producing cow more vulnerable to high ambient temperatures and humidity than animals that are less active metabolically. 'Metabolism and productivity run parallel' (Brody 1945), and therefore, high-producing cows are affected more

Table 1 The effect of heat stress on dairy cows

THI	Stress level	Response of dairy cows
72	None	
72–79	Mild	Dairy cows adjust by seeking shade, increasing respiration rate and dilating the blood vessels. The effect on milk production is minimum
80–89	Moderate	Both saliva production and respiration rate increase. Feed intake may be reduced, and water consumption increases. There is an increase in body temperature. Milk production and reproduction decrease
90–98	Severe	Cows become very uncomfortable due to high body temperature, rapid respiration(panting) and excessive saliva production. Milk production and reproduction decrease significantly
>98	Danger	Potential cow death may occur

Source: Chase (2006)

than low-producing cows as a result of the TNZ downward shift (Coppock et al. 1982).

Heat stress increases loss of body fluid due to sweating and panting. Continued loss of the body fluid during heat can reach a critical level and may pose problem to thermoregulation and cardiovascular integrity (Silanikove 1994). Table 1 shows the responses of animals during heat stress at various levels of THI.

The general homeostatic responses to thermal stress in mammals are reduction in faecal and urinary water losses, reduction in feed intake and production and increased sweating, as well as initial increases in respiratory rates and heart rates. In response to stress, several changes in physical, biochemical and physiological processes occur to counteract the negative effects of heat stress and maintain thermal equilibrium. Most of these adjustments are to dissipate body heat to the environment and reduce the metabolic heat production.

6.1 Water Vaporisation

Evaporation of water is an effective means of cooling the body since it takes about 580 cal to evaporate (latent heat of vaporisation) 1 g of water from the body. At normal temperature and humidity, about 25% of the heat produced in resting animal is lost by evaporation of water from the skin and respiratory passages. Water loss through the insensible perspiration, cutaneous and respiratory is constant under basal conditions. An increased flow of blood through the skin causes an increase in heat loss, but the mechanisms of sweating and panting offer much efficient ways to increase the evaporative heat loss (Andersson and Jonasson 1992). The evaporative heat dissipation is the most effective mean of thermal regulation in the animal's body due to the latent heat of vaporisation of water. Water vaporised from skin surface increases by active sweating and that from respiratory surfaces increases by increased pulmonary ventilation or panting. An inverse relationship exists between sweating and panting and thus between cutaneous and respiratory evaporation in animals (Kibler and Yeck 1959; Mclean 1963).

Evaporative water loss can be calculated if faecal and urinary water losses are known. Water vaporisation loss was about 120 ml/h, at moderate temperature, in cattle (Kibler and Brody 1950). McDowell and Weldy (1960) estimated that water losses from the skin surface and from the respiratory tract increased by 77 and 55%, respectively, and water consumption increased by 110%, when cows acclimated to 21.1°C were exposed to 32.2°C for 2 weeks.

Water turnover rate has been used for comparing heat adaptability of different species and breeds. The water turnover rate based on tritiated water was observed to be the highest in Holstein–Friesians cattle reaching 96%, followed by goats 90%, sheep 77% and buffaloes 72% when ambient temperature of the climatic chamber increased from 18 to 32°C. This indicated that the native fat-tailed sheep and water buffaloes were less affected by heat and 12 h of water deprivation, than H Friesians and goats (Kamal et al. 1982).

In general, buffalo cows' volume of water intake (free water ± water through feedstuffs), metabolic water, water voided (faeces ± urine) and water vaporisation were 20.9, 1.6, 17.3 and 5.3 L in winter season and 36.8, 2.3, 20.0 and 19.2 L, respectively, in summer season. The water input (intake ± metabolic) was observed to be about twofold in summer. Water vaporisation increased by four times in summer as that of winter. Voided water showed slight seasonal difference (Pal et al. 1975). In Friesian cows, the water inputs for free, dietary and metabolic water were 19.05, 0.41 and 1.35 L/day, respectively, under mild climate (16°C and 62% RH) and 31.16, 0.37 and 1.24 L/day, respectively, under hot climate (39°C and 62% RH, for 7 h daily over 11 days). The water outputs for urinary, faecal and total evaporative water losses were 7.07, 7.72 and 6.02 L/day under mild climate, respectively, and were 9.61, 7.61 and 13.33 L/day, respectively, under hot climate. The differences in all components between the two climatic conditions were significant except for faecal water output (Kamal et al. 1982). Particularly, chronic heat exposure of 6- and 12-month-old buffalo calves was accompanied with increases in total body water content (8.52 and 9.63%), free water intake (25.16 and 56.41%), total water intake (28.49 and 48.34%), urine excretion (24.79 and 108.0%) and evaporative water loss (51.15 and 69.37%), respectively. Significant decreases were also observed in 6- and 12-month-old buffalo calves in total body solids content (15.73 and 16.12%), metabolic water (20.84 and 16.81%) and faecal water excretion (36.42 and 8.49%), respectively (Nessim 2004). In order to maintain total body water content at a relatively constant level, it is important that a continuous water supply must be provided.

6.2   Physiological Responses

Responses of the cow to temperatures above the TNZ include raised respiration rates and rectal temperature (Omar et al. 1996), panting, drooling, reduced heart rates, profuse sweating (Blazquez et al. 1994), decreased feed intake (NRC 1981) as well as reduced milk production (Abdel-Bary et al. 1992). Physical responses to heat stress in dairy cows are breed specific (Finch 1986), with the *B. indicus* and other tropical breeds being less responsive to thermal stress than *B. taurus* cattle. The differences in response to heat stress between cattle breeds are attributed to varying levels of adaptability to hot environments. Sharma et al. (1983) showed that, within *B. taurus* dairy cattle breeds, the Jersey was less sensitive to thermal stress than the Holstein–Friesian.

6.3   Sweating and Panting

Sweating plays an important role in heat loss mechanisms of crossbred cattle. Excess heat is dissipated by evaporation of sweat as a protective mechanism of the body against overheating. Though crossbreds have a limited sweating capacity as compared to zebu animal, they are able to increase sweating rate and doubled the rate of evaporation from skin at ambient temperature of 38°C and relative humidity of 59% (Upadhyay and Aggarwal 1997). The efficiency of function and total output of sweat glands can be increased by continuous exposure to hot conditions. Crossbreds also increase pulmonary frequency and ventilation rate in an attempt to maintain a thermal equilibrium during exercise under heat stress (Upadhyay and Aggarwal 1997). In buffaloes, pulmonary functions increase with increase in ambient temperature. Under mild hot conditions, low respiratory frequency of buffaloes and high volume of O_2 consumption and low pulmonary volume may contribute partially to an improved efficiency of energy utilisation, thus affecting thermal equilibrium in buffaloes. Respiratory evaporation has been found to be more important in maintaining thermal homeostasis than sweating under hot conditions in buffaloes (Aggarwal and Upadhyay 1998; Fig. 3).

Increase in water evaporation in response to increased ambient temperature was found in buffaloes and cattle by many workers (Kellaway and

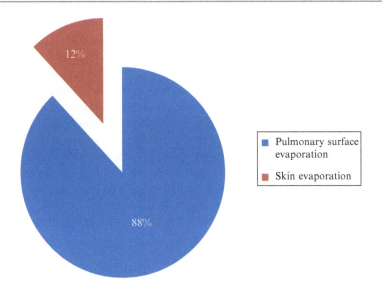

Fig. 3 Relative contribution of sweating and pulmonary evaporative losses in buffaloes (Aggarwal and Upadhyay 1998)

Colditz 1975; Krishana et al. 1975; Saxena and Joshi 1980; Daniel et al. 1981). In buffaloes' body, the water vaporisation loss was 6 and 18 kg/day (Raghavan and Mullick 1961) or 5.3 and 19.2 L/day (Ghosh et al. 1980), during winter and summer, respectively; Raghavan and Mullick (1961) also stated that cattle lost similar amounts although they drank much less water. In Jersey, Brown Swiss and Holstein heifers, water vaporisation increased from 4 to 14 L/day when environmental temperature increased from 15.0 to 35.0°C (Kamal et al. 1962). Water losses from the skin increase sharply. Evaporative loss through pulmonary surface and through skin in indigenous and crossbred cows has been shown in Fig. 4 (Aggarwal and Upadhyay 1998). Skin vaporisation accounts for 16–26% of the dissipation of body heat at air temperature of 10°C and from 40 to 60% at 27°C and over 80% at air temperatures above 38°C, in cattle (Roubicek 1969). Skin water loss was found to be negatively correlated with skin temperature and rectal temperature (Aggarwal and Upadhyay 1998). Water losses through the lungs increase gradually as ambient temperature increases. Pulmonary surface evaporative loss of buffaloes was 64.5±4.1 g/m²/h before exposure to direct solar radiation, and increased by 98% after 4 h of exposure (Aggarwal and Upadhyay 1998). Respiratory vaporisation as g/h/kg body weight

increased from 0.13 to 0.22 in Brahman, from 0.17 to 0.38 in Santa Gertrudis and from 0.28 to 0.42 in Shorthorn, when environmental temperature increased from 20.6 to 37.8°C, respectively (Kibler and Yeck 1959). In lactating Holstein cows, significant increase in respiratory vaporisation was recorded with rising air temperature (Kibler et al. 1965). Kibler et al. (1962) reported breed differences in respiratory vaporisation. Concerning the effect of relative humidity at high ambient temperature on respiratory vaporisation, Kibler and Brody (1953) reported that respiratory vaporisation decreased with rising relative humidity at high ambient temperature. Particularly, the buffalo calf showed little or no evaporative heat loss, and all internal temperatures were the same and rose progressively with time, when high environmental temperature was accompanied by high relative humidity. The buffalo calf has shown little or no evaporative heat loss with no changes in internal temperatures; however, temperature increased progressively with time at high environmental temperature and relative humidity.

At high environmental temperature and low relative humidity, some evaporation of water from the mucosa occurs during inspiration and helps in cooling the mucosa. Air humidity governs the evaporative cooling of the mucosa and determines the extent of water regain by condensation

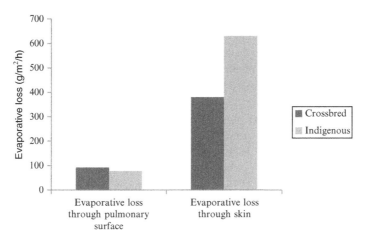

Fig. 4 Evaporative loss through skin and pulmonary surface in indigenous and crossbred cattle (Aggarwal and Upadhyay 1997)

Table 2 Relative contributions (%) of sweating and panting to evaporative heat loss in various domestic animals in a hot–dry environment

| | Relative contribution (%) to evaporative heat loss | | | | | |
	Donkey	Camel	Cow	Sheep/goat	Buffalo	Pig
Sweating	100	95	85	40	12	0
Panting	0	5	15	60	88	100

Source: Modified from Jenkinson (1972)

on cold mucosa upon exhalation. The passive transfer of water through the skin depends partly on the surface temperature of the epidermis and the vapour pressure of the air.

Thermoregulatory sweating is elicited in two ways:

- First by a rise of the CNS temperature and reflexively
- Second by stimulation of skin warm receptors and other parts of the body outside the CNS, that is, blood vessels, abdomen and viscera

Although reflex sweating may occur in absence of an increased central temperature, a high skin temperature cannot elicit full-scale sweating without simultaneous CNS facilitation.

Regarding sweating, the eccrine sweat glands (in humans) are innervated by cholinergic sympathetic nerve fibres, and the apocrine sweat glands (in many domestic animals) develop from hair follicles. The latter glands do not receive nerve supply, but are sensitive to epinephrine carried in the blood stream. The eccrine sweat glands are responsible for thermal sweating, but apocrine sweat glands are important for evaporative heat loss. The relative importance of sweating as a heat-dissipation mechanism varies among species (Table 2). In cows, maximum evaporation from the skin surface amounts to about 150 $g/m^2/h$ at an external temperature of 40°C. The respiratory evaporation under the same condition is only about one-third of that amount. In sheep, sweat secretion is less important than in the cow. Sweat secretion in shorn sheep during heat stress is 32 $g/m^2/h$, which means that sheep may dissipate about 20 kcal/h by sweating. Consequently, evaporative heat loss via respiratory passages is more important in sheep than in the cow. In humans, heat loss from sweating may be as high as 1,000 kcal/h (Andersson and Jonasson 1992).

Panting appears to be the more efficient of the two methods of evaporative cooling, that is, sweating and panting. Both methods use latent heat from the body core, but sweating can also use solar radiation on the body surface. Panting also provides its own airflow over moist surfaces, thus facilitating water evaporation. Salt and electrolytes are not lost, as in sweating, unless the saliva drips out of the mouth. The panting cools the nasal and oral passages as the cool blood flows into the venous sinus bathing the carotid plexus. Thus, the blood supply to the brain can be kept cool, even when the body temperature is rising (Taylor and Lyman 1972). The disadvantages of panting include a risk of respiratory alkalosis, particularly in the goat (Jenkinson 1972), sheep and buffaloes, and the increase in work and therefore heat production by the respiratory muscles. However, much of this work is reduced by the elastic property of the respiratory system, which has its own natural frequency of oscillation. The high respiratory rate associated with panting has the effect of keeping the system oscillating at its own resonant frequency with the minimum of muscular effort. Thus, the thermoregulatory efficiency of panting is high in such species as sheep, which show no increase in total body heat production above normal levels (Hales and Brown 1974).

The evaporation is a very significant process for loss of body water. A fully hydrated camel weighing 260 kg lost 9l of water a day through sweat when standing in the desert sun. This quantity represented a loss of 4% of total body weight, and a loss much in excess of 25% would probably be fatal. Assuming that heat load and therefore evaporation are proportional to body surface, then water loss under hot, desert conditions increases exponentially with decreasing size. There is very little difference in water loss per hour in the camel at 1.0%, and man at 1.5%, but the rate in animals weighing 2.5 kg is nearly 5%. Many animals also have lower lethal limits than the camel (Schmidt-Nielsen 1965). The need to preserve vital functions, as an animal becomes dehydrated, results in a reduction in the rate of evaporative cooling. The sequel to this reduction is either a rise in body temperature or a depression of heat production.

Kibler and Brody (1952) observed similar sweating rates for *B. taurus* and *B. indicus* breeds; however, Allen (1962) showed that *B. indicus* or zebu cattle had significantly higher sweating rates

than breeds from temperate regions and ascribed elevated sweating rates of *B. indicus* zebu cattle to their higher density of sweat glands. Schmidt-Nielsen (1964) reported that as the environmental temperature rose, *B. taurus* cattle showed an appreciable increase in evaporation between 15 and 20°C, with a maximum rate of evaporation being reached before 30°C. On the contrary, Brahman cows (*B. indicus*) had initially lower evaporation rates, but rapid evaporation rates occurred when temperatures were between 25 and 30°C, and continued rising up to 40°C. Cattle of temperate and tropical regions possess the same type of sweat glands, one to each hair follicle (Findlay and Yang 1950). However, Dowling (1955) found that tropical breeds have a higher density of hair follicles (1,698/cm^2 for zebu) than is the case in *B. taurus* breeds (1,064/cm^2 for Shorthorn), and in zebu cattle, sweat glands are located much closer to the skin surface than is the case in temperate breeds of cattle. Blazquez et al. (1994) attributed more significance to the product that these sweat glands produce than to the number of sweat glands per unit area or to their individual size. They measured a fivefold increase in the rate of skin moisture loss (up to 279 g/m^2/h) from the scrotum at 36.2°C in bulls.

The measurement of sweating rate in large animals is difficult due to equipment availability and placement of device. Robertshaw and Vercoe (1980) reported a twofold increase in the rate of skin moisture loss (up to 77 g/m^2/h) from the scrotum after exposure to a temperature of 40°C in bulls (*Bos* sp.). Finch (1986) found that the sweating rates of *B. indicus* increased exponentially with rises in body temperature, whereas in *B. taurus*, the sweating rates tended to plateau after an initial increase. Within *B. taurus* breeds, Singh and Newton (1978) found higher ($P<0.05$) sweating rates in Ayrshire calves than in Guernsey calves and suggested that Ayrshire calves were more capable of acclimation to hot weather than Guernsey calves.

Polypnea or a rapid breathing in cattle, buffalo, dog, sheep, goat, etc. is evoked by heat load and lack of oxygen. Polypneic panting is breathing occurring at a frequency between 200 and 400 breaths/min with open mouth and protrusion of tongue. Panting is often accompanied by increased salivary secretion and may cause a considerable increase in respiratory evaporative cooling. Panting may be initiated in response to a raised environmental temperature and an increase in the temperature of the blood supplying the brain. Panting in cattle and buffalo may be induced due to work and high external temperature by a rise in body temperature or by local warming of the anterior hypothalamus. This indicates that the panting, like sweating, may be initiated both reflexively and centrally. Birds, which have no sweat glands, increase evaporation not only by panting but also by a mechanism named regular flutter. It consists of rapid oscillations of the thin floor of the mouth and the upper part of the throat (Andersson and Jonasson 1992).

The proportion of metabolic heat that is dissipated from the cow by evaporation increases with rising environmental temperatures and a decreasing temperature gradient between the animal and air. The morphology and functioning of the apocrine sweat glands of cattle during hot climatic conditions have been extensively investigated (Montgomery et al. 1984). Blazquez et al. (1994) reported that increased blood flow to the skin is positively correlated to the sweating rate. The correlation between rectal temperature and water loss from skin was observed to be negative in buffaloes which could be due to higher rectal temperature in relation to skin water loss (Aggarwal and Upadhyay 1998).

7   Heat Production

A thermal steady state in an animal exists, when the heat gain and the heat loss are in equilibrium. In livestock species similar to other mammalian homeotherms, the thermoregulatory mechanisms consist of a series of physiological adjustments help to establish a thermal equilibrium under heat stress to maintain equality in heat gain and heat loss. These physiological adjustments are highly dependent on the external temperature and physiological state of animal. Normally the variable insulation, mainly due to circulatory adjustments in the thermoneutral zone of constant metabolism,

is sufficient to maintain a thermal steady state. But, above and below this thermoneutral zone, circulatory adjustments are no longer enough for maintenance of heat balance or body temperature.

Heat production (HP) is a measure of the sum total of energy transformations happening in the animal per unit time (Yousef 1985). Heat production is directly under control of the nervous system (Hammel 1968), by the endocrine system, through modification of appetite and digestive processes, and indirectly by alterations of the activity of respiratory enzymes and protein synthesis (Yousef 1985). The influence of environmental temperature on feed intake, production and thermoregulation in the animal greatly affects the rate of heat production (Brody 1945). Because of their effect on the rate of metabolism, the concentrations of hormones such as thyroxine, triiodothyronine, growth hormone and glucocorticoids are closely related to heat production (Yousef and Johnson 1966). Other factors influencing heat production in mammals include the following: body size (Brody 1945), the environment (Salem et al. 1982) animal species and breed and the availability of feed and water (Graham et al. 1959). Breed differences between Jersey and Holstein cows in the rate of heat production and dissipation have been reported, which may be attributed to differences in body size. The temperature gradient between internal organs and external environment is steeper in the smaller Jerseys than in the larger Holstein cows (Kibler and Brody 1952).

8   Heat Increment

Under warm to hot (ambient temperatures of 15 and >25°C, respectively) conditions, the cow gains heat from solar radiation and from the usual metabolic processes. If the heat gain exceeds heat loss from radiation, convection, evaporation and conduction, heat is stored in the body, and core temperature rises (Finch 1986). At night or at low ambient temperatures, heat stored is dissipated to the environment, and the body temperature may reach an equilibrium or normal body set point. Webster et al. (1976) discussed other heat generation mechanisms including the cost of eating and ruminating, heat produced by fermentation in the rumen and the increased heat produced by the tissues of the liver and the gut as the heat increment (HI). Webster et al. (1975) cited a range in energy costs of eating, 11 J/kJ of ME for grass pellets and 151 J/kJ of ME for fresh grass and concluded that the energy cost of ruminating could be discounted as a contribution to HI. Webster et al. (1975) found no differences in HI due to diet source and estimated the heat of fermentation in sheep to be 285 J/kJ of digestible energy from forage diets; however, heat production increased exponentially with increasing ME intake. At an intake of 34 kJ of ME/kg$^{0.75}$, heat production in tissues of the gut was 7 kJ/kg$^{0.75}$ per 24 h. Fasting heat production of gut tissues in sheep was 4 kJ/kg$^{0.75}$. Therefore, HI due to feeding in the gut was 2.9 kJ/kg$^{0.75}$, of which 1.7 kJ was fermentation heat and 1.2 kJ was aerobic metabolism in gut tissues. The processes of ingestion and digestion account for about 25–30% of total HI (Webster et al. 1976). Armstrong and Blaxter (1957) ascribed most of the variation in total HI to the nature of substrates made available for metabolism.

9   Assessment of Adaptability

Heat adaptability is a complex character that depends on the integrity and coordination of various systems such as the respiratory, circulatory, excretory, nervous, endocrine and enzymatic systems. The coordination of all these systems to maintain the productive potential under thermal stress is variable not only between species but also between breeds and even between individuals within a breed.

Particularly, prediction of the possibility of a breed to fit in a new region can be achieved by constructing climographs by plotting the means of monthly maximum and minimum temperature–humidity indices (THI) in the two localities, using climatic data collected from both the original and new environments. Similarity of each of position, shape and area of the two patterns so formed after joining the twelve points indicates such possibility. However, disease and parasite

criteria, the feed situation, prices of inputs and products and the market situation also have to be considered, in this respect.

The suitable stock for the tropics and subtropics should be morphologically and physiologically equipped to withstand heat and drought. Accordingly, adaptability of the animals to hot climate could be evaluated on morphological and/or physiological basis. This could be achieved according to morphological characteristics that assist to adapt to hot climate and/or according to physiological parameters such as the actual response or adaptability to hot climate after testing the animals under the hot climate conditions. Morphological characteristics of animals suitable to hot climate should include large skin area in relation to unit of live weight, shielded eyes, pigmented skin and eyelids (to lessen susceptibility to eye cancer) and short sleek light-coloured hair. In addition to that, the animals should be able to shed their coats early in spring, walk long distances, use low water intake, tolerate high intake of salts either in drinking water or in forages and poor quality food, afford harsh treatment and resist ticks and other pests. Particularly, animals with long or woolly coats pick up a large number of larval ticks than animals with short sleek coats. With such information in mind, animals with permanent short coats can be used in fairly hot rather humid regions, and those with heavy coat, but shed early and decisively in the spring, can be used in regions that are fairly hot and humid in summer.

Proper and more accurate evaluation could be based on the ability of the animals to maintain expression of their inherited functional potential during their lifetime when raised under hot conditions. This is the typical definition of adaptability or heat tolerance. The obtained parameters may be expressed as either adaptability indices (Marai et al. 2005, 2008) or heat tolerance indices (Marai et al. 2006) and are used either in choosing breeds to be located in new areas different than their original ones or in evaluating the animals for adaptability to the prevailing conditions under which they exist.

Many parameters were suggested to estimate adaptability, using the relative changes in thermal, water and/or nitrogen balances of the animals to the hot climate, and have been summarised in publications (Habeeb et al. 1997; Marai and Habeeb 1998). Marai and Habeeb (1998) and Marai et al. (2008) suggested the 'adaptability index' as a parameter for estimating the heat tolerance to the climatic conditions, using all the traits, that is, thermal, water and/or nitrogen balances, as well as the physiological, productive or reproductive traits of the animals under the conditions which they have to live. The 'adaptability index' equals 100– the average relative deviations in the traits studied regardless of the minus or plus signs. The relative deviation in each trait was estimated as [the difference between the two conditions in the values of the trait, i.e. the hot and mild conditions (normal; control) divided by the value of the trait in the mild conditions] × 100. The selected animal for the hot region should manifest the least changes in most of the parameters estimated (Marai et al. 2005, 2006, 2008). Marai and his group have estimated the buffalo adaptability index to the subtropical environment of Egypt as 89.1%. The average value of the relative deviations as a function to the hot environment was 10.9%. Estimation of the adaptability index using the data of El-Masry and Marai (1991) showed nearly the same values in buffaloes (89.6% as adaptability index and the average relative deviations 10.4%). Adaptability for Friesian cows was estimated as 82.9% in the subtropical environment of Egypt on the same bases. The estimated adaptability index can be used as a simple and valid index for selection of high-productive animals which maintain high milk or meat production under hot climate conditions. The gain in body weight may not be a reliable indicator or mislead about heat adaptability because the increase in body weight may also occur due to increase either in body water or in body protein and fat.

The adaptability of animal to heat needs to be expressed in relation to production as indicated in the recommendations of FAO/IAEA Panel (1974) 'Heat tolerance must be assessed in relation to production and not only in relation to maintenance of thermal balance'. The loss of body solids associated with a standard heat stimulus may be an index. Kamal (1993) developed four heat

tolerance indices (HTI) related to hot summer growth performance of young water buffalo calves, depending on heat-induced changes in some physiological and biochemical parameters. These were the cortisol-HTI, glucose-HTI, evaporative water (EW)-HTI and nitrogen retention (NR)-HTI. The cortisol-HTI and glucose-HTI were applicable to young calves (6 months old). Meanwhile, the EW-HTI and NR-HTI were applicable to older calves (12 months old). Such parameters can be used in screening heat tolerance in terms of growth or milk production under hot climate. This is preferable to screening the mature animals directly for productivity because the latter is time consuming, as it takes 2 years for the calves to reach their mature body weight and 3 years at least to produce milk (Nessim 2004). Such suggestions considered the main two factors controlling productive performance under heat stress are the genetic productive potentials for growth and milk production and the heat tolerance of the animal. The latter characteristic expresses the animal's capability to attain, as much as, the genetic maximum production under heat stress. Other HTI parameters were suggested by other workers (Rhoad 1944; Lee and Phillips 1948). Tunica dartos index (TDI) was developed to estimate the ability of the bull and ram to tolerate elevation of ambient temperature (El-Darawany 1999; Marai et al. 2006).

10   Genetic Adaptations/Cellular Changes During Heat Stress

The genetic adaptations developed in zebu cattle in the process of its evolution are ascribed to the acquisition of genes for thermotolerance. Zebu breeds can regulate their body temperature under vide range than *Bos Taurus* breeds of European origin (McDowell et al. 1953; Cartwright 1955; Allen et al. 1963; Finch 1986; Carvalho et al. 1995; Hammond et al. 1996). Genetic differences in their thermotolerance have been attributed to the cellular functions (Malayer and Hansen 1990; Kamwanja et al. 1994; Paula-Lopes et al. 2003; Hernández-Cerón et al. 2004). Thermal stress triggers a complex programme of gene expression and biochemical adaptive responses (Fujita 1999; Lindquist 1986). Biologically, the ability to survive and adapt to thermal stress is a fundamental requirement of cellular life, as cell stress responses are ubiquitous among both eukaryotes and prokaryotes, and key heat-shock proteins (HSPs) involved in these responses are highly conserved across all species (Lindquist 1986; Parsell and Lindquist 1993). In euthermic species, in which core temperature is very precisely regulated, considerable variations in core temperature can occur during severe environmental stress, exercise and fever. The ability to survive and adapt to severe systemic physiological stress is critically dependent on the ability of cells to mount an appropriate compensatory stress response.

The effects of heat stress on cellular function include:

1. Inhibition of DNA synthesis, transcription, RNA processing and translation
2. Inhibition of progression through the cell cycle
3. Denaturation and misaggregation of proteins
4. Increased degradation of proteins through both proteasomal and lysosomal pathways
5. Disruption of cytoskeletal components
6. Alterations in metabolism that lead to a net reduction in cellular ATP
7. Changes in membrane permeability that lead to an increase in intracellular Na^+, H^+ and Ca^{2+}

In mammalian cells, nonlethal heat shock produces changes in gene expression and in the activity of expressed proteins, resulting in cell stress response (Jaattela 1999; Lindquist 1986). This response increases thermotolerance (i.e. the ability to survive subsequent, more severe heat stresses) that is associated with increased expression of HSPs. A cell stress response leading to HSPs production can be induced by other stressors like toxins, chemicals, pyrogens and heavy metals, and the response initiated by one stressor often leads to cross-tolerance to others (Parsell and Lindquist 1993). At increasingly severe heat exposures, heat shock leads to activation of the apoptotic process and to cellular necrosis (Creagh et al. 2000). The apoptosis of cells exposed to different stressors appears to depend critically on the sequence of exposure (DeMeester et al. 2001). There were no significant differences between

Brahman, Senepol and Angus in the amount of heat-shock protein 70 (HSP70) in heat-shocked lymphocytes (Kamwanja et al. 1994), and the nonsignificant lower HSP70 amounts in Brahman and Senepol may indicate that protein denaturation in response to elevated temperature (one of the signals for HSP70 synthesis; Ananthan et al. 1986) is reduced in Brahman and Senepol. Therefore, magnitude of transcription in response to stressor or high temperature seems to be important for expression of genetic differences in breeds as there are no differences between Brahman and Holstein embryos in resistance to elevated temperature at the two-cell stage (Krininger et al. 2003), a time when the embryonic genome is largely inactive (Memili and First 2000).

11   Genetic Improvement for Adaptation

Acclimation and adaptation are two different processes in response to a stressor. Animals are considered acclimated to a given stressor when body temperature returns to prestress levels (Nienaber et al. 1999). Systemic, tissue and cellular responses associated with acclimation are coordinated, require several days or weeks to occur and are therefore, not homeostatic in nature (Bligh 1976). Furthermore, when stress is removed, these changes decay. Adaptation, on the other hand, requires modifications of the genetic makeup and is a process involving populations and exposure for very long periods.

Genetic improvement is an evolutionary action; evolution should be defined as a continuous process of adaptation of the populations of organisms to the ever-changing geological, biological and climatic conditions (Dobzhansky 1970). Because of the almost infinite number of combinations of environmental factors, organisms must have a great variety of genetic types capable to deal with a range of climatic, nutritional or other conditions. In a word, any population must be genetically heterogeneous – that is, with a great genetic diversity – in order to be able to survive under the challenge of the changing environment. Therefore, any population in a specific ambient is composed by a majority of well-adapted individuals, while a minor number of individuals present genotypes that are not good enough for that environment, but are well suited for different conditions. This is the basis for the livestock genetic improvement. Rhoad (1940) was probably the first to propose the selection of livestock for traits related to heat tolerance. Da Silva (1973) estimated the genetic variation of some traits in Brazilian beef cattle, observing that the increase in the body temperature after exposure of the animals to the sun in the hottest period of the day presented a moderate heritability coefficient (0.443) and a high negative genetic correlation (−0.895) with the average daily weight gain. Da Silva et al. (1988) determined the heritabilities of the sweating rate (0.222), skin pigmentation (0.112), hair coat pigmentation (0.303) and thickness (0.233) and hair length (0.081) of Jersey cattle bred in a tropical region. For Holstein cattle in a similar environment, the heritability of hair length was found as 0.20 by Pinheiro (1996). On the other hand, evidence has been found that supports the existence of a major gene, which is a dominant one and responsible for producing a very short, sleek hair coat in cattle (Olson et al. 2003).

However, little attention has been paid to the genetic aspects of the adaptation of livestock to their environment; it has generally been considered faster and easier to improve production through alterations of the environment, and most part of the research efforts has been oriented to environment modification. Numerous arguments have been used against animal breeding options, but there seems to be no a priori reason why genetic progress for adaptation is not possible. Present programmes of genetic improvement of livestock in tropical countries should take into account not only production traits (milk yield, weight gain, egg or wool production) but also those traits related to the interaction with environmental factors as the solar radiation, wind, air temperature and humidity. Additional research on this subject will likely provide avenues by which livestock production could have significant progresses in the years to come.

12 Genomic Responses During Acclimation

High-producing cattle and buffaloes are vulnerable to environmentally induced hyperthermia because the metabolic heat load is proportionate to their milk production. To counteract an increasing heat load, animals must adapt on a cellular level and to increase heat dissipation and minimise heat production. The information on genomic alterations and associated molecular mechanisms that lead to cellular heat stress acclimation is poorly understood in cattle and buffaloes. Studies on cattle have aimed at characterising the global changes in cellular gene expression and microarray analysis utilising bovine-specific cDNA arrays obtained from the National Bovine Functional Genomics Consortium. Microarray analysis has helped profiling of bovine mammary epithelial cell (BMEC) gene expression in response to acute heat stress using an in vitro system that approximates mammary development and function (Collier et al. 2008). During severe and acute hyperthermia (42°C), the BMEC exhibits morphological changes (regression of ductal branches) and reduced cellular growth. Consistent with these physical alterations in cell behaviour, gene expression associated with protein synthesis and cellular metabolism also decreases. In this model system, Hsp70 gene expression in BMEC remained elevated for 4 h at 42°C and then returned to basal levels after 8 h of exposure, indicating the end of heat tolerance and activation of genes associated with apoptosis (Collier et al. 2008). Another study on lactating dairy cattle evaluated the gene expression profile of liver tissue in response to an extended period (14 days) of the heat stress (Rhoads et al. 2005). The liver's pivotal role in whole-body metabolism via coordination of endogenously and exogenously derived nutrients is most likely altered by heat stress-induced reductions in feed intake and shifts in metabolism.

A large proportion of an animal's mass comprises skeletal muscle, which can have a profound impact on whole-animal energy metabolism and nutrient homeostasis, especially during periods of stress. To better understand how an environmental heat load influences the set points of several metabolic pathways within skeletal muscle, Rhoads et al. (2008) examined heat stress effects on skeletal muscle of beef cattle adaptation to chronic heat stress using microarray analysis. Skeletal muscle (semimembranosus) biopsies were obtained during thermal neutral conditions and again after exposure to heat stress. Data interrogation by pathway analysis identified dramatic changes in the skeletal muscle transcriptional profile revealing that during heat stress, bovine skeletal muscle may experience mitochondrial dysfunction leading to impaired cellular energy status. This may have broad implications for the reduced growth, decreased milk production and heat intolerance observed in ruminants during heat stress especially if skeletal muscle is not able to make necessary contributions to whole-body energy homeostasis. Taken together, the microarray data demonstrate that bovine cells and tissues undergo changes in cellular behaviour, which may be important for individual tissue function, whole-body metabolism and overall physiological acclimation to heat stress.

13   Conclusions

High ambient temperature is a major constraint on animal productivity in tropical climates since it evokes series of drastic changes in the animal's physiological functions that include depression in feed intake efficiency and utilisation; disturbances in metabolism of water, protein, energy and mineral balances, enzymatic reactions, hormonal secretions and blood metabolites. Such changes result in impairment of production and reproduction performance. The effect of heat stress is aggravated when it is accompanied by high ambient humidity. Susceptibility to heat stress depends on many factors, including adaptation, housing and management. Determining nutrient requirements for animals exposed to different climates and effective ambient temperature is a significant challenge that requires consideration of many factors affecting animal's ability to maintain a normal temperature. Heat and cold outside the

thermoneutral zone forces an animal to utilise compensatory mechanism to maintain its body temperature within normal limits. Management, housing, diet and climate affect the energy requirements of the animal. Cattle from zebu breeds are better able to regulate their body temperature in response to heat stress than are cattle of *B. taurus* breeds of European origin. Among the genetic adaptations that have developed in zebu cattle during its evolution have been the acquisition of genes for thermotolerance and disease resistance. Thus, an alternative scheme to crossbreeding for utilising the zebu genotype for livestock production in hot climates is to incorporate those zebu genes that confer thermotolerance into European breeds while avoiding undesirable genes. Heat stress reduces milk production in cows with high genetic merit for milk production. Strategies to alleviate metabolic and environmental heat loads in early lactation need to be developed. Milk production per cow has increased over threefold in response to advances in animal nutrition, in technology and in biotechnology as well as genetic progress for milk production. Using these facts as a basis, it is apparent that genetic selection and other variables enhancing milk production may have resulted in adjustments in factors important to lactation and nutritional physiology of the dairy cow. One such factor, important especially in hot environments, is the thermoregulatory capabilities and the physiological effects of heat stress on high-producing cows. There is a need to further define the physiological and cellular basis for thermotolerance in zebu cattle and buffaloes that associate with their disease resistance. The understanding will help targeting identified genes responsible for better adaptation and to exploit the zebu genome for improving their production under tropical climate conditions under limited resources.

References

Abdel-Bary HH, Mahmoud MM, Zaky HI, Mohamed AA (1992) Effects of season and month of calving on estrous performance, services conception and milk yield of Friesian cows in Egypt. Egypt J Anim Prod 29:229–253

Ablas DS, Titto EAL, Pereira AMF, Titto CG, Leme TMC (2007) Comportamento de bubalinos a pasto frente a disponibilidade de sombra e água para imersão. Ciência Anim Brasileira 8:167–175

Aggarwal A (2004) Effect of environment on hormones, blood metabolites, milk production and composition under two sets of management in cows and buffaloes. PhD thesis submitted to National Dairy Research Institute, Karnal (Haryana), India

Aggarwal A, Singh M (2008) Skin and rectal temperature changes in lactating buffaloes provided with showers and wallowing during hot-dry season. Trop Anim Health Prod 40:223–228

Aggarwal A, Upadhyay RC (1997) Pulmonary and cutaneous evaporative water losses in Sahiwal and Sahiwal × Holstein cattle during solar exposure. Asian Australas J Anim Sci 10:318–323

Aggarwal A, Upadhyay RC (1998) Studies on evaporative heat loss from skin and pulmonary surfaces in male buffaloes exposed to solar radiations. Buffalo J 2:179–187

Al-Haidary A, Spiers DE, Rottinghaus GE, Garner GB, Ellersieck MR (2001) Thermoregulatory ability of beef heifers following intake of endophyte-infected tall fescue during controlled heat challenge. J Anim Sci 79:1780–1788

Allen TE (1962) Responses of Zebu, Jersey, and Zebu × Jersey crossbred heifers to rising temperature, with particular reference to sweating. Aust J Agric Res 13:165–179

Allen TE, Pan YS, Hayman R (1963) The effect of feeding on the evaporative heat loss and body temperature in Zebu and Jersey heifers. Aust J Agric Res 14:580–593

Ananthan J, Goldberg AL, Voellmy R (1986) Abnormal proteins serve as eukaryotic stress signals and trigger the activation of heat shock genes. Science 232:522–524

Andersson BE, Jonasson H (1992) Temperature regulation and environmental physiology. In: Decker BC (ed) Physiology of small and large animals. Hamilton, Philadelphia, pp 886–895

Armstrong DG, Blaxter KL (1957) The heat increment of steam-volatile fatty acids in fasting sheep. Br J Nutr 11:247–273

Ashour GAS (1990) Water balance in bovine as related to heat regulation. PhD dissertation, Faculty of Agriculture, Cairo University, Cairo, Egypt

Asker AA, Ragab MT (1952) Causes of variation in birth weight of Egyptian cattle and buffaloes. Indian J Vet Sci 22:265–272

Berman A (1968) Nycthemeral and seasonal patterns of thermoregulation in cattle. Aust J Agric Res 19:181–188

Berman A, Folman YM, Kaim M, Mamen Z, Herz D, Wolfenson A, Graber Y (1985) Upper critical temperatures and forced ventilation effects for high-yielding dairy cows in a tropical climate. J Dairy Sci 68:488–495

Bitman JA, Lefcourt DL, Stroud B (1984) Circadian and ultradian temperature rhythms of lactating dairy cows. J Dairy Sci 67:1014–1023

Blazquez NB, Long SE, Mayhew TM, Perry GC, Prescott NJ, Wathes CM (1994) Rate of discharge and morphology of sweat glands in the perineal, lumbodorsal and scrotal skin of cattle. Res Vet Sci 57:277–284

Bligh J (1976) Introduction to acclimatory adaptation, including notes on terminology. In: Bligh J, Cloudsley-Thompson JL, McDonald AG (eds) Environmental physiology of animals. Wiley, New York

Bligh JK, Harthoorn AM (1965) Continuous radiotelemetric records of the deep body temperature of some unrestrained African mammals under near-natural conditions. J Physiol 176:145–162

Brody S (1945) Bioenergetics of growth. Reinhold, New York

Cartwright TC (1955) Responses of beef cattle to high ambient temperatures. J Anim Sci 14:350–362

Carvalho FA, Lammoglia MA, Simoes MJ, Randel RD (1995) Breed affects thermoregulation and epithelial morphology in imported and native cattle subjected to heat stress. J Anim Sci 73:3570–3573

Castro AC, Lourenço Júnior JB, Santos NFA, Monteiro EMM, Aviz MAB, Garcia AR (2008) Sistema silvipastoril na Amazônia: ferramenta para elevar o desempenho produtivo de búfalos. Ciência Rural 38:2395–2402

Chase LE (2006) Climate change impacts on dairy cattle. Fact sheets, Climate change and Agriculture. Promoting Practical and Profitable Responses. http://Climate and Farming.org/pdfs/Fact Sheets/III.3cattle.pdf

Choshniak I, McEwan-Jenkinson D, Blatchford DR, Peaker M (1982) Blood flow and catecholamine concentration in bovine and caprine skin during thermal sweating. Comp Biochem Physiol 71:37–42

Cockrell WR (1974) The husbandry and health of the domestic buffalo. F.A.O. Rome, Italy

Collier RJ, Collier JL, Rhoads RP, Baumgard LH (2008) Invited review: genes involved in the bovine heat stress response. J Dairy Sci 91:445–454

Coppock CE, Grant PA, Portzer SJ, Charles DA, Escobosa A (1982) Lactating dairy cow responses to dietary sodium, chloride, and bicarbonate during hot weather. J Dairy Sci 65:566–576

Creagh EM, Sheehan D, Cotter TG (2000) Heat shock proteins—modulators of apoptosis in tumour cells. Leukemia 14:1161–1173

Da Silva RG (1973) Improving tropical beef cattle by simultaneous selection for weight and heat tolerance. Heritabilities and correlations of the traits. J Anim Sci 37:637–642

Da Silva RG, Arantes-Neto JG, Holtz-Filho SV (1988) Genetic aspects of the variation of the sweating rate and coat characteristics of Jersey cattle. Braz J Genet 11:335–347

Daniel SJ, Hasan QZ, Murty VN (1981) Note in water metabolism in lactating crossed cows. Indian J Anim Sci 51:358–360

Das SK, Upadhyay RC, Madan ML (1999) Heat stress in Murrah buffalo calves. Livest Prod Sci 61:71–78

DeMeester SL, Buchman TG, Cobb JP (2001) The heat shock paradox: does NF- B determine cell fate? FASEB J 15:270–274

Dobzhansky T (1970) Genetics of the evolutionary process. Columbia University Press, New York

Dowling DF (1955) The hair follicle and apocrine gland populations of Zebu (*Bos indicus* L.) and Shorthorn (*Bos taurus* L.) cattle skin. Aust J Agric Res 6: 645–654

Du Prezz JH, Hattingh PJ, Giesecke WH, Eisenberg BE (1990) Heat stress in dairy cattle and other livestock under southern African conditions. Monthly temperature-humidity index mean values and their significance in the performance of dairy cattle. Onderstepoort J Vet Res 57:241–248

El-Darawany AA (1999) Tunica dartos thermoregulatory index in bull and ram in Egypt. Indian J Anim Sci 69:560–563

El-Masry KA, Marai FM (1991) Comparison between Friesians and water buffaloes in growth rate, milk production and some blood constituents, during winter and summer conditions. Egypt J Anim Prod 53:39–43

Elvinger F, Natzke RP, Hansen PJ (1992) Interactions of heat stress and bovine somatotropin affecting physiology and immunology of lactating cows. J Dairy Sci 75:449–462

Finch VA (1986) Body temperature in beef cattle: its control and relevance to production in the tropics. J Anim Sci 62:531–542

Findlay JD, Yang SH (1950) The sweat glands of Ayrshire cattle. J Agric Sci 40:126–133

Folk GE (1974) Textbook of environmental physiology. Lea & Febiger, Philadelphia, pp 97–118

Fujita J (1999) Cold shock response in mammalian cells. J Mol Microbiol Biotechnol 1:243–255

Gaughan JB, Mader TL, Holt SM, Josey MJ, Rowan KJ (1999) Heat tolerance of Boran and Tuli crossbred steers. J Anim Sci 77:2398–2405

Ghosh TK, Singh WB, Verma DN, Saxena KK, Ranjhan SK (1980) Measurement of water turnover and water requirements in different species of animals in two seasons. Indian J Anim Sci 50:615–619

Graham NM, Wainman FW, Blaxter KL, Armstrong DG (1959) Environmental temperature, energy metabolism and heat regulation in sheep. I. Energy metabolism in closely clipped sheep. J Agric Sci (Camb) 52:13–24

Habeeb AAM, Marai IFM, Owen JB (1997) Genetic improvement of livestock for heat adaptation in hot climates. In: Proceedings of the international conference on animal production & health, Zagazig University, Zagazig, Egypt

Hafez ESE, Badreldin AL, Shafie MM (1955) Skin structure of Egyptian buffaloes and cattle with particular reference to sweat glands. J Agric Sci 46:19–30

Hales GRS, Brown GD (1974) Net energetic and thermoregulatory efficiency during panting in the sheep. Comp Biochem Physiol 49:413–422

Hammel HT (1968) Regulation of internal body temperature. Annu Rev Physiol 30:641–646

Hammond AC, Olson TA, Chase CC Jr, Bowers EJ, Randel RD, Murphy CN, Vogt DW, Tewolde A (1996) Heat tolerance in two tropically adapted *Bos taurus* breeds, Senepol and Romosinuano, compared with Brahman, Angus, and Hereford cattle in Florida. J Anim Sci 74:295–303

Hansen PJ (1990) Effects of coat colour on physiological and milk production responses to solar radiation in Holsteins. Vet Rec 127:333–334

Hernández-Cerón J, Chase CC Jr, Hansen PJ (2004) Differences in heat tolerance between preimplantation embryos from Brahman, Romosinuano, and Angus Breeds. J Dairy Sci 87:53–58, http://www.climateandfarming.org/pdfs/FactSheets/III.3Cattle.pdf

Hutchinson JCD, Brown GD (1969) Penetrance of cattle coats by radiation. J Appl Physiol 26:454–464

Igono MO, Bjotvedt G, Sanford-Crane HT (1992) Environmental profile and critical temperature effects on milk production of Holstein cows in desert climate. Int J Biometeorol 36:77–87

Jaattela M (1999) Heat shock proteins as cellular lifeguards. Ann Med 31:261–271

Jenkinson DM (1972) Evaporative temperature regulation in domestic animals. Symp Zool Soc (Lond) 31:345–356

Johnson HD (1965) Environmental temperature and lactation (with special reference to cattle). Int J Biometeorol 9:103–116

Johnston JE, Hamblin FB, Schrader GT (1958) Factors concerned in the comparative heat tolerance of Jersey, Holstein, and Red Sindhi-Holstein (F1) cattle. J Anim Sci 17:473–479

Kadzere CT, Murphy MR, Silanikove N, Maltz E (2001) Heat stress in lactating dairy cows: a review. Livest Prod Sci 77:59–91

Kamal TH, Ibrahim II (1969a) The effect of the natural climate of the Sahara and controlled climate on the rectal temperature and cardiorespiratory activities of Friesian cattle and water buffaloes. Int J Biometeorol 13:275–285

Kamal TH, Ibrahim II (1969b) The effect of the natural climate of the Sahara and controlled climate on thyroid gland activity in Friesian cattle and water buffaloes. Int J Biometeorol 13:287–294

Kamal TH, Johnson HD, Ragsdale AC (1962) Metabolic reactions during thermal stress in dairy animals acclimated to 50 and 80°F. Univ Mo Agric Exp Stat Res Bull 785:1–114

Kamal TH, El-Banna IM, Ayad MA, Kotby EA (1978) The effect of hot climatic and management on water requirements and body water in farm animals using tritiated water. Arab J Nucl Sci Appl 11:160–184

Kamal TH, Mehrez AZ, El-Shinnawy MM, Abedel-Samee AM (1982) The role of water metabolism in the heat stress syndrome in Friesian cattle. In: Proceedings of 6th international conference on animal and poultry production, Cairo, Egypt, vol 1, pp 14–26

Kamal TH (1993) Indices of heat tolerance and amelioration of heat stress. In: Prospects of Buffalo Production in the Mediterranean and the Middle East. In: Proceedings of International Symposium Jointly organized by ESAP, EAAP, FAO, CHEAM and DIE. Cairo, Egypt. EAAP Pub No 62. Pudoc Sci Publ Wageningen pp 198–200

Kamwanja LA, Chase CC Jr, Gutierrez JA, Guerriero V Jr, Olson TA, Hammond AC, Hansen PJ (1994) Responses of bovine lymphocytes to heat shock as modified by breed and antioxidant status. J Anim Sci 72:438–444

Kellaway RC, Colditz PJ (1975) The effect of heat stress on growth and nitrogen metabolism in Friesian and Fl Brahman×Friesian heifers. Aust J Agric Res 26:15–22

Kibler HH (1964) Environmental physiology and shelter engineering. LXVII. Thermal effects of various temperature-humidity combinations on Holstein cattle as measured by eight physiological responses. Missouri Agric Exp Stn Res Bull 862

Kibler HH, Brody S (1950) Environmental physiology with special reference to domestic animals. X. Influence of temperature, 5° to 95° F, on evaporative cooling from the respiratory and exterior body surfaces of Jersey and Holstein cows. Missouri Agr Exp Sta Res Bul 461:1–19

Kibler HH, Brody S (1952) Environmental physiology with special reference to domestic animals. XIX. Relative efficiency of surface evaporative, respiratory evaporative, and non-evaporative cooling in relation to heat production in Jersey, Holstein, Brown Swiss and Brahman cattle, 5° to 105° F. Missouri Agr Exp Sta Res Bul 497:1–31

Kibler HH, Brody SS (1953) Influences of humidity on heat exchange and body temperature regulation in Jersey, Holstein

Kibler HH, Yeck GR (1959) Vaporization rates and heat tolerance of growing Shorthorn, Santa Gertrudis and Brahman calves raised at constant 50° and 85 °F temperature. Univ Mo Agric Exp Stat Res Bull 701

Kibler HH, Yeck GR, Erry LL (1962) Vaporization rates in Brown Swiss. Holstein and Jersey calves During Growth and Constant 50° and 85 °F Temperature. Univ Mo Agric Exp Sta Res Bull 792

Kibler HH, Johnson HD, Shanklin MD, Hahn L (1965) Acclimation of Holstein cattle to 84 °F (29°C) temperature changes in heat producing and heat dissipating functions. Univ Mo Agric Exp Sta Res Bull 893

Koga A (1991) Effects of high environmental temperatures on some physicochemical parameters of blood and heat production in swamp buffaloes and Holstein cattle. Anim Sci Technol (Jpn) 62:1022–1028

Koga A (1999) Internal changes of blood compartments and heat distribution in swamp buffaloes in hot condition: comparative study of thermoregulation in buffaloes and Friesian cows. Asian Australas J Anim Sci 12:886–890

Krininger CE III, Block J, Al-Katanani YM, Rivera RM, Chase CC Jr, Hansen PJ (2003) Differences between Brahman and Holstein cows in response to estrous synchronization, superovulation and resistance of embryos to heat shock. Changes in inhibitory effects of arsenic and heat shock on growth of preimplantation bovine embryos. Mol Reprod Dev 63:335–340

Krishana G, Razdan MN, Roy SN (1975) Effects of nutritional and seasonal variations on water metabolism in lactating Zebu cows (*Boss Indicus*) in tropical and subtropical regions of India. Trop Agric Trin 52:35–41

Lee PHK, Phillips RW (1948) Assessment of the adaptability of livestock to climatic stress. J Anim Sci 1:391–426

Lemerle C, Goddard ME (1986) Assessment of heat stress in dairy cattle in Papua New Guinea. Trop Anim Health Prod 18:232–242

Lindquist S (1986) The heat-shock response. Annu Rev Biochem 55:1151–1191

Lough DS, Beede DL, Wilcox CJ (1990) Effects of feed intake and thermal stress on mammary blood flow and other physiological measurements in lactating dairy cows. J Dairy Sci 73:325–332

Magdub A, Johnson HD, Belyea RL (1982) Effect of environmental heat and dietary fiber on thyroid physiology of lactating cows. J Dairy Sci 65:2323–2331

Malayer JR, Hansen PJ (1990) Differences between Brahman and Holstein cows in heat-shock induced alterations of protein secretion by oviducts and uterine endometrium. J Anim Sci 68:266–280

Marai IFM, Habeeb AAM (1998) Adaptability of *Bos taurus* cattle under hot arid conditions. Ann Arid Zone 37:253–281

Marai IFM, Ayyat MS, Abd-El-Monem UM (2001) Growth performance and reproductive traits at first parity of New Zealand White female rabbits as affected by heat stress and its alleviation, under Egyptian conditions. Trop Anim Health Prod 33:1–12

Marai IFM, Habeeb AAM, Gad AE (2005) Tolerance of imported rabbits grown as meat animals to hot climate and saline drinking water in the subtropical environment of Egypt. J Anim Sci 81:115–123

Marai IFM, El-Darawany AA, Abou-Fandoud EI, Abdel-Hafez MAM (2006) Tunica dartos index as a parameter for measurement of adaptability of rams to subtropical conditions of Egypt. Anim Sci J (Jpn) 77:487–494

Marai IFM, Habeeb AAM, Gad AE (2008) Performance of New Zealand White and Californian male weaned rabbits in the subtropical environment of Egypt. Anim Sci J 79:472–480

Marai IFM, Daader AH, Soliman AM, El-Menshawy SMS (2009) Nongenetic factors affecting growth and reproduction traits of buffaloes under dry management housing (in sub-tropical environment) in Egypt. Livest Res Rural Dev 21:1–13

McArthur AJ, Clark JA (1988) Body temperature of homeotherms and the conservation of energy and water. J Therm Biol 3:9–13

McDowell RE (1972) Improvement of livestock production in warm climates. Freeman/San Francisco Press, San Francisco

McDowell RE, Weldy JR (1960) Water exchange of cattle under heat-stress. In: Proceedings of 3rd international biometeorological congress, London. Pergamon Press, New York, pp 414–424

McDowell RE, Lee DHK, Fohrman MH (1953) The relationship of surface area to heat tolerance in Jerseys and Sindhi-Jersey (F1) crossbred cows. J Anim Sci 12:747–756

McDowell RE, Hooven NW, Camoens JK (1976) Effects of climate on performance of Holsteins in first lactation. J Dairy Sci 59:965–973

Mclean JA (1963) The regional distribution of cutaneous moisture vaporization in the Ayrshire calf. J Agric Sci 61:275

Memili E, First NL (2000) Zygotic and embryonic gene expression in cow: a review of timing and mechanisms of early gene expression as compared with other species. Zygote 8:87–96

Montgomery I, Jenkinson DM, Elder HY, Czarnecki D, Mackie RM (1984) The effects of thermal stimulation on the ultra-structure of the human atrichial sweat gland. 1. The fundus. Br J Dermatol 110:319–320

Mullick DN (1960) Effect of humidity and exposure to sun on the pulse rate, respiration rate, rectal temperature and haemoglobin level, in different sexes of cattle and buffaloes. J Agric Sci (Camb) 54:391–394

Nagarcenkar R, Sethi RK (1981) Association of adaptive traits with performance traits in buffaloes. Indian J Anim Sci 51:1121–1123

Nessim MG (2004) Heat-induced biological changes as heat tolerance indices related to growth performance in buffaloes. PhD thesis, Faculty of Agriculture, Ain Shams University, Cairo, Egypt

Nienaber JA, Hahn GL, Eigenberg RA (1999) Quantifying livestock responses for heat stress management: a review. Int J Biometeorol 42:183–188

NRC (1981) Effect of environment on nutrient requirements of domestic animals. National Academic, Washington, DC, pp 75–84

Olson TA, Lucena C, Chase CC Jr, Hammond AC (2003) Evidence of a major gene influencing hair length and heat tolerance in *Bos Taurus* cattle. J Anim Sci 81:80–90

Omar EA, Kirrella AK, Soheir A, Fawzy A, El-Keraby F (1996) Effect of water spray followed by forced ventilation on some physiological status and milk production of post calving Friesian cows. Alex J Agric Res 4:71–81

Pal S, Pal RN, Thomas CK (1975) Effect of restricted access to water supply and shelter on the physiological norms and production of Murrah buffaloes. Indian J Dairy Sci 28:41–48

Pandy MD, Roy A (1969) Variation in cardiorespiratory rates, rectal temperature, blood haematocrit and haemoglobin as measures of adaptability in buffaloes to hot environment. Br Vet J 125:63–70

Parsell DA, Lindquist S (1993) The function of heat-shock proteins in stress tolerance: degradation and reactivation of damaged proteins. Annu Rev Genet 27:437–449

Paula-Lopes FF, Chase CC Jr, Al-Katanani YM, Krininger CE III, Rivera RM, Tekin S, Majewski AC, Ocon OM, Olson TA, Hansen PJ (2003) Genetic divergence in cellular resistance to heat shock in cattle: differences between breeds developed in temperate versus hot climates in responses of preimplantation embryos, reproductive tract tissues and lymphocytes to increased culture temperatures. Reproduction 125:285–294

Perez JH (2000) Parameters for the determination and evaluation of heat stress in dairy cattle in South Africa. J Vet Res 67:263–271

Pinheiro MG (1996) Variação genética de caracteristicas da capa externa de vacas da raça Holandesa em ambiente tropical. PhD, University of São Paulo, Ribeirão Preto

Ragab MT, Asker AA, Ghazy MS (1953) Effect of age on milk yield and length of lactation period in Egyptian buffaloes. Indian J Dairy Sci 6:181–188

Raghavan GV, Mullick DN (1961) Effect of air temperature and humidity on the metabolism of nutrients in buffalo. Ann Biochem Exp Med 21:277

Rhoad AO (1940) A method of assaying genetic differences in the adaptability of cattle to tropical and subtropical climates. Emp J Exp Agric 8:190–198

Rhoad AO (1944) The Iberia heat tolerance test for cattle. Trop Agric 21:162–164

Rhoads RP, Sampson JD, Tempelman RJ, Sipkovsky S, Coussens PM, Lucy MC, Spain JN, Spiers DE (2005) Hepatic gene expression profiling during adaptation to a period of chronic heat stress in lactating dairy cows. FASEB J 19:1673

Rhoads RP, Obrien MD, Greer K, Cole L, Sanders S, Wheelock JB, Baumgard LH (2008) Consequences of heat stress on the profile of skeletal muscle gene expression in beef cattle. FASEB J 22:1165.1

Robertshaw D, Vercoe JE (1980) Scrotal thermoregulation of the bull (*Bos* spp.). Aust J Agric Res 31:401–407

Roubicek CB (1969) Water metabolism (Chapter 19.1n). In: Hafez ESE, Dyer IA (eds) Animal growth and nutrition. Lea and Febiger, Philadelphia, pp 353–373

Salem MH, Yousef MK, El-Sherbiny AA, Khalili MH (1982) Physiology of sheep and goats in the tropics. In: Yousef MK (ed) Animal production in the tropics. Praeger, New York, pp 148–157

Saxena SK, Joshi BC (1980) Dynamics of body water under environmental heat stress. Indian J Anim Sci 50:383–388

Schmidt-Nielsen K (1964) Desert animals. Physiological problems of heat and water. Oxford University Press, Oxford

Schmidt-Nielsen K (1965) Desert animals: physiological problems of heat and water. Clarendon, Oxford

Searle AG (1968) Comparative genetics of colour of mammals. Logos Press, London

Seif SM, Johnson HD, Lippincott AC (1979) The effects of heat exposure (31°C) on Zebu and Scottish Highland cattle. Int J Biometeorol 23:9–14

Shafie MM (1985) Physiological responses and adaptation of water buffalo. In: Yousef MK (ed) Stress physiology in livestock, 2, ungulates. CRC, Boca Raton

Shafie MM (1993) Biological adaptation of buffaloes to climatic conditions. In: Proceedings of prospects of buffalo production in the Mediterranean and the Middle East, Cairo, Egypt. EAAP Publication no. 62. Pudoc Scientific Publishers, Wageningen, pp 176–185

Shafie MM, Abou El-Khair MM (1970) Activity of the sebaceous glands of bovines in hot climate (Egypt). J Anim Prod Cairo, UAR 10:81–98

Sharma AK, Rodriguez LA, Mekonnen G, Wilcox CJ, Bachman KC, Collier RJ (1983) Climatological and genetic effects on milk composition and yield. J Dairy Sci 66:119–126

Shearer JK, Beede DK (1990) Thermoregulation and physio- logical responses of dairy cattle in hot weather. Agri-Pract 11:5–17

Silanikove N (1994) The struggle to maintain hydration and osmoregulation in animals experiencing severe dehydration and rapid rehydration: the story of ruminants. Exp Physiol 79:281–300

Singh SP, Newton WM (1978) Acclimatization of young calves to high temperatures: physiological responses. Am J Vet Res 39:795–797

Spain JN, Spiers D (1998) Effect of fan cooling on thermoregulatory responses of lactating dairy cattle. In: Proceedings of the fourth international dairy housing conference. ASAE, St. Louis, pp 232–238

Stewart RE (1953) Absorption of solar radiation by the hair of cattle. Agric Eng 34:235–238

Stewart RE, Brody S (1954) Environmental physiology and shelter engineering with special reference to domestic animals. XXIX. Effect of radiation intensity on hair and skin temperatures and on respiration rates of Holstein, Jersey, and Brahman cattle at air temperatures 45°, 70°, and 80°F. Mo Res Bull 561

Taylor CR, Lyman CP (1972) Heat storage in running antelopes: independence of brain and body temperatures. Am J Physiol 222:114–117

Titto EAL, Russo HG, Lima CG (1996) Efeito do banho de água sobre o conforto térmico de bubalinos. In: Proceedings of 6th Congresso de Zootecnia, Évora/Portugal. Actas do VI Congresso de Zootecnia, pp 15–18

Tucker CB, Rogers AR, Schütz KE (2008) Effect of solar radiation on dairy cattle behavior, use of shade and body temperature in a pasture-based system. Appl Anim Behav Sci 109:141–154

Upadhyay RC, Aggarwal A (1997) Pulmonary and skin evaporative heat loss during exercise in hot dry conditions in crossbreds. Ind J Anim Sci 67:51–53

Wathes CM, Jones CDR, Webster AJF (1983) Ventilation, air hygiene and animal health. Vet Rec 113:554–559

Webster AJF, Osuji PO, White F, Ingram JF (1975) The influence of food intake on portal blood flow and heat production in the digestive tract of sheep. Br J Nutr 34:125–139

Webster AJF, Osuji PO, Weekes TEC (1976) Origins of the heat increment of feeding in sheep. In: Proceedings 7th symposium energy metabolism, EAAP Publication 19, pp 45–53

West JW (1993) Interactions of energy and bovine somatotropin with heat stress. J Dairy Sci 77:2091–2102

West JW (1999) Nutritional strategies for managing the heat-stressed dairy cow. Am Soc Anim Sci Am Dairy Sci Assoc 2:21–35

West JW (2003) Effects of hot, humid weather on milk temperature, dry matter intake, and milk yield of lactating dairy cows. Am Dairy Sci Assoc 86: 232–242

Yousef MK (1985) Basic principles. Stress physiology in livestock, vol 1. CRC Press, Boca Raton

Yousef MK, Johnson HD (1966) Calorigenesis of dairy cattle as influenced by thyroxine and environmental temperature. J Anim Sci 25:150–156

Zhengkang H, Zhenzhong C, Shaohua Z, Vale WG, Barnabe VH, Mattos JCA (1994) Rumen metabolism, blood cortisol and T3, T4 levels and other physiological parameters of swamp buffalo subjected to solar radiation. In: Proceedings of World Buffalo Congress, San Paulo, Brazil, vol 2, pp 39–40

Zicarelli F, Campanile G, Gasparrini B, Di Palo R, Zicarelli L (2005) Influence of the period and of the space on the milk production and on the consumption of dry matter in the Italian Mediterranean Buffalo. In: Proceedings 3rd Congresso Nazionale sull'Allevamento del Bufalo – 1st Buffalo Symposium Europe and the Americas, Paestum (SA), Italy, pp 75–76

Heat Stress and Hormones

Contents

Abstract

Activation of the hypothalamic–pituitary–adrenal axis and the consequent increase in plasma glucocorticoid concentrations are two of the most important responses of the animals to heat stress. The short- and long-term environmental heat affects endocrine glands and in turn release of hormones, namely, thyroxine, cortisol, growth hormone and catecholamines. Some of them result in initial increase due to acute stressors and a decline in plasma levels after prolonged exposure to stressors has been observed. The relationship of amounts in plasma of these hormones to milk production appears to be related directly for cortisol, growth hormone and prolactin with an inverse relationship with thyroxine. Epinephrine and norepinephrine are elevated with prolonged environmental heat stress. Hormones in plasma are important as potential indicators of the physiological status of a cow and reflect the physiological compensations a cow undergoes at various stages of lactation and exposure to heat stress. The plasma thyroxine (T$_4$) and triiodothyronine (T$_3$) levels have been observed to decline under heat stress as compared to thermoneutral conditions. The decline in thyroid hormones along with decreased plasma growth hormone (GH) level has a synergistic effect to reduce heat production. A reduced secretion of GH is required for survival of the homeotherm during heat stress. The concentration of insulin-like growth factor-1 (IGF-1) has been observed to decrease during

A. Aggarwal and R. Upadhyay, *Heat Stress and Animal Productivity*,
DOI 10.1007/978-81-322-0879-2_2, © Springer India 2013

Table 1 Some endocrine adaptation made during heat acclimation in cattle

Tissue	Response	References
Adrenal	Reduced aldosterone secretion	Collier et al. (1982a)
	Reduced glucocorticoid secretion	Collier et al. (1982b), Ronchi et al. (2001), and Aggarwal (2004)
	Increased epinephrine secretion	Alvarez and Johnson (1973)
	Increased progesterone secretion	Collier et al. (1982b) and Ronchi et al. (2001)
Adipose tissue	Increased leptin secretion	Bernabucci et al. (2006) and Aggarwal et al. (2010)
Pituitary	Increased prolactin secretion	Wetteman and Tucker (1979), Ronchi et al. (2001), and McGuire et al. (1991)
	Decreased somatotropin secretion	
Liver	IGF-1 unchanged or increased	Collier et al. (1982b)
Thyroid	Decreased thyroxine secretion	Collier et al. (1982b), Nardone et al. (1997), and Aggarwal (2004)
Placenta	Decreased oestrone sulphate secretion	Collier et al. (1982b)

Source: Modified from Bernabucci et al. (2010)

the summer months. Aldosterone concentration declines due to a fall in serum K levels and increased excretion in sweat during heat stress. Heat stress has a detrimental effect on animal reproduction partly by disrupting the normal release of gonadotrophin-releasing hormone from the hypothalamus and luteinising hormone (LH) and follicle-stimulating hormone from the anterior pituitary gland. Heat stress reduces the degree of dominance of the selected follicle as reduced steroidogenic capacity of its theca and granulosa cells and a fall in blood oestradiol concentrations. Plasma progesterone levels may be increased or decreased depending on whether the heat stress is acute or chronic and also on the metabolic state of the animal. Insufficient *progesterone* secretion by the corpus luteum during summer is a probable reason of low fertility in cattle and buffalo during summer months in tropical climates.

1   Introduction

Thermal stress impacts physiological, biochemical and productive functions of livestock species. The changes observed during the year in meteorological events are also reflected in production functions. The weather or seasonal changes observed during the year impacts on animal growth, reproduction and lactation. With the increase in average production of dairy cows, the metabolic heat output increases substantially rendering them more susceptible to heat and metabolic stress. These, in turn, alter cooling and housing requirements for livestock. Substantial progress has been made in the last quarter century in delineating the mechanisms by which thermal stress and photoperiod influence productive performance of dairy animals. Acclimation to thermal stress is now identified as a homeorhetic process under endocrine control. The process of acclimation occurs in two phases, that is, acute and chronic, and involves changes in hormone secretion as well as receptor populations in target tissues (Collier et al. 2006a). The time required to complete both phases vary from species to species and may take days to weeks. The opportunity may exist to modify endocrine status of animals and improve their resistance to heat and cold stress. New estimates of genotype x environment interactions support use of molecular and genomics tools to identify the genetic basis of heat stress sensitivity and tolerance. Improved understanding of environmental effects on nutrient requirements has resulted in diets for high merit dairy animals during different weather conditions. Successful cooling strategies for lactating dairy cows are based on maximising available routes of heat exchange, convection, conduction, radiation and evaporation. The endocrine system coordinates metabolic events during thermal stress (Beede and Collier 1986). A summary of endocrine adaptations that occur in mammals and particularly in cattle has been presented in Table 1.

Cows categorised as low, medium and high producers tend to have higher milk temperatures with increasing production (Igono et al. 1988) and concentrations of milk somatotropin decline significantly when THI exceeded 70. The decline probably occurs to reduce metabolic heat production. Reduced concentration of key metabolic hormones observed during heat stress is probably an attempt to reduce metabolic heat production. Scott et al. (1983) reported a negative relationship for plasma thyroxine concentration and rectal temperature, but the initiation of night cooling, at the time when rectal temperature was highest, was most beneficial in maintaining thermoneutral plasma thyroxine concentration, suggesting that strategically cooling the heat-stressed cow could enhance the metabolic potential.

Heat stress is a major contributing factor to the low fertility of dairy cows during summer (Ingraham et al. 1974; Ray et al. 1992; Thompson et al. 1996; Al-Katanani et al. 1999; Khodaei et al. 2006). Effects of heat stress on reproductive hormones and other physiological functions are a direct consequence of the increase in body temperature due to heat stress or of the physiological changes cows undergo to reduce the magnitude of hyperthermia (De Rensis and Scaramuzzi 2003; Khodaei 2003). High ambient temperature adversely affects normal reproduction of cattle (Plasse et al. 1970), swine (Edwards et al. 1968) and sheep (Thwaites 1968), the syndrome being short oestrus, abnormal oestrus cycle, increased proportion of abnormal ova shed, decreased fertilisation rate and increased embryonic and fetal mortality early in gestation (Stott 1972). These physiological manifestations as influenced by environmental heat have been associated with altered endocrine functions (Stott and Wiersma 1971). Hormonal secretion in animals during heat stress including short-term and long-term temperature modification using environmental chambers, seasonal comparisons of hormonal profiles and the use of microclimatic modification during periods of heat stress has been extensively evaluated. Differences in experimental conditions contribute to the disparity in results on hormonal secretions during heat stress. Some of the variations in hormonal responses to heat stress reflect

that ovarian steroid concentrations are dependent not only on rate of secretion from ovarian tissue but also on rate of vascular perfusion of the ovary, on possible adrenal release in case of progesterone, on metabolism in the liver and other organs and on the degree of haemodilution or haemoconcentration (Wise et al. 1988a). The extent to which heat stress affects other physiological characteristics could lead to variable changes in steroid hormone concentrations in peripheral blood. Heat stress can also cause either dilution, concentration or no effect on blood plasma volume (Richards 1985; McGuire et al. 1989; Johnson et al. 1991; Elvinger et al. 1992), and the nature of effect of heat stress on blood volume affects steroid hormone concentrations in blood. Hyperthermia has been shown to decrease ovarian blood flow (Lublin and Wolfenson 1996) and to inhibit angiogenesis (Fajardo et al. 1988). Blood flow and vascular density determine the follicular perfusion rate, which directly influences the rates of nutrient uptake and hormonal release by the follicle.

2 Thyroxine (T$_4$) and Triiodothyronine (T$_3$)

Thyroid hormones are important in an animal's adaptation to a hot environment. The thyroid gland secretes triiodothyronine (T$_3$) and tetraiodothyronine/thyroxine (T$_4$) which provide a major mechanism important for acclimation (Johnson and Van Jonack 1976; Horowitz 2001) and are known indicators of heat stress (Pusta et al. 2003). Both triiodothyronine (T$_3$) and thyroxine (T$_4$) are associated with metabolic homeostasis and are susceptible to climatic changes (Perera et al. 1985). Shade or cooling can, therefore, alter thyroid activity when cattle are exposed to heat stress (Collier et al. 1982a; Aggarwal 2004; Aggarwal and Singh 2009; Table 2). It is well known that heat acclimation decreases endogenous levels of thyroid hormones and that mammals that have adapted to warmer climates follow this pattern (Johnson and Van Jonack 1976; Horowitz 2001). These hormones are the primary determinants of basal metabolic rate and

Table 2 Average thyroxine, triiodothyronine, insulin and cortisol levels in crossbred cows kept with and without mist cooling

| Hormone | Hot–dry season | | | Hot–humid season | | |
	Without cooling	With cooling by mist and fan	Δ change	Without cooling	With cooling by mist and fan	Δ change
Thyroxine (ng/ml)	45.79±0.54	46.26±0.47	+0.47	44.49±0.51	46.21±0.94	+1.72
Triiodothyronine (ng/ml)	1.52±0.05	1.35±0.04	−0.17	1.33±0.33	1.35±0.03	+0.02
Insulin (μU/ml)	9.08±0.40	10.15±0.30	+1.07	6.00±0.29	8.86±0.23	+2.26
Cortisol (ng/ml)	4.17±0.15	2.38±0.07	−1.79	4.35±0.25	2.83±0.04	−1.52

Source: Aggarwal and Singh (2009)

Table 3 Average values of thyroxine, triiodothyronine, insulin and cortisol levels in showering and wallowing groups during hot–dry and hot–humid season

| Parameter | Hot–dry season | | Hot–humid season | |
	Buffaloes under water showers	Wallowing buffaloes	Buffaloes under water showers	Wallowing buffaloes
Thyroxine	50.65±0.50	52.57±0.67	48.25±0.54	50.57±0.61
Triiodothyronine	1.97±0.03	1.88±0.03	1.83±0.04	1.99±0.03
Insulin	8.30±0.26	10.86±0.27	7.86±0.33	9.62±0.30
Cortisol	4.80±0.14	2.60±0.08	4.33±0.16	2.64±0.32

Source: Aggarwal and Singh (2010)

have a positive correlation to weight gain or tissue production (Magdub et al. 1982). The plasma T_4 and T_3 levels decreased under heat stress as compared to thermoneutral conditions (Magdub et al. 1982).

The response of T_4 and T_3 to heat stress is slow and it takes several days to weeks for levels to reach a new steady state (Silanikove 2000). A decline in the plasma concentrations of T_4 from 2.2 to 1.16 ng/ml has been reported by Johnson et al. (1988) whereas a reduced thyroid activity in thermal acclimated cattle has been reported by Gale (1973). This decline in thyroid hormones along with decreased plasma GH level has a synergistic effect to reduce heat production (Yousef and Johnson 1966b).

Thyroid hormones play an important role in growth regulation and are essential for maintenance of the basal metabolic rate (Carlson 1969). Thyroxine (T_4) and triiodothyronine (T_3) are biologically active, and T_3 is several times more active than T_4. Seasonal rhythm occurs in the level of some of these hormones. T_3 level was found to be at maximum in winter, decreased in spring and continued to decline reaching the

lowest value in summer in both buffaloes and Friesians (Kamal and Ibrahim 1969). T_3 was found to take at least 72 h to reach its minimum level after heat exposure (Kamal and Ibrahim 1969). The decline in T_3 in the heat-stressed buffaloes may be responsible for the decline in milk components during the hot July month in Egypt (Habeeb et al. 2000) and the decline ($P<0.01$) in daily body weight gain with elevated temperature (Habeeb et al. 2007). The relatively high total solid (fat, protein and lactose) percentages observed during February can be ascribed to the favourable conditions of the mild climate and abundance of green fodder in winter (Habeeb et al. 2000). In a comparative study in buffaloes kept under water showers and allowed to wallowing, it was found that wallowing buffaloes had higher levels of T_4 (Aggarwal and Singh 2010; Table 3) during hot–dry and hot–humid seasons.

In young and old buffalo calves, acute heat exposure (33–43°C, 40–60 RH%) was observed to decrease plasma T_3 and T_4 levels (Nessim 2004). Plasma concentration of T_3 was found to decrease significantly when ambient temperature rose from 24 to 38°C in the climatic chamber in

Table 4 Effects of environmental heat on T_4 and T_3 in plasma and excretion in milk

	Thermoneutral conditions (17.6°C)	Heat stress conditions (31.2°C)	Δ change
Plasma T_4 (ng/ml)	79.11 ± 2.11	66.07 ± 3.90	13.04
Plasma T_3 (ng/ml)	1.46 ± 0.20	0.62 ± 0.01	0.84
Milk T_4 excretion (μg/day)	30.17 ± 1.40	16.77 ± 1.44	13.40
Milk T_3 excretion (μg/day)	21.00 ± 0.82	9.00 ± 0.79	12.00
Milk yield (kg/day)	19.28 ± 0.42	11.92 ± 1.00	7.36
Rectal temp. (°C)	38.70 ± 0.03	40.20 ± 0.07	1.5

Source: Magdub et al. (1982)

male Friesian calves (Habeeb et al. 2001). Plasma T_3 concentration declined from 151 to 126 ng/dl in Friesian calves which were exposed to direct solar radiation in summer (Yousef et al. 1997). In lactating buffaloes, plasma T_3 concentration was observed to decrease significantly ($P<0.01$) with the increase of ambient temperature from 17.5 to 37.1°C (Habeeb et al. 2000). T_4 exhibited, in general, a similar trend to that of T_3 in buffaloes (Dwaraknath et al. 1984). Thyroxine levels in buffalo calves were 5.5, 5.3 and 5.6 μg/dl at 7, 8 and 9 months of age, respectively (Yousef 1992). Thyroid hormones' changes in response to heat stress are probably some of the attempts to reduce metabolic heat production besides other endocrine and metabolic changes induced (El-Nouty and Hassan 1983; Abdalla et al. 1989, 1991; Johnson et al. 1989). Magdub et al. (1982) reported significant reduction in concentrations of triiodothyronine (T_3) and thyroxine (T_4) in plasma and in milk of lactating cows during heat stress (Table 4).

Blood T_4 and T_3 levels have been negatively correlated with RH, THI, rectal temperature and respiration rate in Friesian heifers during hot–humid and dry winter season (Perez et al. 1997). In the hot–humid season, Perez and Fernandez (1988) classify two groups of animals, with significant differences in T_3 and T_4 levels between the two groups. One of the two groups showed serious impairment of all parameters and was considered non-heat tolerant and the other was less affected and was considered heat tolerant. The results of Chaiyabutr et al. (2008) showed that levels of arterial plasma T_4 were lower in cooled animals than in non-cooled animals at all stages of lactation, but this was not apparent for

arterial plasma T_3 concentrations. It is probable that animals in both groups are able to restore thermal balance without restriction of feed DMI, resulting in unchanged plasma levels of thyroid hormones at all stages of lactation.

A higher level of T_4 due to acute heat stress has previously been observed in lactating cows (Johnson and Van Jonack 1976). The inverse relationship between milk yield and the plasma T_4 level of cooled animals could be due to greater utilisation of plasma T_4 and mammary drain on iodine during high milk production (Johnson and Van Jonack 1976).

3 Cortisol

The hypothalamic–pituitary–adrenal axis gets activated in response to heat or climatic stress that consequently increases plasma glucocorticoid concentrations in blood. Adrenal corticoids, mainly cortisol, elicit physiological adjustments that enable animals to tolerate stress (Christison and Johnson 1972). Blood cortisol levels increase significantly due to increase in ambient temperature in cattle and buffaloes to different levels. Exposure of non-pregnant female buffaloes for 2–3 h to solar radiation at 42.1°C increased plasma cortisol concentration rapidly for 30 min, followed by a gradual fall (Zhengkang et al. 1994). The cooled animals have low plasma cortisol concentration than that of non-cooled animals (Aggarwal 2004; Aggarwal and Singh 2010). The elevated plasma cortisol concentrations of non-cooled animals may reflect the stress due to high temperatures (Chaiyabutr et al. 2008).

Acute versus chronic heat stress – Acute and chronic thermal stress exhibit differing responses on glucocorticoid concentrations and levels are elevated in acute response but not in chronic stress (Collier et al. 1982b). The initial rise in plasma glucocorticoids is due to activation of the adrenocorticotrophin release in the hypothalamus in response to thermoceptors of the skin (Chowers et al. 1966); the subsequent decline to normal even after continuing heat stimulus indicates other response probably a negative glucocorticoid feedback and a decrease in the glucocorticoid binding transcortin (Lindner 1964). The glucocorticoids act as vasodilators to help heat loss and have stimulatory effect on proteolysis and lipolysis, hence providing energy to the animal to help offset the reduction of intake (Cunningham and Klein 2007).

Concentration of cortisol is altered by acute and chronic heat exposure (Christison and Johnson 1972) and by changes in photoperiod (Leining et al. 1980). Acute heat exposure (33–43°C, 40–60 RH%) of young and old buffalo calves has been observed to induce significant increases in plasma cortisol concentration. The values were 408 and 213% in young and old calves, respectively (Nessim 2004). In Friesian calves, plasma cortisol concentration also increased from 11 to 29 ng/ml due to direct solar exposure during heat stress (Yousef et al. 1997) and increased from 3.8 to 6.5 ng/ml when ambient temperature increased from 24 to 38°C in climatic chamber when exposed to heat (Habeeb et al. 2001).

Abilay et al. (1975b) observed that cortisol levels decreased during prolonged heat exposure after a temporary increase at the beginning of heat stress. The decline in cortisol in the heat-stressed lactating cows and buffaloes during hot months may be responsible for decline in milk components. The increase of plasma cortisol level during acute heat stress is attributed to the fact that glucocorticoid hormones have hyperglycaemic action through the gluconeogenesis process, thus enhancing glucose formation in heat-stressed animals. The decline which occurs in chronic heat stress is attributed to the fact that cortisol is thermogenic in animals and, consequently, the reduction of adrenocortical activity

under thermal stress is a thermoregulatory protective mechanism preventing a rise in metabolic heat production in hot environment. This indicates that the role of the adrenal cortex gland in acute and chronic adaptation is not similar and under the adrenal–thyroid axis response to stress (Alvarez and Johnson 1973).

The study by Alvarez and Johnson (1973) reported that glucocorticoids increased by 38% after 1 h and 62% after 2 h of exposure of cattle to hot conditions, reaching a peak of 120% at 4 h, then declined gradually to values not different from normal at 48 h and remained at or below this level for the rest of the exposure duration. In hot conditions, cortisol hormone is considered a thermoregulatory protective mechanism against metabolic heat production under thermal stress (Collier et al. 1982b). Therefore, cortisol plays a role in adaptation to short- and long-term heat stress (Norman and Litwack 1987; Beraidinell et al. 1992). The likely response and contradiction in results observed in the cortisol levels to change in ambient temperature may be due to the differences in animals, their heat tolerance, physiological status, production level, type of production, blood sampling time and duration of exposure to heat stress.

Cortisol and immunity – The studies on the plasma cortisol in response to thermal stress in dairy cows are inconclusive and conflicting. The exposure of cows to high environmental temperature may result in increased glucocorticoid concentrations in the plasma (Wise et al. 1988a; Silanikove 2000). Conversely, other studies reported conflicting results indicating that dairy cattle kept under conditions of high environmental temperature show a reduction in the levels of glucocorticoid secretion (Abilay et al. 1975b; Ronchi et al. 2001). Differences in the experimental conditions have presumably contributed to the disparity in these results. Late pregnancy and nonlactation may be a source of additional stimuli that, in cows exposed to high environmental temperature, activate the hypothalamic–pituitary–adrenal axis causing an increased secretion of cortisol. However, studies in rats also suggested an inhibitory effect of pregnancy on stress-induced immunosuppression and neuroendocrine changes, thereby promoting homeostasis within the neuroendocrine–immune system against stress (Nakamura et al. 1997). The increase in plasma cortisol during summer in calving cows provides at least a partial explanation for the changes in cell-mediated and humoral immunity. Studies in humans and laboratory animals under a large variety of experimental conditions have also demonstrated that intense adrenocortical activity depress T-cell activity and elevate B-cell activity (MacMurray et al. 1983; Kok et al. 1995; Elenkov and Chrousos 1999; Moynihan 2003). Therefore, under stressful conditions, glucocorticoids may suppress cellular immunity and boost humoral immunity via an effect on T helper 1 (Th1)/Th2 cells and type 1/type 2 cytokine production (Elenkov and Chrousos 1999). Webster et al. (2002) have indicated that glucocorticoids induce a shift from a Th1 (cellular) to a Th2 (humoral) pattern of immunity and that this is mainly to downregulate the Th1 cytokines for allowing dominant expression of the Th2 cytokines (Franchimont et al. 1998; Agarwal and Marshall 2001).

4   Insulin

Plasma insulin decreases due to exposure to elevated temperature in cattle. The decreased values were 30, 54 and 33% in Friesian calves (Habeeb 1987), Friesian heifers (Sejrsen et al. 1980) and Holstein cows, respectively (Abdel-Samee et al. 1989). A decrease in insulin level is likely to facilitate decrease in heat production. However, other researchers either did not detect significant difference between the stressed and unstressed Friesian bulls in plasma insulin concentration (McVeign and Tarrent 1982) or reported an increase in insulin levels during exposure to elevated temperature (Chaiyabuter et al. 1987), and the increase in insulin levels during exposure to elevated temperature was correlated with a marked increase in plasma glucose level. Such increase may be related to the action of glucocorticoids.

Nutrition plays a major role in controlling reproductive processes. However, the physiological mechanisms through which nutrition mediates its effects are not well understood. Clearly, changes in the availability of nutrients are perceived by the hypothalamus and influence gonadotrophin secretion via effects on hypothalamic GnRH release (Foster et al. 1989; Ebling et al. 1990). Insulin and leptin are hypothesised to be the adiposity signal for the brain to regulate body weight. Jonsson et al. (1997) suggested that the reduction in dry matter intake of heat-stressed animals results into a negative energy balance that may prolong the postpartum period and decrease fertility in dairy cows. Negative energy balance leads to decreased plasma concentrations of insulin and glucose that affect follicle development. Lower plasma insulin and glucose lead to impaired follicular development and delayed ovulation. Insulin also appears to play a role in regulating synthesis and secretion of leptin. Adipose tissue fragments cultured in the presence of insulin have been observed to increase both synthesis and secretion of leptin (Barr et al. 1997). A positive correlation between insulin and leptin levels has been observed in cows (Block et al. 2001, 2003; Aggarwal et al. 2010). The depression of insulin (Habeeb 1987; Aggarwal 2004), thyroxine (El-Mastry and Habeeb 1989; Aggarwal 2004; Aggarwal and Singh 2009) and cortisol (Kamal et al. 1989; Aggarwal 2004) may contribute to a decrease in milk synthesis and milk yield, and changes in milk composition may be observed.

5   Insulin-Like Growth Hormone 1 (IGF-1)

IGF-1 is synthesised by most of body tissues and may act in an autocrine or paracrine manner (Cohick and Clemmons 1993). The liver is the primary source of circulating IGF-1, which may directly affect target tissues, such as the mammary gland of pregnant and lactating animal to stimulate the synthesis of milk components. Most of the effects of growth hormone are mediated by IGF-1. The concentration of IGF-1 has been found to decrease during summer months (Ingraham et al. 1982; Butler and Smith 1989;

Hamilton et al. 1999). Since plasma levels of IGF-1 are directly related to energy status and IGF-1 is critical to ovarian follicular development (Beam and Butler 1999), it is likely that IGF-1 is associated with poor reproductive performance of cows during hot summer. These effects may be mediated through reduced feed intake as cows in negative energy balance have lower IGF-1 levels (Webb et al. 1999; Bousquet et al. 2004) and an increase in dietary intake improves LH pulse frequency (Spicer et al. 1990) and the diameter of the dominant follicle in heifers (Diskin et al. 1999).

6 Growth Hormone

Growth hormone is a calorigenic hormone which exerts its effects on almost all tissues of the body. The plasma GH levels decline at thermoneutrality in heat-stressed Jersey cows (Mitra et al. 1972). Igono et al. (1988) observed that GH content in milk of low, medium and high production groups declined when THI exceeded 70. Plasma growth hormone reductions that occurred with heat-stressed cows did not occur in thermoneutral conditions for cows fed restricted intakes that were similar to those consumed during heat stress. The decreased GH leads to less calorigenesis aimed in the maintenance of heat in the body (Bauman and Currie 1980). In addition to calorigenesis, GH also enhances heat production by stimulating thyroid activity (Yousef and Johnson 1966a). Therefore, a reduced secretion of GH may help survival of the homeotherm under high ambient temperatures.

7 Aldosterone

Plasma aldosterone concentration in *B. taurus* decreases significantly due to high environmental temperature (El-Nouty et al. 1980; Niles et al. 1980; Aboul-Naga 1987) and prolonged heat exposure reduce mineralocorticoids. The decrease in aldosterone and parathyroid hormone secretion during heat stress in cattle is associated with a rise in body fluids and water turnover rate. In addition, the increase of glucocorticoid hormone

leads to tissue destruction, thereby eliminating minerals from the body in the urine and faeces. The relationship between plasma aldosterone concentration and urine electrolyte excretion during heat stress in bovines has been documented by El-Nouty et al. (1980). Plasma aldosterone levels during the first few hours of prolonged heat exposure decline (40% lower), and thereafter, the decline occurs rapidly. This decline in aldosterone level has been observed due to a fall in serum K levels mainly because of its increased excretion in sweat (El-Nouty et al. 1980). The differences in the ruminants and nonruminants are likely to occur with respect to Na and K loss during thermal stress, as nonruminants produce sweat high in Na and low in K concentrations (Lippsett et al. 1961) contrary to ruminants.

8 Adiponectin and Leptin

The transition dairy cows undergo large metabolic adaptations in glucose, fatty acid and mineral metabolism to support lactation and avoid metabolic dysfunction (Overton and Waldron 2004). But the level of feeding of cows during the last month before calving has been postulated to have only small effects on the cows' metabolic and hormonal status (Roche et al. 2005). Since leptin is thought to play a critical role in regulating energy metabolism in mammals (Block et al. 2003; Aggarwal et al. 2010), nutritional changes induced by heat stress may influence leptin. Adipose tissue performs its regulatory function by secreting biologically active molecules called adipocytokines (Trayhurn 2005). Among these, adiponectin and leptin are expressed almost exclusively in differentiated adipocytes and are crucial for the regulation of energy balance and carbohydrate/lipid metabolism (Havel 2002, 2004). Adiponectin is one of the most important and abundant adipocytokines and have been observed to exert anti-diabetic, anti-atherogenic and anti-inflammatory roles (Goldstein and Scalia 2004; Pittas et al. 2004). However, leptin is a modulator of the appetite and the energetic balance and may function as an endocrine, a paracrine as well as an autocrine factor (Friedman 2002; Hall et al. 2002). The upregulation of leptin expression in adipose cells from heat shock indicates that leptin helps in the thermoregulatory processes to limit body hyperthermia by a central action that is responsible for the decrease in feed intake, energy metabolism and body fat (Houseknecht et al. 1998). Adipocytes have been observed to express leptin receptors indicating that leptin may act directly on adipocytes for regulating energy and metabolism. This indicates that two regulatory systems 'short-loop' leptinergic system and hypothalamic 'long-loop' energy regulatory systems independently exist (Wang et al. 2005). The leptin in severe heat-shocked cells induces adipocyte apoptosis (Ambati et al. 2007). Leptin changes in severe heat-shocked 3T3-L1 adipocytes might indicate a proapoptotic signal of leptin (Bernabucci et al. 2009). The increase of leptin expression may be a consequence of a possible adaptative response to heat shock and downregulation of adiponectin might be a cell heat-shock response that is accompanied by a reduction in protein synthesis to favour induction of heat-shock response (Linquist 1986; Collier et al. 2006b).

Adiponectin has a potent anti-inflammatory effect and leptin exerts a proinflammatory role (Fantuzzi 2005). The upregulation of leptin and downregulation of adiponectin determine the alteration of energy and lipid metabolism observed in cattle exposed to hot environment (Bernabucci et al. 2009; Ronchi et al. 1999). The downregulation of adiponectin and upregulation of leptin are comparable with the regulation of adipokines observed in humans having obesity, atherosclerosis, diabetes type 2 and metabolic syndrome (Kamigaki et al. 2006; Lafontan and Viguerie 2006). Altered adipokine levels also occur in a variety of inflammatory conditions (Fantuzzi 2005). Hsps have role in the inflammatory response and participate in cytokine signal transduction and in the control of cytokine gene expression (Moseley 1998). Therefore, changes of adipokine gene and protein expression and *Hspa2* gene expression observed in heat-shocked adipocytes might be responsible for a proinflammatory response associated to insulin resistance (Ailhaud 2006) and CVD (Fantuzzi 2005) in heat-stressed humans. Studies need to be undertaken to verify whether heat-shock-induced impairment of adipokines are similar to humans and in livestock and affect energy balance in animals during metabolic diseases and heat stress. In vivo studies are necessary in high-producing cows to confirm the association between adipokine expression and heat stress.

Reduced leptin and/or Ob-Rb expression in heat-shocked lymphocytes may represent an adaptive mechanism to high environmental temperatures, which may contribute to explain the immunosuppression that may take place in cows suffering from severe heat stress (Lacetera et al. 2009).

9   Reproductive Hormones

9.1   FSH and LH

Ovarian activity is primarily regulated by gonadotrophins released from pituitary under the influence of gonadotrophin-releasing hormone from the hypothalamus. The release of luteinising hormone (LH) and follicle-stimulating hormone from the anterior pituitary gland is disrupted by heat stress. Therefore, heat stress has a detrimental effect on reproduction partly by disrupting the normal release of these hormones (Dobson et al. 2003). The effect of heat stress on LH concentrations in peripheral blood has been observed to be inconsistent. Gwazdauskas et al. (1981) and Gauthier (1986) reported unchanged concentrations, while Roman-Ponce et al. (1981) reported increased concentrations and some authors reported decreased concentrations (Madan and Johnson 1973; Wise et al. 1988a) in heat-stressed cows. These discrepancies may be associated with differences in sampling frequency, which varied from once a day to once every 3 h and depended on whether heat stress was acute or chronic. Regarding the pattern of LH secretion in heat-stressed cows, Gilad et al. (1993) found lower LH basal concentrations and lower LH amplitude in heat-stressed cows with low plasma oestradiol, and Wise et al. (1988a) found lower LH pulse frequency in the heat-stressed cows

Table 5 Progesterone concentrations in peripheral plasma of dairy cows exposed to various types of heat stress

Type of heat exposure	Response	References
Short term, acute (hot chamber)	Increased	Wilson et al. (1998b), Trout et al. (1998), and Gwazdauskas et al. (1981)
	No change	
Short term, acute (solar radiation)	No change	Roman-Ponce et al. (1981) and Roth et al. (2000)
	Increased	
Long term, chronic (summer heat stress)	Decreased	Howell et al. (1994), Jonsson et al. (1997), Wolfenson et al. (1988, 2002), Younas et al. (1993), and Wise et al. (1988a)
	No change	

compared to the cows under cooling. Conflicting results have been reported regarding the preovulatory LH surge in heat-stressed cows. Madan and Johnson (1973) reported a reduction in the endogenous LH surge caused by heat stress in heifers, and some authors reported that it was unchanged in cows (Gwazdauskas et al. 1981; Rosenberg et al. 1982; Gauthier 1986). Gilad et al. (1993) have suggested that these differences are related to preovulatory oestradiol levels because heat stress had no effect on tonic LH secretion or GnRH-induced LH release in cows with high concentrations of plasma oestradiol and heat stress depressed LH concentrations in cows with low concentrations of plasma oestradiol. Because most studies report that LH levels are decreased by heat stress, it can be concluded that in summer, the dominant follicle develops in a low LH environment and these results in reduced oestradiol secretion from the dominant follicle leading to poor expression of oestrus and low fertility (De Rensis and Scaramuzzi 2003). Plasma inhibin concentrations in summer are low in heat-stressed cows (Wolfenson et al. 1993) and in cyclic buffaloes (Palta et al. 1997), perhaps reflecting reduced folliculogenesis since a significant proportion of plasma inhibin comes from small and medium size follicles.

Gilad et al. (1993) reported low concentrations of FSH in acute and chronic heat-stressed cows which also had lower concentrations of oestradiol while no alterations in concentrations of FSH were observed in cows which had normal concentrations of oestradiol. Conversely, Ronchi et al. (2001) reported no differences in frequency, amplitude of FSH pulses and baseline concentrations of FSH between cows exposed and unexposed to high ambient temperatures. However,

Roth et al. (2000) observed high plasma concentrations of FSH in heat-stressed cows than in cooled cows. Increased concentration of FSH in heat-stressed cows has been attributed to the concentration of inhibin (Roth et al. 2000; De Rensis and Scaramuzzi 2003). Increased concentrations of FSH in heat may be due to decreased plasma inhibin production by compromised follicles as inhibin is an important factor in the regulation of FSH secretion. A negative relationship between plasma FSH and immunoreactive inhibin concentrations has been observed (Findlay 1993; Kaneko et al. 1995, 1997).

9.2   Progesterone

The studies on the effect of heat stress on plasma progesterone concentrations report variable results (Table 5). Wilson et al. (1998a, b) observed that heat stress had no effect on plasma progesterone levels in lactating cows and dairy heifers during the second half of the oestrus cycle exposed to heat in a climatic chamber and luteolysis was delayed. In another study, Roth et al. (2000) observed that plasma progesterone during the oestrus cycle in cows were almost similar in heat-stressed cows and cooled cows and did not differ between groups and during the subsequent cycle. However, Wolfenson et al. (1988) and Wise et al. (1988b) found that plasma progesterone concentrations were decreased in heat-stressed cows. Rosenberg et al. (1982) found that plasma progesterone concentrations measured during the oestrus cycle before the first insemination were higher during winter than the summer in multiparous cows. Jonsson et al. (1997) also reported that plasma progesterone concentrations during the life of the second corpus luteum after calving were lower during summer than the winter and that THI during the first 14 days after calving was negatively correlated with progesterone production. Younas et al. (1993) measured plasma progesterone concentrations in cooled and non-cooled cows during summer and found that plasma progesterone concentrations were lower in non-cooled cows compared to cooled cows. Ronchi et al. (2001) also reported that plasma progesterone concentrations were lower in heat-stressed Holstein heifers as compared to thermoneutral Holstein heifers. In contrast, exposure of heifers to heat stress for two successive cycles has been reported to result in increased plasma progesterone concentrations on day 2–19 of the first cycle and on day 2–8 of the second cycle (Abilay et al. 1975a), and Trout et al. (1998) found higher progesterone concentrations in heat-stressed cows until day 19 of oestrus cycle.

Wise et al. (1988a) found that plasma progesterone in non-cooled cows tended to be higher on day 3–5 of the oestrus cycle when compared with cooled cows. However, plasma progesterone concentrations depend on its rate of production by the corpus luteum, possible adrenal release of progesterone, the degree of haemodilution and haemoconcentration, metabolic clearance rate, hepatic blood flow and feed intake (Vasconcelos et al. 2003). An increase in hepatic blood flow occurs following feed intake (Sangsritavong et al. 2002) and more than 90% of progesterone in hepatic portal blood is metabolised during the first pass through the liver (Parr et al. 1993). Insufficient progesterone secretion by the corpus luteum is a possible cause of low fertility of cows during summer. Studies also report higher progesterone concentrations in summer or lower or similar to that in winter (Wolfenson et al. 1997). These variations among the findings on plasma progesterone concentrations have been attributed to several factors associated with low luteal cell blood perfusion, progesterone metabolism in the liver, blood volume changes, adrenal release of progesterone, level or degrees of hyperthermia, extent and duration of heat exposure and differences in physiology of the animal, nutritional level and stage of lactation of a cow and other factors that affect plasma progesterone (Wolfenson et al. 2000; De Rensis and Scaramuzzi 2003).

However, heat stress may also directly alter progesterone production by the corpus luteum (Wolfenson et al. 1997). Chronic heat stress possibly impairs follicle and corpus luteum, and luteinised theca cells are more susceptible to heat stress than luteinised granulosa cells (Wolfenson et al. 2002). Plasma progesterone levels are significantly higher in winter than in summer (Wolfenson et al. 2002). Low progesterone prior to AI is related to enhanced uterine $PGF_{2\alpha}$ secretion, to alterations in the growth pattern of ovarian follicles and to their steroidogenic capacity (Mann and Lamming 2001; Santos et al. 2004).

Low progesterone concentrations in the circulation of cows have been associated with compromised reproductive function and reduced pregnancy rates (Butler et al. 1996; Lamming et al. 1989; Mann et al. 1995, 2001). Wolfenson et al. (2002) analysed progesterone production in vitro by theca and granulosa cells obtained from cows in cool and hot seasons as well as progesterone concentrations in general circulation. Under chronic summer stress conditions, progesterone production was markedly low and heat stress-induced damage to follicular functions may be carried over to the subsequently formed corpus luteum. Low plasma progesterone concentrations during the luteal phase of the pre-conception oestrus cycle can compromise follicular development leading to abnormal oocyte maturation and early embryonic death (Ahmad et al. 1995). At the time of conception, low progesterone concentrations may lead to the failure of implantation (Mann et al. 1999; Lamming and Royal 2001). The influence on conception occurs most probably due to the need for synchronous development of the embryo and corpus luteum as the delayed or advanced development of the corpus luteum may lead to implantation failure (Lamming and Royal 2001). The pattern of the postovulatory progesterone may alter fertility (Darwash et al. 1999). However, the use of exogenous progesterone post-insemination to supplement endogenous progesterone is likely to support pregnancy and improve chances of establishment (Robinson et al. 1989), but treatment may not be beneficial in all cases (Breuel et al. 1990). An early atresia of bovine follicles is characterised by a decrease in androgen production by thecal cells (McNatty et al. 1984) and early atresia in medium-sized follicles because heat stress effects could be associated with low oestradiol production by granulosa cells and an increased progesterone concentration in the follicular fluid of heat-stressed cows (Roth et al. 2000).

9.3   Oestrogen Concentrations

The effect of heat stress on oestradiol in peripheral blood has been inconclusive similar to progesterone. The blood oestradiol concentrations remain unaffected (Roman-Ponce et al. 1981) or increased (Rosenberg et al. 1982) and may decline (Gwazdauskas et al. 1981). An effect that is consistent is a decreased concentration of luteinising hormone (LH) and reduced dominance of the selected follicle (Rosenberg et al. 1982). Lactating dairy cows housed in a climate-controlled chamber at the ambient temperature of 29°C and 60% relative humidity and dairy heifers housed at ambient temperature of 33°C and 60% relative humidity had a low concentration of oestradiol in plasma between days 11 and 21 of oestrus cycle and were associated with small follicular size (Wilson et al. 1998a, b). Likewise, Wolfenson et al. (1995) also observed that oestradiol concentrations were lower in heat-stressed cows during days 4 and 8 of the first follicular wave than in non-heat-stressed cows. Pronounced reduction in concentrations of plasma oestradiol in cows during summer and experience of exposure to heat stress was observed to influence levels (Badinga et al. 1993). Exposure to heat stress for longer periods may severely impair follicular function resulting in a reduction in oestradiol production. A lower oestradiol concentration in the follicular fluid of dominant follicles on day 8 of the oestrus cycle in late summer compared to early summer has been observed (Badinga et al.

1993). Wolfenson et al. (1997) reported that the oestradiol concentrations in follicular fluid were low in autumn and summer and high in winter. In contrast, Badinga et al. (1994) reported highest concentration of oestradiol in the hottest month compared to the cooler months in Florida in a field study.

Oestradiol concentration in the follicular fluid and androstenedione production by thecal cells have been found to be both lower in dominant follicles collected in autumn than in those collected in winter (Wolfenson et al. 1997). Hyperthermia decrease ovarian blood flow (Lublin and Wolfenson 1996) and inhibit angiogenesis (Fajardo et al. 1988). Cardiac output redistribution during hyperthermia in favour of skin reduces blood flow and follicular perfusion rate, which directly influences the rates of nutrient uptake and hormonal release by the follicle.

Exposure of cows to heat stress results in impaired steroidogenesis 20 and 26 days later, in medium-sized and preovulatory follicles, respectively (Roth et al. 2001). The delayed effect is expressed in different ways in granulosa and thecal cells. Granulosa cells produce low oestradiol in medium-sized follicles and affect viability in the preovulatory follicles. The steroid production in thecal cells is susceptible to heat stress and androgen production in follicle is affected. Therefore, heat stress influences follicular steroidogenic capacity, follicular dynamics, oocyte quality and embryo development (Roth et al. 1999, 2000) that may lead to the low fertility of cows during summer.

During follicular development, the dominant follicles acquire high oestrogenic activity compared with other follicles. Follicular oestrogenic activity depends on the capacity for oestrogen production of the granulosa cells (Fortune 1994). Oestrogen production is stimulated by the binding of follicle-stimulating hormone (FSH) to its receptors on the granulosa cell membrane, activating the aromatase enzyme that converts testosterone to oestradiol (Dorrington et al. 1975; Erickson and Hsueh 1978). The synthesised oestradiol, in turn, contributes to the maintenance and expression of receptors for luteinising hormone

(LH) and FSH (LH-R and FSH-R, respectively) on granulosa cells (Hsueh et al. 1994; Bodensteiner et al. 1995). Heat stress suppresses aromatase activity in granulosa cells and decreases the oestradiol concentrations in the follicular fluid and plasma of dairy cows (Badinga et al. 1993). In addition, in both lactating cows and heifers, heat stress decreased serum oestradiol concentration between days 11 and 21 of the oestrus cycle (Wilson et al. 1998a, b). Seasonal and acute heat stresses were observed to affect steroid production in the dominant follicles of cows (Wolfenson et al. 1997). Thus, heat stress affects the viability of granulosa cells and the follicular oestrogenic activity. Follicular cells in atretic follicles are normally eliminated by apoptosis (Tilly 1993; Palumbo and Yeh 1994), as can be observed in granulosa cells during follicular development (Tilly 1996).

Apoptosis of granulosa cells due to heat stress
Apoptosis of granulosa cells is provoked by the lack of survival factors such as FSH and oestradiol (Billing et al. 1993; Chun et al. 1996). Expression of genes encoding apoptosis-regulating proteins in the follicle has been reported. Two members of the *bcl-2* family, *bcl-2* and *bax*, have been shown to regulate apoptosis in the ovary (Tilly et al. 1995). In the rat ovary, Bax appears to antagonise the apoptosis-suppressive effects of Bcl-2, and the fate of the cell appears to be decided by the balance between these two regulatory proteins (Tilly et al. 1995). The ability of gonadotrophins to prevent apoptosis and atresia in ovarian follicles may be linked to a shift in the ratio of bcl-2 to bax gene expression (Tilly et al. 1995). FSH receptors are confined to the granulosa cells of healthy, developing follicles (Camp et al. 1991). Oestradiol contributes to the maintenance and expression of both LH-R and FSH-R in granulosa cells (Hsueh et al. 1994; Bodensteiner et al. 1995). FSH-R expression in the granulosa cells is also regulated by oestradiol-independent mechanisms (Shimizu et al. 2005).

The PMSG administration causes a marked increase of FSH-R mRNA expression and FSH-binding sites (Nakamura et al. 1991; LaPolt et al. 1992). Locally produced intraovarian growth factors, like transforming growth factor-beta

(TGF-ß) and insulin-like growth factor-1 (IGF-1), induce FSH-R expression in granulosa cells (Dunkel et al. 1994; Zhou et al. 1997). Heat stress suppresses FSH-R expression in granulosa cells of follicles at the early antral, antral and preovulatory stages after PMSG treatment. The increase in atresia caused by heat stress (Shimizu et al. 2000) may be induced by the inhibition of FSH-R expression in granulosa cells.

One key group of intracellular factors regulating apoptosis is the Bcl-2 family of proteins (Adams and Cory 1998). The members of this family can be subdivided into antiapoptotic proteins (such as Bcl-2 and Bcl-xL) and proapoptotic proteins (such as Bax and BAD). Anti- and proapoptotic proteins regulate cell death by binding to each other and forming heterodimers (Oltvai et al. 1993; Yang et al. 1995). Therefore, a delicate balance between anti- and proapoptotic Bcl-2 family members exists in each cell, and the relative concentrations of these two groups of proteins determine cell death or survival. The levels of *bcl-2* and *bax* mRNA and the ratio of *bax* to *bcl-2* mRNA are the deciding factor for apoptotic signal. However, the expression of *bax* mRNA in granulosa cells of follicles from heat-stressed animals has been found to increase significantly after the 12-h culture period suggesting that the amount of *bax* present in granulosa cells that maintain a constitutive level of *bcl-2* expression may be linked to the induction of granulosa cell apoptosis and follicular atresia caused by heat stress. Another apoptotic signalling cascade, the Fas/Fas ligand (FasL) system, can also lead to granulosa cell apoptosis (Kim et al. 1998). Thus, it is likely that heat stress stimulates the Fas/FasL cascade in the granulosa cells of growing follicles. Heat stress inhibits the FSH-R signalling pathway in the granulosa cells of growing follicles that decrease oestrogen levels and induce follicular atresia. Thus, the regulators of FSH-R expression may be potential targets for therapeutic intervention for improving summer fertility (Shimizu et al. 2005).

The mechanisms by which heat stress alters the reproduction in susceptible mammals are through concentrations of circulating reproductive hormones through adrenal–gonadal axis and involve adrenocorticotrophic hormone (ACTH). Heat stress-induced cortisol secretion (Roman-Ponce et al. 1981; Wise et al. 1988a; Elvinger et al. 1992) may block oestradiol-induced sexual behaviour (Hein and Allrich 1992). Increased corticosteroid secretion (Roman-Ponce et al. 1981) can inhibit GnRH and thus LH secretion (Gilad et al. 1993). Heat stress inhibited the secretion of gonadotrophins in cows with low plasma concentrations of oestradiol compared to those with high concentrations (Gilad et al. 1993). However, the high concentrations of oestradiol can counteract the effect of heat stress. Therefore, the response of neuroendocrine mechanism controlling gonadotrophin secretion is different and more sensitive to heat stress when concentrations of plasma oestradiol are low in cows. The heat stress may also act directly on the ovary to decrease its sensitivity to gonadotrophin stimulation (Wolfenson et al. 1997). Also the somatic cells within the follicles (theca and granulosa cells) may also be damaged by heat stress. In terms of steroid production, the thecal cells and granulosa cells were observed to be susceptible to heat stress (Roth et al. 2001). Any alteration in the secretory activity of the follicle and the corpus luteum induced by heat stress would be important factors in summer infertility in vulnerable cows particularly high producing.

9.4 Gonadotrophins and Corticosteroids

Heat stress affects the hypothalamic–pituitary–adrenal axis and the sympathoadrenal system (Tilbrook et al. 2000) to initiate and modulate most of the activities. Stimulation of the hypothalamic–pituitary–adrenal axis is characterised by activation of corticotrophin-releasing factor (CRF) and arginine vasopressin (AVP) neurones in the paraventricular nucleus and secretion of these neuropeptides into the hypophyseal portal system to stimulate the corticotrophs of the anterior pituitary gland (Tilbrook et al. 2000). The corticotrophs produce a variety of peptides derived from pro-opiomelanocortin, including ACTH, endorphin and melanocyte-

stimulating hormone, all of which are released in response to heat stress (Engler et al. 1989). In terms of the response to heat stress, ACTH acts on the adrenal cortex to stimulate the synthesis and secretion of glucocorticoids like cortisol. The higher concentrations of cortisol in heat stressed cows influence LH release and cortisol has been implicated to inhibit LH in the bovine species (Gangwar et al. 1965). Various studies have shown that administration of natural or synthetic glucocorticoids can inhibit the secretion of the gonadotrophins in sheep (Juniewicz et al. 1987) and in dairy cattle (Thibier and Rolland 1976). Nevertheless, increased secretion of glucocorticoids is not always associated with decreased secretion of the gonadotrophins, particularly in cases of acute stress. Suppression of reproduction is more likely under conditions of chronic stress and may involve action at the hypothalamus or pituitary. Furthermore, there may be species differences in the extent to which glucocorticoids inhibit the secretion of LH and FSH (Tilbrook et al. 2000). Under prolonged heat stress conditions, secretion of the gonadotrophins may be suppressed and reproduction may also be inhibited.

9.5 Prostaglandin

Heat stress compromises uterine environment with decreased blood flow to the uterus and increased uterine temperature which may lead to implantation failure and embryonic mortality. These effects are associated with the production of heat-shock proteins by the endometrium during heat stress and reduced production of interferon-tau by the conceptus. Moreover, heat stress can affect endometrial prostaglandin secretion and activate luteolysis leading to premature luteolysis and embryo loss. Heat shock of 42 and 43°C have been observed to increase output of prostaglandins by cultured endometrium collected at day 17 of the oestrus cycle (Putney et al. 1988; Malayer and Hansen 1990). However, heat stress on day 17 of pregnancy increases uterine production in response to oxytocin (Wolfenson et al. 1993).

9.6 Androstenedione

Thecal cells at 40.5°C reduce androstenedione production but generally had no effect on oestradiol-17 production from cultured granulosa cells (Wolfenson et al. 1997), but the mechanism by which heat stress induces a decrease in androstenedione production in thecal cells is not clear. Analyses of mRNA content for LH receptor in thecal cells obtained from preovulatory follicles did not provide any evidence for alterations of mRNA content related to previous heat exposure (Roth et al. 2000). However, the significant decrease of LH-stimulated androstenedione production by thecal cells may indicate that heat exposure induces impairment of LH receptor function.

10 Catecholamines

The epinephrine and norepinephrine are predominant catecholamine hormones involved in stress response. They consist of a 6-carbon ring with a carbon side chain. The type of side chain determines the type of catecholamine and provides biological specificity. The catecholamines bind to specific membrane-bound G-protein receptors that initiate an intracellular cAMP signalling pathway to rapidly activate cellular responses. The speed at which these responses are activated provides the foundation for many of the catecholamine effects. The responses mediated by epinephrine and norepinephrine are commonly called the 'fight-or-flight response' because they have immediate effects on increasing the readiness and activity of the animal. Upon detection of a stressor, epinephrine and norepinephrine are released by both the adrenal medulla and nerve terminals of the sympathetic nervous system. These hormones are synthesized beforehand and stored in secretory vesicles and release occurs rapidly in response to a stressor. Epinephrine and norepinephrine activate organism-level responses within seconds of a stressor response. These catecholamines activate a number of responses in different organs, including decreasing visceral activity and inhibiting digestion, increasing visual acuity, increasing brain blood flow and arousal, increasing gas exchange efficiency in the lungs, breaking down glycogen to release glucose stores, inducing vasodilation in muscles, inducing vasoconstriction in the periphery, increasing heart rate and inducing piloerection. The classic fight-or-flight response is designed to help the animal survive an acute threat such as an attack by a predator or conspecific competitor. Thus, catecholamines not only activate beneficial responses such as increasing alertness and providing energy to muscles but also inhibit processes, such as digestion, during an acute emergency (McEwen and Goodman 2001; Nelson 2005; Norman and Litwack 1987; Norris 2007; Sapolsky et al. 2000).

The concentration of catecholamines is elevated during both acute and chronic thermal stress. Alvarez and Johnson (1973) have reported an average increase of 45 and 42% in short-term and 91 and 70% in long-term heat exposures for epinephrine and norepinephrine, respectively. Allen and Bligh (1969) have reported that catecholamines activate sweat glands of cattle and are involved in regulating sweat gland activity. Plasma epinephrine levels varied between 285 and 1,575 ng/ml and norepinephrine levels ranged from 1,500 to 2,525 ng/ml during summer in crossbred cattle (Aggarwal et al. 2005).

11 Prolactin

Circulating prolactin levels are observed to increase during thermal stress in mammals (Collier et al. 1982a; Ronchi et al. 2001; Roy and Prakash 2007). Reduced nutrient intake decreases circulating prolactin concentrations in ruminants (Bocquier et al. 1998), and direct effect of heat stress on serum prolactin levels has also been observed (Ronchi et al. 2001). Increase in prolactin in lactating dairy cows, besides prolactin's well-known role in maintaining galactopoiesis in some mammalian species and lactogenesis in ruminants, may play an important role in acclimation through improved insensible heat loss and sweat gland function (Beede and Collier 1986). Bromocriptine, a prolactin inhibitor, affects sweat gland function by preventing increased sweat gland discharge (Kaufman et al. 1988). Prolactin levels also differ with season and it may be involved in acclimation (Leining et al. 1979).

12   Conclusions

The biological mechanism by which heat stress impacts animal production and reproduction is both direct and indirect. The decrease in feed intake and also because of an altered endocrine status, reduction in nutrient absorption and increased maintenance requirements result in a net decrease in nutrient/energy availability. Heat stress acclimation is accomplished by changes in homeostatic responses and may include homeorhetic processes involving an altered endocrine status that ultimately affects target tissue responsiveness to environmental stimuli. Stress hormones are also implicated in the acclimatory response to heat stress and they primarily include thyroid hormones, prolactin, growth hormone, glucocorticoids and mineralocorticoids. The thyroid hormones, T_4 and T_3, provide a major mechanism important for acclimation and have received considerable research attention in animal production system. It is established that heat acclimation decreases endogenous levels of thyroid hormones to reduce endogenous heat production. The decrease in thyroid hormones during heat stress is to facilitate decrease in basal metabolic rate and muscle activity by decreasing heat production. Under heat stress, cortisol hormone is another protective mechanism preventing metabolic heat production. In other words, it is thermogenic and consequently reduces adrenocortical activity, under thermal stress. Therefore, cortisol is involved in adaptation to short- and long-term heat stress. Plasma levels of LH and oestradiol are reduced during summer in cows affecting fertility. Heat stress may affect the secretion of the gonadotrophins through mechanisms that modify the synthesis or the secretion of GnRH, the responsiveness of the gonadotrophs to the actions of GnRH or the feedback actions of gonadal hormones. Prolonged or chronic heat stress results in suppressed gonadotrophin secretion and inhibition of reproduction but, when the duration of the stress response is transient or acute, the effects are less clear.

Reduced leptin and/or Ob-Rb expression in heat-shocked lymphocytes may represent an adaptive mechanism to high environmental temperatures, which may limitedly explain the immunosuppression mechanisms observed in cows during summer. Therefore, a greater understanding is required to understand mechanism associated with immune suppression and hormonal changes during high milk production in heat stressed cows.

Appendices

Stressor: A threatening or unpredictable stimulus that causes a stress response.

Stress response: The physiological, hormonal and behavioural changes that result from exposure to a stressor.

Chronic stress: A state that an organism enters when repetitive or long-term exposure to a stressor has exceeded an organism's regulatory capacities.

Context of a stressor: The physical and psychological conditions present when a stressor appears.

Acclimation: After repeated or chronic exposure to a single stressor, an animal no longer perceives the stressor to be threatening and reduces its physiological stress response. The decrease in stress response is specific to that stressor and does not generalise to other stressors as long as the animal is capable of distinguishing between them.

Sensitisation: When acclimation to one stressor increases subsequent stress responses to novel stressors.

'Stress hormones': A generic and non-scientific term for hormones whose concentrations change in response to stressors and are indicative of a stress response. They are divided in two main types: catecholamines (e.g. epinephrine/adrenaline, norepinephrine/noradrenaline) and glucocorticoid–steroid hormones (e.g. cortisol, corticosterone). Some hormones (e.g. cortisol) have been traditionally used as indicative of stress. However, they may exhaust under repetitive stimuli and may not reflect chronic stress.

Steroid hormones: A class of hormones (including testosterone, oestradiol and cortisol) typified by a four-ring structure.

(Romero 2004)

Hormone	Increase/decrease	Effect	References
Prolactin	Increase	Elevated PRL is involved in meeting increased water and electrolyte demands of heat-stressed cows	Collier et al. (1982b) and Roy and Prakash (2007)
Growth hormone	Decrease	The decreased GH leads to less calorigenesis aimed in maintenance of heat in the body (Bauman and Currie 1980). In addition to calorigenesis, GH also enhances heat production by stimulating thyroid activity (Yousef and Johnson 1966a). Therefore, a reduced secretion of this hormone is all the more necessary for survival of the homeotherm in high ambient temperatures	Igono et al. (1988)
Thyroxine	Decrease	This decline in thyroid hormones along with decreased plasma GH level has a synergistic effect to reduce heat production (Yousef and Johnson 1966b)	Johnson et al.(1988), Aggarwal (2004), Aggarwal and Singh (2009), and Aggarwal and Singh (2010)
Glucocorticoid	Increase and then decrease	The initial rise in plasma glucocorticoids is due to activation of the adrenocorticotrophin (ACTH)-releasing mechanism in the hypothalamus by thermoceptors of the skin (Chowers et al. 1966), whereas the later decline to normal, in spite of continuing heat stimulus, indicates a negative glucocorticoid feedback and a decrease in the glucocorticoid-binding transcortin (Lindner 1964). The glucocorticoids work as vasodilators to help heat loss and have stimulatory effect on proteolysis and lipolysis, hence providing energy to the animal to help offset the reduction of intake (Cunningham, and Klein 2007)	Alvarez and Johnson (1973) and Aggarwal (2004)
Aldosterone	With prolonged exposure, it was 40% lower and declined rapidly during later hours of exposure	This decline in aldosterone concentration is due to a fall in serum K levels because of its increased excretion in sweat and is explained on the basis of a major difference between ruminants and nonruminants with respect to location of Na and K loss during thermal stress. Nonruminants produce sweat high in Na and low in K concentrations (Lippsett et al. 1961), but this is vice versa for ruminants	El-Nouty et al. (1980)
Catecholamines	Increase during both acute and chronic thermal stress	Catecholamines activate sweat glands of cattle and are involved in regulating sweat gland activity	Alvarez and Johnson (1973), Allen and Bligh (1969), and Aggarwal et al. (2005)
ADH	Increase	To conserve water and increase water intake (El-Nouty et al. 1980)	Cunningham and Klein (2007)
Progesterone	Increase/decrease	Low progesterone prior to AI is related to enhanced uterine $PGF_{2\alpha}$ secretion, to alterations in the growth pattern of ovarian follicles and to their steroidogenic capacity (Mann and Lamming 2001; Santos et al. 2004)	Wolfenson et al. (1988) Wise et al. (1988b), Younas et al. (1993), and Ronchi et al. (2001)
Triiodothyronine	Decrease	Decrease in thyroid hormones during heat stress is due to the decrease in basal metabolic rate and muscle activity in order to decrease heat production	Johnson et al. (1988), Pusta et al. (2003), and Aggarwal 2004

Effect of shade, no shade and shade + spray with fan on plasma hormones in cows during summer

Hormone	Shade	No shade	Shade + spray + fan	References
Plasma growth hormone (ng/ml)	4.6 ± 0.21	–	5.7 ± 0.31	Igono et al. (1987)
Plasma prolactin (ng/ml)	59.9 ± 1.79	–	44.9 ± 0.94	Igono et al. (1987)
Progesterone (ng/ml)	1.56	1.91	–	Roman-Ponce et al. (1981)
Oestradiol (pg/ml)	7.75	8.50	–	Roman-Ponce et al. (1981)
LH (ng/ml)	3.87	5.45	–	Roman-Ponce et al. (1981)
Corticoids (ng/ml)	8.72	13.04	–	Roman-Ponce et al. (1981)

Serum concentrations of oestradiol and cortisol during the oestrus cycle of cooled and control cows

	Oestradiol			Cortisol		
Days of oestrus cycle	Without cooling	With cooling	Δ change	Without cooling	With cooling	Δ change
Day 3	9.1 ± 2.4	7.9 ± 1.9	−1.2	10.6 ± 1.0	5.9 ± 1.3	−4.7
Day 5	5.8 ± 1.7	6.4 ± 1.2	+0.6	6.1 ± 0.7	3.8 ± 0.4	−2.3
Day 10	8.8 ± 0.5	4.3 ± 0.7	−4.5	6.2 ± 0.9	3.0 ± 0.5	−3.2
Day 12	6.9 ± 1.4	4.7 ± 1.5	−2.2	4.7 ± 0.5	1.9 ± 0.3	−2.8
Day 14	6.7 ± 1.1	7.7 ± 1.0	+1.1	11.8 ± 1.7	7.5 ± 1.1	−4.3
Day 17	11.0 ± 1.0	13.4 ± 0.4	+2.4	15.6 ± 1.6	6.6 ± 0.6	−9.0

Source: Wise et al. (1988b)

References

Abdalla EB, Kotby EA, DJohnson H (1989) Environmental heat effects on metabolic and hormonal functions and milk yield of lactating ewes. J Dairy Sci 72:471 (Abstract)

Abdalla EB, Johnson HD, Kotby EA (1991) Hormonal adjustments during heat exposure in pregnant and lactating ewes. J Dairy Sci 74:145 (Abstract)

Abdel-Samee AM, Habeeb AAM, Kamal TH, Abdel-Razik MA (1989) The role of urea and mineral mixture supplementation in improving productivity of heat-stressed Friesian calves in the sub-tropics. In: Proceedings of 3rd Egyptian–British conference on animal, fish and poultry production, Alexandria University, Alexandria, vol 2, pp 637–641

Abilay TA, Johnson HD, Madan M (1975a) Influence of environmental heat on peripheral plasma progesterone and cortisol during the bovine estrous cycle. J Dairy Sci 58:1836–1840

Abilay TA, Mitra R, Johnson HD (1975b) Plasma cortisol and total progesterone levels in Holstein steers during acute exposure to high environmental temperature (42°C) conditions. J Anim Sci 41:113–118

Aboul-Naga AI (1987) The role of aldosterone in improving productivity of heat stressed farm animals with different technique. PhD thesis, Faculty of Agriculture, Zagazig University, Zagazig

Adams JM, Cory S (1998) The Bcl-2 protein family: arbiters of cell survival. Science 281:1322–1326

Agarwal SK, Marshall GD Jr (2001) Dexamethasone promotes type 2 cytokine production primarily through inhibition of type 1 cytokines. J Interferon Cytokine Res 21:147–155

Aggarwal A (2004) Effect of environment on hormones, blood metabolites, milk production and composition under two sets of management in cows and buffaloes. PhD thesis submitted to National Dairy Research Institute, Karnal (Haryana), India

Aggarwal A, Singh M (2009) Changes in hormonal levels during early lactation in summer calving cows kept under mist cooling system. Indian J Anim Nutr 26:337–340

Aggarwal A, Singh M (2010) Hormonal changes in heat stressed Murrah buffaloes under two different cooling systems. Buffalo Bull 29:1–6

Aggarwal A, Upadhyay RC, Singh SV, Kumar P (2005) Adrenal-thyroid pineal interaction and effect of exogenous melatonin during summer in crossbred cattle. Indian J Anim Sci 75:915–921

Aggarwal A, Ashutosh, Kaur H, Mani V (2010) Effect of vitamin E supplementation on leptin and insulin hormones in crossbred and indigenous cows. Annual report. National Dairy Research Institute, Karnal, Haryana

Ahmad N, Schrick FN, Butcher RL, Inskeep EK (1995) Effect of persistent follicles on early embryonic losses in beef cows. Biol Reprod 52:1129–1135

Ailhaud G (2006) Adipose tissue as a secretory organ: from adipogenesis to the metabolic syndrome. Comptes Rendus Biologies 329:570–577

Al-Katanani YM, Webb DW, Hansen PJ (1999) Factors affecting seasonal variation in non-return rate to first service in lactating Holstein cow in a hot climate. J Dairy Sci 82:2611–2616

Allen TE, Bligh J (1969) A comparative study of the temporal patterns of cutaneous water vapor loss from some domesticated mammals with epithelial sweat glands. Comp Biochem Physiol 31:347

Alvarez MB, Johnson HD (1973) Environmental heat exposure on cattle plasma catecholamine and glucocorticoids. J Dairy Sci 56:189–194

Ambati S, Kim HK, Yang JY, Lin J, Della-Fera MA, Baile CA (2007) Effects of leptin on apoptosis and adipogenesis in 3T3-L1 adipocytes. Biochem Pharmacol 73:378–384

Badinga L, Thatcher WW, Diaz T, Drost M, Wolfenson D (1993) Effect of environmental heat stress on follicular development and steroidogenesis in lactating Holstein cows. Theriogenology 39:797–810

Badinga L, Thatcher WW, Wilcox CJ, Morris G, Entwistle K, Wolfenson D (1994) Effect of season on follicular dynamics and plasma concentrations of estradiol-17ß, progesterone and luteinizing hormone in lactating Holstein cows. Theriogenology 42:1263–1274

Barr VA, Malide D, Zarnowski MJ, Taylor SI, Cushman SW (1997) Insulin stimulates both leptin secretion and production by rat white adipose tissue. Endocrinology 138:4463–4472

Bauman DE, Currie WB (1980) Partitioning of nutrients during pregnancy and lactation. A review of mechanisms involving homeostasis and homeorhesis. J Dairy Sci 63:1514–1545

Beam SW, Butler WR (1999) Effects of energy balance on follicular development and first ovulation in postpartum dairy cows. J Reprod Fertil 54:411–424

Beede DK, Collier RJ (1986) Potential nutritional strategies for intensively managed cattle during thermal stress. J Anim Sci 62:543–554

Beraidinell LJG, Godfrey RW, Adair R, Lunstra DD, Byerley DJ, Gardenas H, Randel RD (1992) Cortisol and prolactin concentrations during different seasons in relocated Brahma and Hereford bulls. Theriogenology 37:641–654

Bernabucci U, Lacetera N, Basirico L, Ronchi B, Morera P, Seren E, Nardone A (2006) Hot season and BCS affect leptin secretion of periparturient dairy cows. J Dairy Sci 89:348–349

Bernabucci U, Basirico L, Morera P, Lacetera N, Ronchi B, Nardone A (2009) Heat shock modulates adipokines expression in 3T3-L1 adipocytes. J Mol Endocrinol 42:139–147

Bernabucci U, Lacetera N, Baumgard LH, Rhoads RP, Ronchi B, Nardone A (2010) Metabolic and hormonal acclimation to heat stress in domesticated ruminants. Animal 4:1167–1183 & The Animal Consortium, doi:10.1017/S175173111000090X

Billing H, Furuta I, Hsueh AJW (1993) Estrogens inhibit and androgens enhance ovarian granulosa cell apoptosis. Endocrinology 133:2204–2212

Block SS, Butler WR, Ehrhardt RA, Bell AW, Van Amburgh ME, Boisclair YR (2001) Decreased concentration of plasma leptin in periparturient dairy cows is caused by negative energy balance. J Endocrinol 171:339–348

Block SS, Smith JM, Ehrhardt RA, Diaz MC, Rhoads RP, Van Amburgh ME, Boisclair YR (2003) Nutritional and developmental regulation of plasma leptin in dairy cattle. J Dairy Sci 86:3206–3214

Bocquier F, Bonnet M, Faulconnier Y, Guerre-Millo M, Martin P, Chilliard Y (1998) Effects of photoperiod and feeding level on adipose tissue metabolic activity and leptin synthesis in the ovariectomized ewe. Reprod Nutr Dev 38:489–498

Bodensteiner KJ, Wiltbanl MC, Bergfelt DR, Ginther OJ (1995) Alternations in follicular estradiol and gonadotropin receptors during development of bovine antral follicles. Theriogenology 44:499–512

Bousquet D, Bouchard E, DuTremblay D (2004) Decreasing fertility in dairy cows: myth or reality? In: Proceedings 23rd World Buiatrics Congress, Que'bec, 11–16 July

Breuel KF, Spitzer JC, Thompson CE, Breuel J (1990) First-service pregnancy rate in beef heifers as influenced by human chorionic gonadotrophin administration before and/or after breeding. Theriogenology 34:139–145

Butler WR, Smith RD (1989) Interrelationship between energy balance on postpartum reproductive function in dairy cattle. J Dairy Sci 7:767–783

Butler WR, Calaman JJ, Beam SW (1996) Plasma and milk urea nitrogen in relation to pregnancy rate in lactating dairy cattle. J Anim Sci 74:858–865

Camp TA, Rahal JO, Mayo KE (1991) Cellular localization and hormonal regulation of follicle-stimulating hormone and luteinizing hormone receptor messenger RNAs in the rat ovary. Mol Endocrinol 5:1405–1417

Carlson JR (1969) Growth regulators. In: Hafez ESE, Dyer IA (eds) Animal growth and nutrition. Lea and Febiger, Philadelphia, pp 138–155

Chaiyabuter N, Buranakarl C, Muangcharoen V, Loypetjra P, Pichaicharnarong A (1987) Effects of acute heat stress on changes in the rate of liquid flow from the rumen and turnover of body water of swamp buffalo (Bubalus bubalis). J Agric Sci (Camb) 108:549–553

Chaiyabutr N, Chanpongsang S, Suadsong S (2008) Effects of evaporative cooling on the regulation of body water and milk production in crossbred Holstein cattle in a tropical environment. Int J Biometeorol 52:575–585. doi:10.1007/s00484-008-0151-x

Chowers I, Hammel HT, Eisenman J, Abrams RM, McCann SM (1966) Comparison of effect of environmental and preoptic heating and pyrogen on plasma cortisol. Am J Physiol 210:606

Christison GI, Johnson HD (1972) Cortisol turnover in heat stressed cows. J Anim Sci 35:1005–1010

Chun SY, Eisenhauer KM, Minami S, Billig H, Perlas E, Hsueh AJ (1996) Hormonal regulation of apoptosis in early antral follicles: follicle-stimulating hormone as a major survival factor. Endocrinology 137:1447–1456

Cohick WS, Clemmons DR (1993) The insulin like growth factor. Annu Rev Physiol 55:131–153

Collier RJ, Beede DK, Thatcher WW, Israel LA, Wilcox CJ (1982a) Influences of environment and its modification on dairy animal health and production. J Dairy Sci 65:2213–2227

Collier RJ, Doelger SG, Head HH, Thatcher WW, Wilcox CJ (1982b) Effects of heat stress on maternal hormone concentrations, calf birth weight, and postpartum milk yield of Holstein cows. J Anim Sci 54:309–319

Collier RJ, Dahl GE, VanBaale MJ (2006a) Major advances associated with environmental effects on dairy cattle. J Dairy Sci 89:1244–1253

Collier RJ, Stiening CM, Pollard BC, VanBaale MJ, Baumgard LH, Gentry PC, Coussens PM (2006b) Use of gene expression microarrays for evaluating environmental stress tolerance at the cellular level in cattle. J Anim Sci 84:1–13

Cunningham JG, Klein BG (2007) Veterinary physiology, 4th edn. Saunders Elsevier, Missouri

Darwash AO, Lamming GE, Woolliams JA (1999) The potential for identifying heritable endocrine parameters associated with fertility in post-partum dairy cows. Anim Reprod Sci 68:333–347

De Rensis F, Scaramuzzi RJ (2003) Heat stress and seasonal effects on reproduction in dairy cow-a review. Theriogenology 60:1139–1151

Diskin MG, Stagg K, Mackey DR, Roche JF, Sreenan JM (1999) Nutrition and oestrus and ovarian cycles in cattle, http://www.teagasc.ie/research/reports/beef/4009/eopr-4009.pdf. ISBN No. 1841700894

Dobson H, Ghuman S, Prabhakar S, Smith R (2003) A conceptual model of the influence of stress on female reproduction. Reproduction 125:151–163

Dorrington JH, Moon YS, Armstrong DT (1975) Estradiol-17b biosynthesis in cultured granulosa cells from hypophysectomized immature rats: stimulation by follicle-stimulating hormone. Endocrinology 97:1328–1331

Dunkel L, Tilly JL, Shikone T, Nishimori S, Hsueh AJW (1994) Follicle-stimulating hormone receptor expression in the rat ovary: increases during prepubertal development and regulation by the opposing actions of transforming growth factors b and a. Biol Reprod 50:940–948

Dwaraknath PIC, Agarwal SP, Agarwal VK, Dixit NIC, Sharma IJ (1984) Hormonal profiles in buffalo bulls. In: The use of nuclear techniques to improve domestic buffalo production in Asia, Proceedings of isotope & radiation applications of agricultural development, Manila

Ebling FJP, Wood RI, Karsch FJ, Vannerson LA, Suttie JM, Bucholtz DC, Scball RE, Foster DL (1990) Metabolic interfaces between growth and reproduction. 111. Central mechanisms controlling pulsatile luteinizing hormone secretion in the nutritionally growth-limited female lamb. Endocrinology 126:2719–2727

Edwards RI, Omtvedt IT, Turman EJ, Stephens DE, Mahoney GWA (1968) Reproductive performance of gilts following heat stress prior to breeding and in early lactation. J Anim Sci 27:1634

Elenkov IJ, Chrousos GP (1999) Stress hormones, Th1/Th2 patterns, pro/anti-inflammatory cytokines and susceptibility to disease. Trends Endocr Metab 10:359–368

El-Mastry KA, Habeeb AA (1989) Thyroid function in lactating Friesian cows and water buffaloes in winter and summer Egyptian conditions. In: 3rd Egyptian – British conference on animal, fish and poultry production, Alexandria University, Cairo, pp 613–620

El-Nouty FD, Hassan GA (1983) Thyroid hormone status and water metabolism in Hereford cows exposed to high ambient temperature and water deprivation. Indian J Anim Sci 53:807–812

El-Nouty FD, El-Banna IM, Daandon IP, Johnson HD (1980) Aldosterone and ADH response to heat and dehydration in cattle. J Appl Physiol 48:249–255

Elvinger F, Natzke RP, Hansen PJ (1992) Interactions of heat stress and bovine somatotropin affecting physiology and immunology of lactating cows. J Dairy Sci 75:449–462

Engler D, Pham T, Fullerton MJ, Ooi G, Funder JW, Clarke IJ (1989) Studies of the secretion of corticotrophin-releasing factor and arginine vasopressin into the hypophysial-portal circulation of the conscious sheep. Neuroendocrinology 49:367–381

Erickson GF, Hsueh AJW (1978) Stimulation of aromatase activity by follicle-stimulating hormone in rat granulosa cells in vivo and in vitro. Endocrinology 102:1275–1282

Fajardo LF, Prionas SD, Kowalski J, Kwan HH (1988) Hyperthermia inhibits angiogenesis. Radiat Res 114:297–306

Fantuzzi G (2005) Adipose tissue, adipokines and inflammation. J Allergy Clin Immunol 115:911–919

Findlay JK (1993) An update on the role of inhibin, activin, and follistatin as local regulators of folliculogenesis. Biol Reprod 48:15–23

Fortune JE (1994) Ovarian follicular growth and development in mammals. Biol Reprod 50:225–232

Foster DL, Ebling FJP, Micka AF, Vannerson LA, Bucholtz DC, Wood RI, Suttie JM, Fenner DE (1989) Metabolic interfaces between growth and reproduction. I. Nutritional modulation of gonadotropin, prolactin and growth hormone secretion in the growth-limited female lamb. Endocrinology 125:342–350

Franchimont D, Louis E, Dewe W, Martens H, Vrindts-Gevaert Y, de Groote D, Belaiche J, Greenen V (1998) Effects of dexamethasone on the profile of cytokine secretion in human whole blood cell cultures. Regul Pept 73:59–65

Friedman JM (2002) The function of leptin in nutrition, weight and physiology. Nutr Rev 60:1–14

Gale CC (1973) Neuroendocrine aspects of thermoregulation. Annu Rev Physiol 35:391

Gangwar PC, Branton C, Evans DL (1965) Reproductive and physiological response of Holstein heifers to controlled and natural climatic conditions. J Dairy Sci 48:222–227

Gauthier D (1986) The influence of season and shade on oestrous behaviour, timing of preovulatory LH surge and the pattern of progesterone secretion in FFPN and Creole heifers in a tropical climate. Reprod Nutr Dev 26:767–775

Gilad E, Meidan R, Berman A, Graber Y, Wolfenson D (1993) Effect of heat stress on tonic and GnRH-induced gonadotropin secretion in relation to concentration of estradiol in plasma of cyclic cows. J Reprod Fertil 99:315–321

Goldstein BJ, Scalia R (2004) Adiponectin: a novel adipokine linking adipocytes and vascular function. J Clin Endocrinol Metabol 89:2563–2568

Gwazdauskas FC, Thatcher WW, Kiddy CA, Paper MJ, Wilcox CJ (1981) Hormonal pattern during heat stress following PGF2alpha-tham salt induced luteal regression in heifers. Theriogenology 16:271–285

Habeeb AAM (1987) The role of insulin in improving productivity of heat stressed farm animals with different techniques. PhD thesis, Faculty of Agriculture, Zagazig University, Zagazig

Habeeb AAM, Ibrahim MKH, Yousef HM (2000) Blood and milk contents of triiodothyronine (T3) and cortisol in lactating buffaloes and changes in milk yield and composition as a function of lactation number and ambient temperature. Arab J Nucl Sci Appl 33:313–322

Habeeb AAM, Aboulnaga AJ, Kamal TH (2001) Heat-induced changes in body water concentration, Ts, cortisol, glucose and cholesterol levels and their relationships with thermoneutral bodyweight gain in Friesian calves. In: Proceedings of 2nd international conference on animal production and health in semi-arid areas, El-Arish, North Sinai, pp 97–108

Habeeb AAM, Fatma FIT, Osman SF (2007) Detection of heat adaptability using heat shock proteins and some hormones in Egyptian buffalo calves. Egypt J Appl Sci 22:28–53

Hall JE, Crook ED, Jones DW, Wofford MR, Dubbert PM (2002) Mechanisms of obesity-associated cardiovascular and renal disease. Am J Med Sci 324:127–137

Hamilton TD, Vizcarra JA, Wettermann RP, Keefer BE, Spicer LJ (1999) Ovarian function in nutritionally induced anoestrus cows: effect of exogenous gonadotrophin-releasing hormone in vivo and effect of insulin and insulin-like growth factor I in vitro. J Reprod Fertil 117:179–187

Havel PJ (2002) Control of energy homeostasis and insulin action by adipocyte hormones: leptin, acylation stimulating protein, and adiponectin. Curr Opin Lipidol 13:51–59

Havel PJ (2004) Update on adipocytes hormones. Regulation of energy balance and carbohydrate/lipid metabolism. Diabetes 53:143–151

Hein KG, Allrich RD (1992) Influence of exogenous adrenocorticotropic hormone on estrous behavior in cattle. J Anim Sci 70:243–247

Horowitz M (2001) Heat acclimation: phenotypic plasticity and cues to the underlying molecular mechanism. J Therm Biol 26:357–363

Houseknecht KL, Baile CA, Matteri RL, Spurlock ME (1998) The biology of leptin: a review. J Anim Sci 76:1405–1420

Howell JL, Fuquay JW, Smith AE (1994) Corpus luteum growth and function in lactating Holstein cows during spring and summer. J Dairy Sci 77:735–739

Hsueh AJW, Billig H, Tsafriri A (1994) Ovarian follicle atresia: a hormonally controlled apoptotic process. Endocr Rev 15:707–724

Igono MO, Johnson HD, Steevens BJ, Krause GF, Shanklin MD (1987) Physiological, productive, and economic benefits of shade, spray, and fan system versus shade for Holstein cows during summer heat. J Dairy Sci 70:1069–1079

Igono MO, Johnson HD, Steevens BJ, Hainen WA, Shanklin MD (1988) Effect of season on milk temperature, milk growth hormone, prolactin, and somatic cell counts of lactating cattle. Int J Biometeorol 32:194–200

Ingraham RH, Gillette DD, Wagner WD (1974) Relationship of temperature and humidity to conception of Holstein cows in tropical climate. J Dairy Sci 54:476–481

Ingraham RH, Kappel LC, Morgan EB, Babcock DK (1982) Temperature-humidity vs seasonal effects on concentrations of blood constituents of dairy cows during the pre and post-calving periods: relationship to lactation level and reproductive functions. In: Proceedings of 2nd international livestock and environment symposium, American Society of Agricultural Engineers, St. Joseph, pp 565–570

Johnson HD, Van Jonack WJ (1976) Effects of environmental and other stressors on blood hormone patterns in lactating animals. J Dairy Sci 59:1603–1617

Johnson HD, Katti PS, Hahn L, Shanklin MD (1988) Short-term heat acclimation effects on hormonal profile of lactating cows. University of Missouri Research Bulletin no 1061, Columbia

Johnson HD, Shanklin MD, Hahn L (1989) Productive adaptability indices of Holstein cattle to environmental heat. In: Agriculture and forest and meteorology conference, pp 291–297, Boston, USA; American Meteorological Society

Johnson HD, Li R, Manalu W, Spencer-Johnson KJ (1991) Effects of somatotropin on milk yield and physiological responses during summer farm and hot laboratory conditions. J Dairy Sci 74:1250–1262

Jonsson NN, McGowan MR, Mcguigan K, Davison TM, Hussian AM, Kafi M, Matschoss A (1997) Relationships among calving season, heat load, energy balance and postpartum ovulation of dairy cows in a subtropical environment. Anim Reprod Sci 47:315–326

Juniewicz PE, Johnson BH, Bolt DJ (1987) Effect of adrenal steroids on testosterone and luteinizing hormone secretion in the ram. J Androl 8:190–196

Kamal TH, Ibrahim II (1969) The effect of the natural climate of the Sahara and controlled climate on thyroid gland activity in Friesian cattle and water buffaloes. Int J Biometeorol 13:287–294

Kamal TH, Habeeb AA, Abdel-Samee AM, Abdel-Razik MA (1989) Supplementation of heat-stressed Friesian cows with urea and mineral mixture and its effect on milk production in the subtropics. In: Proceedings of symposium on ruminant production in the dry subtropics: constraints and potentials, Cairo. EAAP

Publication no. 38, Pudoc Science Publication, Wageningen, pp 183–185

Kamigaki M, Sakaue S, Tsujino I, Ohira H, Ikeda D, Itoh N, Ishimaru S, Ohtsuka Y, Nishimura M (2006) Oxidative stress provokes atherogenic changes in adipokine gene expression in 3T3-L1 adipocytes. Biochem Biophys Res Commun 339:624–632

Kaneko H, Nakanishi Y, Akagi S, Arai K, Watanabe G, Sasamoto S, Hasegawa S (1995) Immunoneutralization of inhibin and estradiol during the follicular phase of the estrus cycle in cows. Biol Reprod 53:931–939

Kaneko H, Taya K, Watanabe G, Noguchi J, Kikuchi K, Shimada A, Hasegawa S (1997) Inhibin is involved in suppression of FSH secretion in the growth phase of the dominant follicle during the early luteal phase in cows. Domest Anim Endocrinol 14:263–271

Kaufman FL, Mills DE, Hughson RL, Peake GT (1988) Effects of bromocriptine on sweat gland function during heat acclimatization. Horm Res 29:31–38

Khodaei MM (2003) Study of some effective factors in reproductive performance of Iranian Holstein Cows. Thesis, Animal Physiology, Tehran University, Tehran

Khodaei M, Roohani MZ, Zare A, Moradi Share Babak M (2006) Study of some effective factors in reproductive performance in sharif abad Ghazvin company and Shir-o-dame Boniad. In: Proceedings of 2nd congress animal science, Tehran

Kim JM, Boone DL, Auyeung A, Tsang BK (1998) Granulosa cell apoptosis induced at penultimate stage of follicular development is associated with increased levels of Fas and Fas ligand in the rat ovary. Biol Reprod 58:1170–1176

Kok FW, Heijnen CJ, Bruijn JA, Westenberg HGM, Van Ree JM (1995) Immunoglobulin production in vitro in major depression: a pilot study on the modulating action of endogenous cortisol. Biol Psychiatry 38:217–226

Lacetera N, Bernabucci U, Basirico L, Morera P, Nardone A (2009) Heat shock impairs DNA synthesis and down-regulates gene expression for leptin and Ob-Rb receptor in concanavalin A-stimulated bovine peripheral blood mononuclear cells. Vet Immunol Immunopathol 127:190–194

Lafontan M, Viguerie N (2006) Role of adipokines in the control of energy metabolism: focus on adiponectin. Curr Opin Pharmacol 6:1–6

Lamming GE, Royal MD (2001) Ovarian hormone patterns and subfertility in dairy cows. In: Diskin MG (ed) Fertility in the high producing dairy cow, BSAS Edinburgh: Occasional Publication 26, pp 105–118

Lamming GE, Darwash A, Black HL (1989) Corpus luteum function in dairy cows and embryo mortality. J Reprod Fertil 37:245–252

LaPolt PS, Tilly JL, Aihara T, Nishimori K, Hsueh AJ (1992) Gonadotropin-induced up- and down-regulation of ovarian follicle-stimulating hormone (FSH) receptor gene expression in immature rats: effects of pregnant mare's serum gonadotropin, human chorionic gonadotropin, and recombinant FSH. Endocrinology 130:1289–1295

Leining KB, Bourne RA, Tucker HA (1979) Prolactin response to duration and wavelength of light in prepubertal bulls. Endocrinology 104:289–294

Leining DB, Tucker HA, Kesner JS (1980) Growth hormone, glucocorticoids and thyroxine response to duration, intensity and wave length of light in pre-pubertal bulls. J Anim Sci 51:932–942

Lindner HR (1964) Comparative aspects of cortisol transport. Lack of firm binding to plasma proteins in domestic ruminants. J Endocrinol 28:301

Linquist S (1986) The heat–shock response. Annu Rev Biochem 55:1151–1191

Lippsett MB, Schwartz TL, Thon NA (1961) Hormonal control of sodium, potassium, chloride and water metabolism. In: Comar CL, Bonner F (eds) Mineral metabolism, lBth edn. Academic, New York

Lublin A, Wolfenson D (1996) Effect of blood flow to mammary and reproductive systems in heat-stressed rabbits. Comp Biochem Physiol 115:277–285. Sinauer Associates, MA

MacMurray JP, Barker JP, Armstrong JD, Bozzetti LP, Kuhn IN (1983) Circannual changes in immune function. Life Sci 32:2363–2370

Madan ML, Johnson HD (1973) Environmental heat effects on bovine luteinizing hormone. J Dairy Sci 56:1420–1423

Magdub A, Johnson HD, Belvea RL (1982) Effect of environmental heat and dietary fiber on thyroid physiology of lactating cows. J Dairy Sci 65:2323

Malayer JR, Hansen PJ (1990) Differences between Brahman and Holstein cows in heat-shock induced alterations of protein secretion by oviducts and uterine endometrium. J Anim Sci 68:266–280

Mann GE, Lamming GE (2001) Relationship between maternal endocrine environment, early embryo development and inhibition of leutolytic mechanism in cows. Reproduction 121:175–180

Mann GE, Lamming GE, Fray MD (1995) Plasma estradiol and progesterone during early pregnancy in the cow and the effects of treatment with buserelin. Anim Reprod Sci 37:121–131

Mann GE, Lamming GE, Robinson RS, Wathes DC (1999) The regulation of interferon-tau production and uterine hormone receptors during early pregnancy. J Reprod Fertil 54:317–328

Mann GE, Merson P, Fray MD, Lamming GE (2001) Conception rate following progesterone supplementation after second insemination in dairy cows. Vet J 162:161–162

McEwen BS, Goodman HM (eds) (2001) Handbook of physiology; section 7: The endocrine system, vol IV: Coping with the environment: neural and endocrine mechanisms. Oxford University Press, New York

McGuire MA, Beede DK, DeLorenzo MA, Wilcox CJ, Huntington GB, Reynolds CK, Collier RJ (1989) Effects of thermal stress and level of feed intake on portal plasma flow and net fluxes of metabolites in lactating Holstein cows. J Anim Sci 67:1050–1060

McGuire MA, Beede DK, Collier RJ, Buonomo FC, Delorenzo MA, Wilcox CJ, Huntington GB, Reynolds

CK (1991) Effect of acute thermal stress and amount of feed intake on concentrations of somatotropin, insulin-like growth factor I (IGF-I) and IGF-II and thyroid hormones in plasma of lactating dairy cows. J Anim Sci 69:2050–2056

McNatty KP, Heath DA, Henderson KM, Lun S, Hurst PR, Ellis LM, WMntgomery G, Morrison L, Hurly DC (1984) Some aspects of thecal and granulosa cell function during follicular development in the bovine ovary. J Reprod Fertil 72:39–53

McVeign JM, Tarrent PV (1982) Behavioral stress and skeletal muscle glycogen metabolism in young bulls. J Anim Sci 54:790

Mitra R, Christison GI, Johnson HD (1972) Effect of prolonged thermal exposure on growth hormone (GH) secretion in cattle. J Anim Sci 34:776–779

Moseley PL (1998) Heat shock proteins and the inflammatory response. Ann N Y Acad Sci 856:206–213

Moynihan JA (2003) Mechanisms of stress-induced modulation of immunity. Brain Behav Immun 17:11–16

Nakamura K, Minegishi T, Takakura Y, Miyamoto K, Hasegawa Y, Ibuki Y, Igarashi M (1991) Hormonal regulation of gonadotropin receptor mRNA in rat ovary during follicular growth and luteinization. Mol Cell Endocrinol 82:259–263

Nakamura H, Seto T, Nagase H, Yoshida M, Dan S, Ogino K (1997) Inhibitory effect of pregnancy on stress-induced immunosuppression through corticotropin releasing hormone (CRH) and dopaminergic systems. J Neuroimmunol 75:1–8

Nardone A, Lacetera NG, Bernabucci U, Ronchi B (1997) Composition of colostrum from dairy heifers exposed to high air temperatures during late pregnancy and early postpartum period. J Dairy Sci 80:838–844

Nelson RJ (2005) An introduction to behavioral endocrinology, 3rd edn. Sinauer Associates, Sunderland

Nessim MG (2004) Heat-induced biological changes as heat tolerance indices related to growth performance in buffaloes. PhD thesis, Faculty of Agriculture, Ain Shams University, Cairo

Niles MA, Collier RJ, Croom WJ (1980) Effect of heat stress on rumen and plasma metabolite and plasma hormone concentration of Holstein cows. J Anim Sci (Suppl) 152 (Abstract)

Norman AW, Litwack G (1987) Hormones. Academic, New York, pp 408–415

Norris DO (2007) Vertebrate endocrinology, 4th edn. Academic, Boston

Oltvai ZN, Milliman CL, Korsmeyer SJ (1993) Bcl-2 heterodimerizes in vivo with a conserved homolog, Bax, that accelerates programmed cell death. Cell 74:609–619

Overton TR, Waldron MR (2004) Nutritional management of transition dairy cows: strategies to optimize metabolic health. J Dairy Sci 87(13 suppl):105–199

Palta P, Mondal S, Prakas BS, Madan ML (1997) Peripheral inhibin levels in relation to climatic variations and stage of estrous cycle in buffalo (Bubalus bubalis). Theriogenology 47:898–995

Palumbo A, Yeh J (1994) In situ localization of apoptosis in the rat ovary during follicular atresia. Biol Reprod 51:888–895

Parr RA, Davis IF, Miles MA, Squires TJ (1993) Liver blood flow and metabolic clearance rate of progesterone in sheep. Res Vet Sci 55:311–316

Perera KS, Gwazdauskas FC, Akers RM, Pearson RE (1985) Seasonal and lactational effects on response to thyrotropin releasing hormone injection in Holstein cows. Domest Anim Endocrinol 2:43–52

Perez JH, Fernandez O (1988) Thyroid hormone levels in heat tolerant and non-tolerant Friesian heifers. Revista de Salud Anim 10:121–130

Perez H, Mendoza E, Alvarez JL, Fernandez O (1997) Effect of the temperature humidity index on secretion of thyroid hormones in Holstein heifers. Revista de Salud Anim 19(2):131–135

Pittas AG, Joseph NA, Greenberg AS (2004) Adipocytokines and insulin resistance. J Clin Endocrinol Metabol 89:447–452

Plasse D, Warnick AC, Koger J (1970) Reproductive behavior of Bos indicus females in a subtropical environment. IV. Length of estrous cycle, duration of estrus, time of ovulation, fertilization and embryo survival in grade Brahman heifers. J Anim Sci 30:63

Pusta D, Odagiu A, Ersek A, Pascal I (2003) The variation of triiodothyronine (T3) level in milking cows exposed to direct solar radiation. J Cent Euro Agric 4:308–312

Putney DJ, Malayer JR, Gross TS, Thatcher WW, Hansen PJ, Drost M (1988) Heat-stress induced alterations in the synthesis and secretion of proteins and prostaglandins by cultured bovine conceptuses and uterine endometrium. Biol Reprod 39:717–728

Ray DE, Halbach TJ, Armstrong DV (1992) Season and lactation number effects on milk production and reproduction in dairy cattle in Arizona. J Dairy Sci 75:2976–2983

Richards JI (1985) Effect of high daytime temperature on the intake and utilization of water in lactating Friesian cows. Trop Anim Health Prod 17:209–217

Robinson NA, Leslie KE, Walton JS (1989) Effect of treatment with progesterone on pregnancy rate and plasma concentrations of progesterone in Holstein cows. J Dairy Sci 72:202–207

Roche JR, Kolver ES, Kay JK (2005) Influence of precalving feed allowance on periparturient metabolic and hormonal responses and milk production in grazing dairy cows. J Dairy Sci 88:677–689

Roman-Ponce H, Thatcher WW, Wilcox CJ (1981) Hormonal interrelationships and physiological responses of lactating dairy cows to a shade management system in a subtropical environment. Theriogenology 16:139–154

Romero LM (2004) Physiological stress in ecology: lessons from biomedical research. Trends Ecol Evol 19:249–255

Ronchi B, Bernabucci U, Lacetera N, Verini Supplizi A, Nardone A (1999) Distinct and common effects of heat stress and restricted feeding on metabolic status of Holstein heifers. Zootecnica e Nutrizione Animale 25:11–20

Ronchi B, Stradaioli G, Verini Supplizi A, Bernabucci U, Lacetera N, Accorsi PA, Nardone A, Seren E (2001) Influence of heat stress or feed restriction on plasma progesterone, oestradiol–17b, LH, FSH, prolactin and cortisol in Holstein heifers. Livest Prod Sci 68:231–241

Rosenberg M, Folman Y, Herz Z, Flamenbaum I, Berman A, Kaim M (1982) Effect of climatic conditions on peri0pheral concentrations of LH, progesterone and oestradiol-17b in high milk-yielding cows. J Reprod Fertil 66:13–146

Roth Z, Arav A, Bor A, Zeron Y, Ocheretny A, Wolfenson D (1999) Enhanced removal of impaired follicles improves the quality of oocytes collected in the autumn from summer heat-stressed cows. J Reprod Fertil Abstr Ser 23:78 (Abstract)

Roth Z, Meidan R, Braw-Tal R, Wolfenson D (2000) Immediate and delayed effect of heat stress on follicular development and its association with plasma FSH and inhibin concentration in cows. J Reprod Fertil 120:83–90

Roth Z, Meidan R, Shaham-Albalancy A, Braw-Tal R, Wolfenson D (2001) Delayed effect of heat stress on steroid production in medium-sized and preovulatory bovine follicles. Anim Reprod Sci 121:745–751

Roy KS, Prakash BS (2007) Seasonal variation and circadian rhythmicity of the prolactin profile during the summer months in repeat-breeding Murrah buffalo heifers. Reprod Fertil Dev 19:569–575

Sangsritavong S, Combs DK, Satori R, Armentano LE, Wiltbank MC (2002) High feed intake increase liver blood flow and metabolism of progesterone and estradiol 17 beta in dairy cattle. J Dairy Sci 85:2831–2842

Santos JEP, Thatcher WW, Chebela RC, Cerria RLA, Galvãoa KN (2004) The effect of embryonic death rates in cattle on the efficacy of estrus synchronization programs. Anim Reprod Sci 82–83:513–535

Sapolsky RM, Romero LM, Munck AU (2000) How do glucocorticoids influence stress-responses? Integrating permissive, suppressive, stimulatory, and adaptive actions. Endocr Rev 21:55–89

Scott IM, Johnson HD, Hahn GL (1983) Effect of programmed diurnal temperature cycles on plasma thyroxine level, body temperature, and feed intake of Holstein dairy cows. Int J Biometeorol 27:47–62

Sejrsen K, Fitzgerald EM, Tucker HA, Huber JT (1980) Effect of plane of nutrition on serum prolactin and insulin in pre-and post-pubertal heifers. J Dairy Sci 53(suppl 1):326–327, Abstract

Shimizu T, Ohshima I, Kanai Y (2000) Effect of heat stress on follicular development in PMSG-treated immature rats. Anim Sci J 71:32–37

Shimizu T, Izumi O, Manabu O, Satoko T, Atsushi T, Masayuki S, Hitoshi M, Yukio K (2005) Heat stress diminishes gonadotropin receptor expression and enhances susceptibility to apoptosis of rat granulosa cells. Reproduction 129:463–472

Silanikove N (2000) Effects of heat stress on the welfare of extensively managed domestic ruminants. Livest Prod Sci 67:1–18

Spicer LJ, Tucker WB, Adams GD (1990) Insulin-like growth factor-I in dairy cows: relationships among energy balance, body condition, ovarian activity, and estrous behavior. J Dairy Sci 13:929–931

Stott GH (1972) Climatic thermal stress. A cause of hormone depression and low fertility in bovine. Biometeorogy 5:113 (Abstract)

Stott GH, Wiersma F (1971) Plasma corticoids as an index of acute and chronic environmental stress. Presented at First conference on biometeorology and tenth conference on agricultural meteorology, Columbia

Thibier M, Rolland O (1976) The effect of dexamethasone (DXM) on circulating testosterone (T) and luteinizing hormone (LH) in young postpubertal bulls. Theriogenology 5:53–60

Thompson JA, Magee DD, Tomaszewski MA, Wilks DL, Fourdraine RH (1996) Management of summer infertility in Texas Holstein dairy cattle. Theriogenology 46:547–558

Thwaites CJ (1968) The influence of age of ewe on embryo mortality under heat stress conditions. Int J Biometeor 12:29

Tilbrook AJ, Turner AI, Clarke IJ (2000) Effects of heat stress on reproduction in non-rodent mammals: the role of glucocorticoids and sex differences. J Reprod Fertil 5:105–113

Tilly JL (1993) Ovarian follicular atresia: a model to study the mechanisms of physiological cell death. Endocr J 1:67–72

Tilly JL (1996) Apoptosis and ovarian function. Rev Reprod 1:162–172

Tilly JL, Tilly KI, Kenton ML, Johnson AL (1995) Expression of members of the bcl-2 gene family in the immature rat ovary: equine chorionic gonadotropin-mediated inhibition of granulosa cell apoptosis is associated with decreased bax and constitutive bcl-2 and bcl-xlong messenger ribonucleic acid levels. Endocrinology 136:232–241

Trayhurn P (2005) Endocrine and signalling role of adipose tissue: new perspectives on fat. Acta Physiol Scand 184:285–293

Trout JP, Mcdowell LR, Hansen PJ (1998) Characteristics of the estrus cycle and antioxidant status of lactating Holstein cows exposed to stress. J Dairy Sci 81:1244–1250

Vasconcelos JLM, Sangsritavong S, Tsai SJ, Wiltbank MC (2003) Acute reduction in serum progesterone concentrations after feed intake in dairy cows. Theriogenology 60:795–807

Wang M, Orci L, Ravazzola M, Unger RH (2005) Fat storage in adipocytes requires inactivation of leptin's paracrine activity: implication for treatment of human obesity. PNAS 102:18011–18016

Webb R, Garnsworthy PC, Gong JG, Robinson RS, Wathes DC (1999) Consequences for reproductive function of metabolic adaptation to load. British Society of Animal Science, Occasional Publication 24, pp 99–112

Webster JI, Tonelli L, Sternberg EM (2002) Neuroendocrine regulation of immunity. Annu Rev Immunol 20: 125–163

Wetteman RP, Tucker HA (1979) Relationship of ambient temperature to serum prolactin in heifers. In: Proceedings of the Society for Experimental Biology and Medicine 146. Academic, New York, pp 909–911

Wilson SJ, Kirby CJ, Koenigsfeld AT, Keisler DH, Lucy MC (1998a) Effects of controlled heat stress on ovarian function of dairy cattle. II. Heifers. J Dairy Sci 81:2132–2138

Wilson SJ, Marion RS, Spain JN, Spiers DE, Keisler DH, Lucy MC (1998b) Effects of controlled heat stress on ovarian function of dairy cattle. I. Lactating cows. J Dairy Sci 81:2124–2131

Wise ME, Armstrong DV, Huber JT, Hunter R, Wiersma F (1988a) Hormonal alteration in the lactating dairy cow in response to thermal stress. J Dairy Sci 71: 2480–2485

Wise ME, Rodriguez RE, Armstrong DV, Huber JT, Wiersma F, Hunter R (1988b) Fertility and hormonal response to thermal relief of heat stress in lactating dairy cows. Theriogenology 29:1027–1035

Wolfenson D, Flamenbaum I, Berman A (1988) Hyperthermia and body energy store effects on oestrus behaviour, conception rate, and corpus luteum function in dairy cows. J Dairy Sci 71:3497–3504

Wolfenson D, Bartol FF, Badinga L, Barros CM, Marple DN, Cummings K, Wolfe D, Lucy MC, Spencer TE, Thatcher WW (1993) Secretion of PGF2a and oxytocin during hyperthermia in cyclic and pregnant heifers. Theriogenology 39:1129–1141

Wolfenson D, Thatcher WW, Badinga L, Savio JD, Meidan R, Lew BJ, Braw-Tai R, Berman R (1995) Effect of heat stress on follicular development during the estrus cycle in lactating dairy cattle. Biol Reprod 52:1106–1113

Wolfenson D, Lew BJ, Thatcher WW, Graber Y, Meidan R (1997) Seasonal and acute heat stress effects on steroid production by dominant follicles in cow. Anim Reprod Sci 47:9–19

Wolfenson D, Roth Z, Meidan R (2000) Impaired reproduction in heat stressed cattle: basic and applied aspects. Anim Reprod Sci 60–61:537–547

Wolfenson D, Sonego H, Bloch A, Shaham-Albalancy A, Kaim M, Folman Y, Meidan R (2002) Seasonal differences in progesterone production by luteinized bovine thecal and granulosa cells. Domest Anim Endocrinol 2:81–90

Yang E, Zha J, Jockel J, Boise LH, Thompson CB, Korsmeyer SJ (1995) Bad, a heterodimeric partner for Bcl-XL and Bcl-2, displaces Bax and promotes cell death. Cell 80:285–291

Younas M, Fuquay JW, Smith AE, Moore AB (1993) Estrus and endocrine responses of lactating Holstein to forced ventilation during summer. J Dairy Sci 76:430–434

Yousef MMM (1992) Growth patterns of calves in relation to rumen development. PhD thesis, Faculty of Agriculture, Cairo University, Giza

Yousef MK, Johnson HD (1966a) Blood thyroxine degradation rate in cattle as influenced by temperature and feed intake. Life Sci 5:1349

Yousef MK, Johnson HD (1966b) Calorigenesis of cattle as influenced by growth hormone and environmental temperature. J Anim Sci 25:1076

Yousef JLM, Habeeb AA, EL-Kousey H (1997) Body weight gain and some physiological changes in Friesian calves protected with wood or reinforced concrete sheds during hot summer season of Egypt. Egypt J Anim Prod 34:89–101

Zhengkang H, Zhenzhong C, Shaohua Z, Vale WG, Barnabe VH, Mattos JCA (1994) Rumen metabolism, blood cortisol and T3, T4 levels and other physiological parameters of swamp buffalo subjected to solar radiation. In: Proceedings of world buffalo Congress, San Paulo, vol 2, pp 39–40

Zhou J, Kumar RT, Matzuk MM, Bondy C (1997) Insulin-like growth factor I regulates gonadotropin responsiveness in the murine ovary. Mol Endocrinol 11:1924–1933

Heat Stress and Milk Production

Contents

Abstract

Heat stress at the initiation of lactation negatively impacts the total milk production. Climatic conditions appeared to have maximum influence during the first 60 days of lactation when high-producing cows are in negative energy balance and make up for the deficit by mobilising body reserves. High-yielding cows are affected more than low yielding ones because the upper critical temperature shifts downwards as milk production, feed intake and heat production increase. High humidity seems to affect buffaloes less than cattle, since buffaloes may be superior to cattle in humid areas if shade or wallows are available. Heat stress causes the rostral cooling centre of the hypothalamus to stimulate the medial satiety centre which inhibits the lateral appetite centre and consequently lowers milk production. The heat-stressed cows depend on glucose for body energy needs; therefore, less glucose is directed towards the mammary gland, and there is a decline in milk production. Negative energy balance is associated with various metabolic changes that are implemented to support the dominant physiological condition of lactation. Marked alterations in both carbohydrate and lipid metabolism ensure partitioning of dietary-derived and tissue-originating nutrients towards the mammary gland, and many of these changes are mediated by endogenous somatotropin which is naturally increased during periods of negative energy balance. Milk constituents are greatly

A. Aggarwal and R. Upadhyay, *Heat Stress and Animal Productivity*,
DOI 10.1007/978-81-322-0879-2_3, © Springer India 2013

affected by hyperthermia. The ability to use powerful new tools in genomics, proteomics and metabolomics to evaluate genetic differences between animals in their response to thermal stress will yield important new information in the next quarter-century and will permit the selection of cattle for resistance to thermal stress.

1   Introduction

Heat stress has adverse effects on milk production and reproduction of dairy cattle (Kadzere et al. 2002; Hansen 2007). The problem of heat stress is growing because increases in milk yield result in greater metabolic heat production and because of anticipated changes in the global climate (Hansen 2007). A decline in milk yield of lactating cows and buffaloes is observed during hot–dry and hot–humid seasons due to high temperature and humidity. High THI (temperature humidity index) negatively impacts milk yield that occurs due to increase in body temperature (Upadhyay et al. 2009). Therefore, a major challenge for high-producing cow and buffaloes under tropical environment is heat stress. It was expected that regulation of body temperature during hyperthermia would be decreased as milk yield increased because of the metabolic heat output associated with lactation. This was demonstrated experimentally in dairy cows in Israel (Berman et al. 1985) and theoretically using mathematical models of heat balance in dairy cows (Berman 2005). Summer depression in fertility in dairy cows was more pronounced for cows with greater milk yield (Al-Katanani et al. 1999). It was traditionally thought that milk synthesis begins to decrease when the THI exceeds 72 (Armstrong 1994), but with increasing milk production, it has been observed that high-yielding dairy cows reduce milk yield at a THI of approximately 68 (Zimbelman et al. 2009). The increasing concern with the thermal comfort of dairy cows is justifiable not only for countries occupying tropical zones but also for nations in temperate zones in which high ambient temperatures are becoming an issue (Nardone et al. 2010).

Improving milk production is, therefore, an important tool for improving the quality of life particularly for rural people in developing countries. The hot–dry environment is relatively less harmful than hot–humid. In hot and humid climates, high ambient temperature and humidity and high direct and indirect solar radiation, wind speed and humidity are the main environmental stressing factors that impose stress on livestock species (Silanikove 2000; Kadzere et al. 2002).

2   Effect of Climatic Variables on Milk Yield

Climatic factors such as air temperature, solar radiation, relative humidity, air flow and their interactions often limit animal performance (Sharma et al. 1983). Quantifying direct environmental effects on milk production is difficult as milk production is also strongly affected by other non-environmental and environmental factors such as nutritional and management (Fuquay 1981). Thatcher (1974) and Johnson (1976) reported decline in the milk and fat production due to direct effect of high temperature on the synthetic and secretory activity in mammary gland of cows (Silanikove 1992). McDowell et al. (1976) also reported decline in milk production by 15%, accompanied by a 35% decrease in the efficiency of energy utilisation for productive purposes in lactating Holstein cow at high temperature on shifting them from an air temperature of 18–30°C. Milk fat, solids-not-fat and milk protein percentage was also observed to decrease by 39.7, 18.9 and 16.9%, respectively. Johnson (1976) also attributed 3–10% of the variance in lactation milk production to climatic factors. Differences in the physiological responses of cattle to the form and duration of heat stress have been reported, and differences have also been noted in productive responses. Bianca (1965) measured a 33% reduction in milk production by different breeds of cows from temperate climates exposed continuously to high (35°C) ambient temperatures. However, cows maintained under similar temperatures during the day but at <25°C at night did not decrease milk production beyond that normally expected under temperate conditions (Richards 1985).

The point on the lactation curve at which the cow experiences heat stress is also important for the total lactation yield. High-producing cows are less able to cope with heat stress during early lactation. Heat stress at the initiation of lactation negatively impacts the total milk production. Climatic conditions appeared to have maximum influence during the first 60 days of lactation when high-producing cows are in negative energy balance and make up for the deficit by mobilising body reserves (Sharma et al. 1983).

Nutrient intake of high-producing cows is closely related to the amount of milk production. The metabolising nutrients generate heat, which may contribute to body temperature maintenance in a cold environment. However, in a warm climate, this heat needs to be dissipated to maintain body temperature and normal physiological functions. This complex interplay between physical and environmental effects influences the physiological functions of cows and affects not only their milk production but also the efficiency of production in turn cow's profitability. High-producing cows have a higher metabolic rate than most other domestic ruminants and a poor water retention mechanism in the body (Silanikove 2000). This may be a consequence of aggressive selection for milk production over the last few decades; cows in Israel and the USA produce 40–70 l of milk per day, compared with 10 l/day or less in their ancestors. Each 10 l/day of milk yield roughly doubles the metabolisable energy requirement of cows, and ~35% of this energy is dissipated as heat (Kadzere et al. 2002). High-yielding cows are affected more than low yielding ones (Barash et al. 2001) because the upper critical temperature shifts downwards as milk production, feed intake and heat production increase (Silanikove 2000). The average body weight due to selection for high production has increased over time. Large cows have larger gastrointestinal tracts that allow them to consume and digest more feed which in turn provides more substrates for milk synthesis. Changes in the physical and genetic constitution of cows may have affected their thermoregulatory capability

as well as how they cope with heat stress (Kadzere et al. 2002).

Maximum temperature and minimum relative humidity are the most critical variables to quantify heat stress, and both variables are combined into THI. Heat stress in lactating cow starts at a THI of 72, which corresponds to 22°C at 100% RH (or 25°C at 50% RH or other combinations of the two parameters). Reduction in milk yield was estimated to be between 0.2 and 0.32 kg per unit increase in THI (Ingraham et al. 1979; Ravagnolo et al. 2000), and milk yield and TDN intake declined by 1.8 and 1.4 kg for each 0.55°C increase in rectal temperature (Johnson et al. 1963). Umphrey et al. (2001) reported that the partial correlation between milk yield and rectal temperature for cows in Alabama was −0.135. West (2003) found that changes in cow body temperature (measured as milk temperature) were most sensitive to same-day climatic factors. The variable having the greatest influence on cow morning milk temperature was the current day minimum air temperature, while cow afternoon milk temperature was most influenced by the current day mean air temperature. DMI and milk yield of cows were most affected by climatic variables, not by body temperature of cow. West (2003) reported that of the environmental variables studied during hot weather, the mean THI two days earlier had the greatest effect on milk yield, while DMI was most sensitive to the mean air temperature two days earlier. Thus, the full impact of climatic variables on production is delayed and may be related to changes in feed intake, delay between intake and utilisation of consumed nutrients or alterations in the endocrine status of the cow. Milk yield of Holstein cows declined 0.88 kg per THI unit increase for the 2-day lag of mean THI, and DMI declined 0.85 kg for each degree (°C) increase in the mean air temperature.

At the peak of lactation, a cow of 700-kg body weight (BW) with a milk yield of 60 kg/day produces about 44,171 kcal/day (25,782 kcal/day at the end of lactation, with a milk yield of 20 kg/day). Exposure to a hot environment may negatively affect growth of young calves. Lower wither height, oblique trunk length, hip width (−35, −26,

−29%, respectively) and body condition score (0.0 vs. +0.4 points) were found in six 5-month-old female Holstein–Friesian calves exposed to hot conditions as compared with a control group, kept under thermoneutrality conditions (Lacetera et al. 1994). Decrease in body growth and body reserves between birth and puberty, especially during the first few months, can be detrimental for milk production during lactation and can increase the replacement rate later (Chillard 1991). High temperatures during late pregnancy and the early postpartum period markedly modify colostrum composition. Holstein–Friesian heifers kept in a climatic chamber and exposed during late pregnancy and the early postpartum period to 82 THI (daytime) and to 76 THI (night-time), when compared to a counterpart maintained under thermoneutrality conditions (65 THI), showed lower colostrum net energy due to a reduction in lactose, fat and protein content. In addition, the analysis of protein fractions showed a reduction in percentages of casein, lactalbumin, IgG and IgA (Nardone et al. 1997) explaining a lower concentration of circulating Ig in summer calves (Stott et al. 1976). An investigation on Israelian Holstein cows in their third and fourth lactation periods showed the antagonistic effect of heat (−0.38 kg of milk/1°C) and photoperiod (+1.157 kg of milk/per 1 h more of light) (Barash et al. 2001). On comparing milk production during summer and spring in a dairy herd located in central Italy, a lower milk yield (−10%), and also lower casein percentages and casein number in summer (2.18 vs. 2.58% and 72.4 vs. 77.7% respectively), was found (Bernabucci et al. 2002). The fall in casein was due to the reduction in α_s-casein and β-casein percentages. However, no differences were observed between the two seasons for κ-casein, α-lactalbumin and β-lactoglobulin. The serum protein contents were higher in summer than in spring. Although performance of dairy cows is better during winter than during heat stress, animals experiencing cold stress also reduce milk yield. The drop starts around −4°C, and marked yield depression occurs at −23°C. The lower critical temperature is −40 or −45°C with a daily milk yield of 36 kg (Broucek et al. 1991).

3   Behaviour of Cows During Heat Stress

Several behaviour modifications in response to heat stressing conditions have been observed in dairy cattle. Dairy cattle change their behaviour to reduce heat load in hot weather. Cows seek shade, reduce feed intake, spend more time standing, spend more time near the water trough and increase respiration rate as ambient conditions become warmer. Animal behaviour can provide insights into how and when to cool dairy cows. In free stall barns, cows stay inside during the hottest part of the day to obtain shelter from intense solar radiation, while during the night, cows go outside (Arave and Albright 1981). High-producing dairy cows showed more eating behaviour than low-producing cows during all times of the day. The eating behaviour of low and high-producing dairy cows decreased after 10:00 h and through the day. High ambient temperature decreases eating behaviour. The best recognised effect of heat stress is an adaptive depression of metabolic rate associated with reduced appetite (Silanikove 2000). Heat stress causes the rostral cooling centre of the hypothalamus to stimulate the medial satiety centre which inhibits the lateral appetite centre and consequently lowers milk production (Kadzere et al. 2002). The decrease (26.2%) in milk production of high-yielding cows is higher than that (15.2%) of low-producing dairy cows during summer. Heat stress in high-producing lactating dairy cows results in a dramatic reduction in rumination (Collier et al. 1982). High-producing cows ruminated less compared to low-producing cows. The underlying mechanism in lower percentage of ruminating behaviour can be explained that high-producing dairy cows coped with the adverse effect of heat stress by ruminating less so that there is lower production of metabolic heat (Kadzere et al. 2002).

High-producing cows tried to lower their body temperature by not lying on the concrete floor where heat from the sun was absorbed during daytime. They preferred standing to minimise surface area contact with the ground (escaping from conduction) so that the distance between the blood vessels and the surface is much greater (Kadzere et al. 2002). Furthermore, it is uncomfortable or even more painful for the cow to be lying with a filled udder, since there is an external pressure and heat from the floor on the udder when lying (Oszterman and Redbo 2001).

4   Effect of Heat Stress on Bovine Somatotropin (bST)-Administered Cows

The feeding and management recommendations for a cow treated with bovine somatotropin (bST) are considered similar to that of high-producing cow. Nutrient intake should be increased with special attention to energy and water needs. The cows should be provided with cooling systems. Israeli research indicated that the response to bST under high environmental temperatures may have been slightly reduced. Work conducted at the University of Missouri showed that heat-stressed injected cows responded to bST with increased milk production and dry matter intake. Hot temperatures did not reduce the effects of bST. The effects of high milk yield have been demonstrated by West et al. (1990, 1991) who reported that milk temperature was greater for cows administered with bST compared with controls in a hot and humid climate; low yielding cows were more responsive to bST than high-yielding cows, possibly because of the higher body temperature associated with greater milk yield. Cows administered with bST exhibited significantly greater heat production in both thermoneutral and hot environments, though cows were apparently able to dissipate the greater heat produced, evidenced by greater total evaporative heat losses and cooling heat loss for the bST-treated cows which enabled cows to maintain normal body temperatures (Manalu et al. 1991). Administration of bST to both lactating and non-lactating cows in a hot, humid climate (Florida) resulted in elevated body temperature and respiratory rate for both groups of cattle, suggesting that the greater heat strain was not due solely to increased milk yield (Cole and Hansen 1993). Either greater heat production or interference with heat loss could explain greater strain in non-lactating cattle and that although bST use is efficacious in hot climates, its use should be coupled with methods to reduce the magnitude of heat stress during summer months.

5   Metabolic Heat Production

Heat production of metabolic functions accounts for approximately 31% of intake energy by a 600-kg cow producing 40 kg of milk containing 4% fat (Coppock 1985). Physical activity increases the amount of heat produced by skeletal muscles and body tissues. Maintenance expenditures at 35°C increase by 20% over thermoneutral conditions (NRC 1981), thus increasing the cow's energy expenditure, often at the expense of milk yield. Body heat production associated with milk yield increases as metabolic processes, feed intake and digestive requirements increase. The heat load accumulated by the cow subjected to heat stress is the sum of heat accumulated from the environment and the failure to dissipate heat associated with metabolic processes. Therefore, with similar body size and surface area, the lactating cow has significantly more heat to dissipate than a non-lactating cow and have greater difficulty dissipating the heat during hot and hot–humid conditions. A comparison of low (18.5 kg/day) or high (31.6 kg/day) milk-producing cows indicated that low and high-yielding cows generated 27 and 48% more heat than non-lactating cows despite having lower body weight (Purwanto et al. 1990). Berman et al. (1985) reported that rectal temperature of cows increased by 0.02°C/kg FCM for cows producing >24 kg/day, and greater heat production can explain the increasing rate of decline in milk yield for cows as production increased from 13.6–18.1 to 22.7 kg of milk per day and THI increased from 72 to 81 (Johnson et al. 1963). High production greatly accentuates heat stress in the lactating cow particularly unable to dissipate extra heat.

6   Role of Acclimation During Heat Stress

Acclimation is a phenotypic response developed by the subject or an animal to an individual stressor within the environment (Fregley 1996). Acclimatization is the process by which an animal adapts to several stressors within its natural environment (Bligh 1976). Acclimation and acclimatization are therefore not evolutionary adaptation or natural selection, which are defined as changes allowing for preferential selection of an animal's phenotype and are based on a genetic component passed to the next generation. The altered phenotype of acclimated animals returns to normal if environmental stressors are removed. This is not the case in animals which are genetically adapted to their environment. Acclimatization is a process that takes several days to weeks to occur via homeorhetic and not homeostatic mechanism. There are three functional differences between acclimatory responses and homeostatic or 'reflex responses':

1. The acclimatory response takes much longer to occur (days or weeks vs. seconds or minutes).
2. The acclimatory responses generally have a hormonal link in the pathway from the central nervous system to the effector cell.
3. The acclimatory effect usually alters the ability of an effector cell or organ to respond to environmental change (Bligh 1976).

These acclimatory responses are characteristic of homeorhetic mechanisms in a species, and the net effect is to coordinate metabolism to achieve a new physiological state. Thus, the seasonally adapted animal is different metabolically during winter than during summer.

Acclimation involves an altered expression of pre-existing features and is a process driven by the endocrine system with the goal of maintaining animal well-being regardless of environmental challenges. The long-term acclimation to heat was classically referred to as acclamatory homeostasis (Horowitz 2001), but has been proposed to be a homeorhetic mechanism (Collier et al. 2005), because it alters the set points of homeostatic-related systems (i.e. basal and stimulated carbohydrate metabolism). Understanding this process

will lead to improved genetic selection of heat stress resistant genotypes. Hormones known to be homeorhetic regulators are also implicated in acclimatory responses to thermal stress and altered photoperiod. These include thyroid hormones, prolactin (PRL), somatotropin (ST), glucocorticoids and mineralocorticoids. One example of acclimatory change in an endocrine regulator is the seasonal rhythm in prolactin concentration when animals acclimate to seasonal changes in temperature and day length. The hypothalamic–pituitary–adrenal axes including corticotropin-releasing hormone, adrenocorticotropic hormone (corticotrophin), cortisol and aldosterone are also altered by thermal stress and are involved in acclimatory responses to thermal stress (Maloyan and Horowitz 2002). Corticotropin-releasing hormone stimulates somatostatin release from the hypothalamus, which can inhibit secretion of ST and thyroid-stimulating hormone from the pituitary and downregulate the thermogenic effects of both ST and thyroid hormones. In dairy cattle, the glucocorticoids decrease during acclimation at 35°C and are lower in thermally acclimated animals compared with controls.

Most of these examples deal with decreased heat production within the body and the cow's increased ability to dissipate heat obtained from the environment. In lactating cows, when long-term heat acclimation occurs, the low-producing cow's milk output will reach a level comparable to what the cow should produce while not under heat stress, but in the higher producing cow, it could still be below the milk production possible in a thermoneutral environment (Johnson and Vanjonack 1976). Some environmental stresses, such as dehydration (Silanikove 1994; Silanikove and Tadmor 1989) and acute heat stress (Maltz et al. 1994; Silanikove 2000), may take cows beyond their current acclimatised-adaptive range quickly (within 24–48 h) which necessitates the induction of emergency physiological responses to avoid lethal effects or death. Such immediate measures include an acute reduction in milk yield, because milk production, particularly in high-yielding dairy cows, intensifies the effects of these external stresses (Silanikove 1994, 2000).

7 Metabolic Adaptations to Heat Stress

Estimating energy balance of cows during heat stress introduces problems independent of those that are inherent to normal energy balance estimations (Vicini et al. 2002). A large number of studies indicate that increased maintenance costs (7–25%) are associated with heat stress (NRC 2002a); however, due to complexities involved in predicting upper critical temperatures for different breeds of cows, no universal equation is available to adjust for this increase in maintenance (Fox and Tylutki 1998). Maintenance requirements of cows are increased, as there is a large energetic cost of dissipating stored heat. If a heat stress correction factor is not incorporated, it results in overestimating energy balance, thereby inaccurately predicting energy status. Due to the reductions in feed intake and increased maintenance costs, and despite the decrease in milk yield, heat-stressed cows enter into a state of negative energy balance (Moore et al. 2005b) and remain in negative energy balance (~4–5 Mcal/day) for long duration of heat stress (Wheelock et al. 2006). Heat stress-induced negative energy balance does not result in elevated plasma NEFA similar to thermoneutral energy balance. Studies on IV glucose tolerance test have demonstrated that glucose disposal (rate of cellular glucose entry) is greater in heat-stressed cows compared to thermal-neutral pair-fed cows (Wheelock et al. 2006). Cows under heat stress also have a greater insulin response to glucose challenge than underfed cows. The changes in plasma NEFA and metabolic/hormonal adjustments in response to glucose challenge can be due to increased insulin effectiveness as insulin is a potent antilipolytic signal (blocks fat breakdown) and the primary driver of cellular glucose entry into the cells. The heat-stressed cows depend on glucose for body energy needs; therefore, less glucose is directed towards the mammary gland and milk production declines.

A typical lactating dairy cow has a maintenance requirement of 9.7 Mcal/day (or 0.08 Mcal/kg BW 0.75; NRC 2001a). Ruminants primarily oxidise (burn) acetate (a rumen-produced VFA) as their principal energy source. The apparent switch in metabolism and the increase in insulin sensitivity is probably a mechanism by which cows decrease metabolic heat production, and oxidising glucose is more efficient (Baldwin et al. 1980). In vivo oxidation of glucose yields 38 ATP or 472 kcal of energy (compared to 637 kcal in a bomb calorimeter) and in vivo fatty acid oxidation (stearic acid) produces 146 ATP or 1814 kcal of energy (compared to 2697 kcal in a bomb calorimeter). Despite having a much greater energy content, due to differences in the efficiencies of capturing ATP, oxidising fatty acids generates more metabolic heat (~2 kcal/g or 13% on an energetic basis) compared to glucose. Therefore, preventing or blocking adipose mobilisation/breakdown and increasing utilization of glucose are presumably strategies to minimise metabolic heat production during heat stress. The mammary gland uses glucose for synthesising milk lactose which is the primary osmoregulator and thus determines milk yield of animals. The mammary cells may not receive adequate amounts of glucose for mammary lactose production, and hence, milk yield is reduced. This may be the primary mechanism which accounts for the additional reductions in milk yield that cannot be explained by decreased feed intake (Bauman and Rhoads 2007). The heat stress also downregulates calcium channels of many T cells that affect milk synthesis (Silanikove et al. 2009).

In addition, heat-stressed cows require special attention with regard to heat abatement and other dietary considerations (i.e. concentrate to forage ratio, HCO_3^-, etc.) having an extra requirement for dietary or rumen-derived glucose precursors. Of the three main rumen-produced volatile fatty acids, propionate is the one primarily converted into glucose by the liver. Highly fermentable starches such as grains increase rumen propionate production, and although propionate is the primary glucose precursor, feeding additional grains is not recommended as this may lead to rumen acidosis.

Metabolic Adaptations to Reduced Nutrient Intake: Due to the reduced feed intake as a result of heat stress and the heat associated with fermenting forages, energy density of the ration is increased. Due to the hyperventilation-induced decrease in blood CO_2, the kidney secretes HCO_3^- to maintain this ratio. This reduces the amount of HCO_3^- that can be used (via saliva) to buffer and maintain a healthy rumen pH. In addition, panting cows drool, and drooling reduces the quantity of saliva that would have normally been deposited in the rumen. Furthermore, heat-stressed cows ruminate less due to reduced feed intake and therefore generate less saliva. The reductions in the amount of saliva produced and salivary HCO_3^- content and the decreased amount of saliva entering the rumen make the heat-stressed cow much more susceptible to subclinical and acute rumen acidosis (Kadzere et al. 2002). When cows begin to accumulate heat, there is a redistribution of blood to the extremities in an attempt to dissipate internal energy. As a consequence, there is reduced blood flow to the gastrointestinal tract, and as a result nutrient uptake may be compromised (McGuire et al. 1989). Therefore, accumulation of end products of fermentation (VFAs) contributes to the reduced pH and indirectly enhances the risk of negative side effects of an unhealthy rumen (i.e. laminitis, milk fat depression, etc.) (Baumgard and Rhoads 2007).

A prerequisite of understanding the metabolic adaptations which occur with heat stress is an appreciation of the physiological and metabolic adaptations to thermal-neutral negative energy balance (i.e. underfeeding or during the transition period). Cows in early lactation are classic examples of when nutrient intake is less than necessary to meet maintenance and milk production costs, and cows typically enter negative energy balance (Moore et al. 2005a). Negative energy balance is associated with various metabolic changes that are implemented to support the dominant physiological condition of lactation (Bauman and Currie 1980). Marked alterations in both carbohydrate and lipid metabolism ensure partitioning of dietary-derived and tissue-originating nutrients towards the mammary gland, and many of these changes are mediated by endogenous somatotropin which is increased during periods of negative energy balance (Bauman and Currie 1980). There is reduction in circulating insulin coupled with a reduction in systemic insulin sensitivity. The reduction in insulin action allows for adipose lipolysis and mobilisation of nonesterified fatty acids (NEFA; Bauman and Currie 1980). Increased circulating NEFA is typical in 'transition' cows and is an important source of energy (and precursor for milk fat synthesis) for cows during negative energy balance. Postabsorptive carbohydrate metabolism is also altered by the reduced insulin action during negative energy balance with the net effect of reduced glucose uptake by systemic tissues (i.e. muscle and adipose). Reduction in nutrient uptake coupled with the net release of nutrients (i.e. amino acids and NEFA) by systemic tissues are key homeorhetic (an acclimated response vs. an acute/homeostatic response) mechanisms implemented by cows in negative energy balance in order to support lactation (Bauman and Currie 1980).

8 Mechanism by Which Heat Stress Reduces Milk Yield

Heat stress reduces both the feed intake and milk yield of cows. The decline in nutrient intake has been identified as a major cause of reduced milk synthesis (Fuquay 1981). A reduction in energy intake combined with increased energy expenditure for maintenance lowers energy balance and partially explains why lactating cattle lose significant amounts of body weight during severe heat stress (Rhoads et al. 2009; Shwartz et al. 2009). Heat-stressed cows have been observed to respond immediately and an immediate reduction (~5 kg/day) in dry matter intake (DMI) with the decrease reaching a peak at ~day 4 and remaining stable thereafter. Thermal-neutral pair-fed cows had a feed intake pattern similar to heat-stressed cows. Heat stress reduced milk yield by ~14 kg/day with production steadily declining for the first 7 days and then reach a plateau. Thermal-neutral pair-fed cows also had a reduction in milk yield of approximately 6 kg/day, but milk production reached its nadir at day 2 and remained relatively stable thereafter. This indicates the reduction in DMI can only account for ~40–50% of the decrease in production when cows are heat stressed and that ~50–60% can be explained by other heat-stressed-induced changes (Rhoads et al. 2007).

The negative energy balance associated with the early postpartum period is coupled with increased risk of metabolic disorders and health problems (Goff and Horst 1997; Drackley 1999), decreased milk yield and reduced reproductive performance (Lucy et al. 1992; Beam and Butler 1999; Baumgard et al. 2002, 2006). It is likely that many of the negative effects of heat stress on production, animal health and reproduction indices are mediated by the reduction in energy balance. However, it is not clear how much of the reduction in performance (yield and reproduction) can be attributed or accounted for by the biological parameters effected by heat stress (i.e. reduced feed intake vs. increased maintenance costs). Seasonal differences in milk production are caused by periodic changes of environment over the year, which has (1) a direct effect on animal's milk production through decreased DMI and (2) an indirect effect through fluctuation in quantity and quality of feed. Process of lactation is a physiological process that presents a substantial challenge to the homeostasis of the cardiovascular and fluid secretory system (Silanikove 1994, 2000; Maltz and Silanikove 1996). The acute and large decrease in milk secretion may, therefore, be considered as having vital importance that makes it necessary to enable the cows to survive under heat stress. The initial reactions to acute heat stress may be emotional responses (Silanikove 2000), but prolonged challenges impact milk yields and composition; milk protein fat contents were found to be reduced. These reductions may be a part of their adaptive response (Kadzere et al. 2002; Collier et al. 2006; Igono et al. 1992).

9   Mechanism of Regulation of Milk Secretion and Mammary Function

Milk secretion and mammary function are regulated acutely by local autocrine feedback mechanisms that involve milk-borne factors which are sensitive to the frequency and efficiency of milking of cows (Daly et al. 1993; Wilde and Peaker 1990). Sustained changes in the frequency of milking and milk secretion are associated with metabolic adaptations (Shennan and McNeillie 1994) and with long-term adaptations in the degree of differentiation and the number of mammary epithelial cells (Liu et al. 1997; Quarrie et al. 1998). The fast modulation of milk secretion in response to external factors, such as emotional stress, heat stress and water deprivation also depends on a negative feedback regulatory system, required for survival of a species increases (Silanikove et al. 2006). The negative feedback system has been shown to comprise an endogenous milk enzymatic system, the plasminogen activator (PA)-plasminogen (PG)-plasmin (PL), that specifically forms a β-casein (CN) fragment (f) (1–28) from β-CN, which acts as the negative control signal by closing potassium channels on the apical membrane of the epithelial cells of the mammary gland (Silanikove et al. 2000, 2006). Inhibition of milk secretion occurs due to downregulation of these channels by inducing undefined inwardly directed cellular signals. A further activation of the PA-PG-PL system, coupled with more extensive degradation of casein-induced involution of the mammary gland in lactating goats and cows, has been observed that forcefully activate the innate immune system (Silanikove et al. 2005, 2006). The concept that PA-PG-PL-β-CN f (1–28) is involved in milk-borne negative feedback regulation of milk secretion was supported experimentally under conditions that simulated stress (intramammary treatment with dexamethasone) (Silanikove et al. 2000; Shamay et al. 2000) and by exposing the cows to dehydration (Silanikove et al. 2000).

The involvement of PA-PG-PL system in regulation of milk secretion (Shamay et al. 2003) and the induction of mammary gland involution (Silanikove et al. 2005) are well documented, and the effects of the PA-PG-PL system are related to enhanced degradation of the extracellular matrix (Lund et al. 2000). The PA-PG-PL system works in mammary secretion by increasing casein degradation and liberation of active components and that β-CN f (1–28) is a principal casein degradation product that is involved in negative control of milk secretion in cows under heat stress.

The acute phase in the regulatory inhibition of milk secretion in cows subjected to heat stress is related to upregulation of the local PA-PG-PL-β-CN f (1–28) peptide in milk and that this peptide in turn downregulates the activity of K^+ channels on apical membranes-derived vesicles. The putative apical K^+ channels belong to the family of voltage-gated channels and that β-CN f (1–28) causes membrane depolarisation, explaining its milk downregulatory effect (Silanikove et al. 2009). To understand the exact mechanism on milk synthesis, further research is needed to determine the nature of the interaction of β-CN f (1–28) with regulatory elements in the apical membrane of mammary gland epithelial cells and to identify these channels and the components of the inward signal transduction. Decrease in Na^+ concentration, increase in K^+ concentration and the consequent decrease in the Na/K ratio are sensitive indicators of the disruption of the tight junction of the mammary gland epithelial cells, which relates to differing ion contents in milk and blood plasma (Stelwagen et al. 1998).

10   Heat Stress Effects on Heifers

Heifers have been observed to generate far less metabolic heat than cows and have greater surface area relative to internal body mass and, therefore, are expected to suffer less from heat stress. However, it has been observed that Holstein females raised at latitudes less than 34 °N weighed 6–10% less at birth and average approximately 16% lower body weight at maturity than those in more northern latitudes, even when sired by the same bulls (NRC 1981). Thus, there are several factors contributing to slower growth and smaller body size, including greater maintenance requirements during hot weather, poor appetite and lower quality forages that are influenced by the same environmental conditions that lead to slow growth of cattle.

During hot weather, reduced feed intake is common, but increased maintenance costs reduce efficiency of feed conversion. In a study from the 1950s where Holsteins, Brown Swiss and Jersey heifers were raised from 1 to 13 months of age in environmental chambers with constant temperatures of 10 or 26.7°C, Holstein heifers raised in the 26.7°C environment were lighter than heifers in the cool environment by 8.2 kg at 3 months and 30.4 kg at 11 months of age. It took Holsteins in the warm environment 11/2 months longer to reach 299-kg body weight (Johnson and Ragsdale 1959). Although the temperature was constant with no diurnal variation, 26.7°C is not extremely hot. In Australia, Friesians, Brahman × Friesian F_1 crosses and Brahmans were exposed to 17.2 and 37.8°C temperatures (Colditz and Kellaway 1972). Comparing the hot versus the cool temperature environments, rectal temperature and respiration rates increased more for Friesians as compared to Brahmans. Intake declined about 17% for Friesians, 1.4% for F_1 crosses and 12% for Brahmans, but initial intake was greater for Friesians, and thus a greater decline would be expected. Gains for Friesians were greatest during cool temperatures but were the least of the three groups when exposed to high temperatures.

Because heifers generate less body heat and can dissipate heat more readily than lactating cows, do heifers require additional cooling? In Egypt, heifers were exposed to winter conditions (17.3°C, 54.5% RH), summer conditions (36°C, 47% RH) and summer conditions with water spraying and an oral diaphoretic (Marai et al. 1995). A diaphoretic (e.g. ammonium acetate) is a compound fed orally to cattle to increase perspiration. Heifers were sprayed with water seven times daily during the hottest period of the day. Heifers that received cooling had lower rectal temperature and respiratory rate, and gain was improved by 26.1% with cooling during summer, even though heifers were only sprayed during the hottest part of the day without the benefit of fans.

11   Genetic Factors Regulating Response to Heat Stress

Development of genomics tools has permitted a much better evaluation of the genotype × environment interactions (G×E). Estimation of G×E effects in dairy cattle has indicated that these effects are larger, and genetic differences in heat tolerance between animals appear to be exacerbated under high temperature conditions. Preimplantation embryos from *B. indicus* cattle are better able to withstand thermal stress as compared to embryos from *B. taurus* cattle. Thus, identifying genetic causes of differences between animals in their response to the environment has potential for improving productivity of animals in adverse environments such as heat stress. The ability to use powerful new tools in genomics, proteomics and metabolomics to evaluate genetic differences between animals in their response to thermal stress will yield important new information in the next quarter-century and will permit the selection of cattle for resistance to thermal stress.

There is genetic variation in heat loss via tissue conductance, non-evaporative heat loss and evaporative heat loss, but more efficient heat loss occurred for Brahman and Brahman cross cattle than with Shorthorn cattle (Finch 1985). Using Brahman, Friesian and Brahman×Friesian F_1 cross heifers, the Brahman×Friesian crosses had superior gains at 38°C but gains for Friesians were greater at 17°C (Colditz and Kellaway 1972). Brahmans gained more slowly at 38°C thereby indicating to be benefits from hybrid vigour under heat stress conditions. Hair colour influences the susceptibility of the cow to heat stress because coat colour is related to the amount of heat absorbed from solar radiation. In *B. Indicus* cattle, the inward flow of heat at the skin of black steers was 16% greater than for brown steers and 58% greater than for white steers (Finch 1986). *B. Taurus* cattle with dark coats exhibited greater heat transfer to the skin, higher body temperature and sharply reduced weight gains than those with white coats, with increasing woolliness of the coat accentuating the effect (Finch 1986). When dairy cows from an Arizona herd were categorised into white (less than 40% black), mixed (40–60% black) or greater than 60% black, no production traits were different (perhaps because cows were cooled for the first 130 days of lactation), but white cows calving in February and March required fewer services per conception and had fewer open days than mixed and black cows (King et al. 1988). Heritability

of coat colour was 0.22. In a study using cows characterised as greater than 70% white or greater than 70% black, white cows had slightly lower body temperatures and greater milk yield, regardless of whether they were in shade or no shade conditions (Hansen 1990).

Because genetic variation exists for traits important to thermoregulation in livestock species, the potential to select sires that can transmit important traits must be considered. However, when bulls were evaluated for genotype environment interactions using daughters in California, New York and Wisconsin, there was no sire region interaction for milk or fat yield (Carabaño et al. 1990). Because the genetic correlation between production and heat tolerance was approximately −0.3, the continued selection for production by ignoring heat tolerance results in decrease in heat tolerance. The correlation being small, a combined selection for production and heat tolerance is likely possible. Functional genomics establishes a verifiable link between gene expression and phenotype. Gene expression arrays in particular allow global analysis of gene expression responses to environmental change. Stress is defined as an external event or condition that produces a 'strain' in a biological system (Lee 1965). When the stress is environmental, the strain is measured as a change in body temperature, metabolic rate, productivity or heat conservation and/or dissipation mechanisms. At the cellular level, acute environmental change initiates the 'heat-shock' or cellular stress response. Changes in gene expression associated with a reaction to an environmental stressor involve acute responses at the cellular level (in most if not all cells) as well as changes in gene expression across a variety of organs and tissues which associated with the acclimation response.

Early work by Guerriero and Raynes (1990) demonstrated elevated heat-shock proteins in response to thermal stress in bovine blood leucocytes. Moderate heat shock (41°C) causes increased heat-shock protein synthesis, decreased protein synthesis, mitochondrial swelling and movement of organelles away from the plasma membrane associated with cytoskeletal reorganisation in the early bovine embryo (Edwards and

Hansen 1997; Edwards et al. 1997; Rivera and Hansen 2001; Rivera et al. 2003). Thermal stress triggers a dramatic and complex programme of altered gene expression in bovine mammary epithelial cells (BMEC) similar to patterns reported in other cell types exposed to thermal stress. As reported by Sonna et al. (2002), these changes include (1) inhibition of DNA synthesis, transcription, RNA processing and translation; (2) inhibition of progression through the cell cycle; (3) denaturation and misaggregation of proteins; (4) increased degradation of proteins through proteasomal and lysosomal pathways; (5) disruption of cytoskeletal components; (6) alterations to metabolism that lead to a net reduction in cellular ATP; and (7) changes in membrane permeability that lead to an increase in intracellular Na^+, H^+ and Ca^{2+} concentrations. Thermal stress induced the changes in gene expression along with rapid regression of BMEC ductal structures. Transcriptional activity indicated a downregulation of a number of genes associated with branching morphogenesis and microtubule activity, thereby suggesting a repression of the genomic signals responsible for promoting ductal growth and networking (Collier et al. 2006). Overall, the transcriptome profile indicated downregulation of genes involved in cell structure, metabolism, biosynthesis and intracellular transport. The upregulated genes during heat shock in BMEC mainly were involved in cellular repair, protein repair and degradation and apoptosis after loss of thermotolerance when HSP-70 gene expression fell to basal levels (Edwards et al. 1997). These data indicate that morphogenic activity in the mammary epithelium might depend upon the expression profile of a core set of genes and that structural assembly is under a positive mode of regulation (i.e. morphogenesis is 'on' by default). In turn, the transition from structural assembly to disassembly might be controlled at the genomic level by simply shutting down cellular biosynthesis and core morphogenic genes. In contrast, transcription of genes encoding repair enzymes and apoptotic proteins is kept off until the cell requires them and removes the inhibition, which suggests a negative mode of regulation.

An additional group of genes dominated by patterns of downregulation were those involved in BMEC differentiation and milk synthesis. This suggests that (1) even during growth and morphogenesis, BMEC expresses detectable mRNA levels of some lactogenic genes; and (2) heat-induced BMEC regression includes transcriptional repression of genes involved in milk synthesis. This strongly implies that milk yield losses in lactating dairy cows exposed to thermal stress are due in part to direct repression of genes associated with milk synthesis.

Thermotolerance in BMEC was found to be lost after 8 h of exposure to thermal stress when HSP-70 gene expression returned to basal levels, which was associated with increased expression of genes in the apoptotic pathways, indicating these cells were in the process of undergoing apoptosis. These studies were carried out using BMEC from non-adapted and non-acclimated cattle (Collier et al. 2006). Thus, a portion of the loss in milk yield during acute thermal stress is associated with direct effects of thermal stress on BMEC. Acute thermal stress of growing bovine mammary epithelial cells directly reduces cellular growth and ductal branching and downregulates genes associated with protein synthesis and cellular metabolism. Chronic thermal stress would likely reduce mammary growth during pregnancy. Furthermore, negative effects of thermal stress on expression of milk protein genes indicate that thermal stress likely has direct negative effects on milk yield.

12   Reducing Heat Stress for Improving Milk Production

The body temperatures of the cow should be maintained below 102.5°F (39.2°C) and respiration rates below 80 per minute. Heat stress should be evaluated at the cows' nose level both lying down and standing at the bunk and in the holding pen. Often, considerable heat stress occurs in the holding area while cows are waiting to be milked. Igono et al. (1992) found that despite high ambient temperatures during the day, a cool period of less than 21°C for 3–6 h minimises the decline in milk yield.

These findings suggest that it is critical to minimise increase in cow's body temperature

Table 1 Effect of water cooling on milk yield during summer

Season	Increase in milk yield (kg/day)	References
Cows		
Hot–dry	4.0	Armstrong (1994)
Hot–dry	2.12	Aggarwal (2004)
Hot–humid	1.9	Flamenbaum et al. (1995)
Hot–humid	1.5	Lin et al. (1998)
Hot–humid	0.85	Aggarwal (2004)
Buffaloes		
Hot–dry	1.05	Gangwar (1985)
Hot–dry	1.0	Aggarwal and Singh (2008)
Hot–humid	0.58	Gangwar (1985)
Hot–humid	0.92	Aggarwal and Singh (2008)

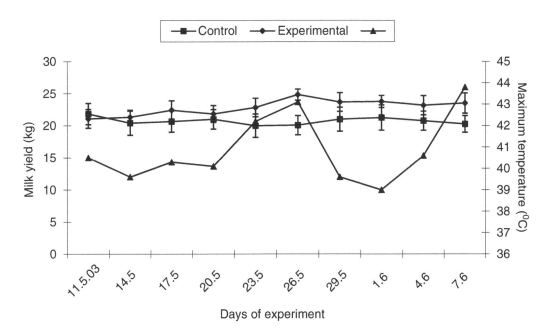

Fig. 1 Milk yield of cows of control group and experimental group (provided with mist and fan system) during hot–dry season

during the hot daylight hours and to find methods for cooling cows during heat stress. Ventilation is very critical. Natural ventilation using side-wall curtains works very well. Fans are required, especially over the feed bunk and in the holding pen. The air speed over the cow should be 400–600 ft (122–183 m) per minute. Usually this requires at least one 36-in. (91 cm) fan (with airflow of 11,000 cfm) for every 30 ft (9.1 m). Fans should be angled downwards. Shade should be provided to the cows. Misters are also helpful for enhancing evaporative cooling. Table 1 shows the effect of water cooling on milk yield in cows and buffaloes during summer. In a study, it was found that when cows in early lactation were provided with mist and fan cooling system during hot–dry and hot–humid seasons from 11.00 A.M. to 4 P.M., the milk yield was significantly increased (Aggarwal 2004; Figs. 1 and 2). Various methods for cooling

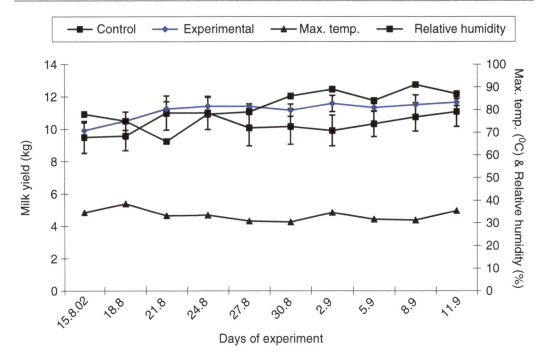

Fig. 2 Milk yield of cows of control group and experimental group (provided with mist and fan system) during hot–humid season

cows and buffaloes during heat stress have been discussed in Chap. 7.

12.1 Nutritional Management to Reduce Heat Stress

There are several key areas related to nutritional management which should be considered while feeding in hot weather.

12.1.1 Water Intake

Water is the most important nutrient for the livestock species. Water intake is closely related to DMI and milk yield of cows, but minimum temperature was the second variable to enter a stepwise regression equation after DMI, indicating the influence of ambient temperature on water consumption (Murphy et al. 1983). Water intake has been observed to increase by 1.2 kg/°C increase in minimum ambient temperature. Intake of dry matter declines during heat stress; therefore, nutrient density of the diet must be increased

in order to increase dietary proteins in relation to requirements, but there is an energetic cost associated with feeding excess proteins. Excess nitrogen above requirements reduces metabolisable energy by 7.2 kcal/g of nitrogen (Tyrrell et al. 1970). Feeding 19 and 23% crude protein diets reduced milk yield by over 1.4 kg (Danfaer et al. 1980), and the energy cost associated with synthesising and excreting urea accounted for the reduced milk yield (Oldham 1984). Dietary protein degradability is critical under heat stress conditions. As diets with low (31.2% of CP) and high (39.2% of CP) rumen undegradable protein fed during hot weather had no effect on DMI, however, milk yield increased (Belibasakis et al. 1995). Cooling the cow may affect the response of the cow to protein supplementation. Although the interaction of protein quality and environment is not significant, the greater response to high quality protein for cows in the cooled environment was attributed to the reduced amount of protein metabolised for energy reduced, and less energy was used in converting NH_3 to urea. When cows are subject to hot weather conditions, rumen degradable protein (RDP) should not exceed 61% of dietary CP, and total protein should not exceed NRC recommendations by greater than 100 g N/day (Huber et al. 1994). One hundred grams N is equivalent to about 3.1% CP in the diet, assuming 20 kg DMI/day. High dietary lysine (241 g/day, 1% of DM) increased milk yield by 3 kg over diets containing 137 g/day lysine (0.6% of DM) (Huber et al. 1994).

Metabolic heat production, though advantageous during cold weather, is a liability during hot weather due to the difficulty in maintaining body heat balance. Heat production for a 600-kg cow yielding 40 kg of 4% fat milk amounted to 31.1% of consumed energy, which was second to faecal energy losses of 35.3% (Coppock 1985). While maintenance was responsible for 23.5% of the heat produced, greater milk yield also increases heat production. Cows at high (31.6 kg/day) and medium (18.5 kg/day) milk yield had 48.5 and 27.3% greater heat production than dry cows (Purwanto et al. 1990). Use of some dietary ingredients may contribute less to heat increment of the diet, thus reducing total heat production of the cow. Lower efficiency for use of acetate may account for the low net energy of high-fibre feeds (Moe 1981) and supports the feeding of low-fibre diets during hot weather.

Particularly, chronic heat exposure of 6- and 12-month-old buffalo calves was accompanied with highly significant ($P<0.01$) increases in total water intake (28.5 and 48.3%), total body water (TBW) content (8. 5 and 9.6%), free water intake (25.2 and 56.4%), urine excretion (24.8 and 108.0%) and evaporative water loss (51.2 and 69.4%). Significant respective decreases were recorded in 6- and 12-month-old buffalo calves in metabolic (which is derived from oxidation of fats, carbohydrates and proteins) (20.8 and 16.8%) and faecal water excretion (36.4 and 8.5%). Dietary water intake decreased (16.3%) due to chronic heat exposure in 12-month-old calves (Nessim 2004).

The consumed water may replace the lost TBS by heat stress, since it was found that a net total body solids loss of 10 kg, in 3 days of elevated heat exposure, was replaced by extra body water retained during these 3 days without a significant change in body weight, in cattle (Kamal and Johnson 1971). Ambient relative humidity showed no significant effect on water consumption in cattle (Mullick 1964). However, Mishra et al. (1963) showed that a drop in dietary water intake occurred as ambient temperature increases in buffalo cows. Particularly, a significant positive correlation between temperature and water consumption and nonsignificant negative correlation between relative humidity and water consumption were found when temperature was held constant in lactating and non-lactating dairy cows (Harbin et al. 1958). Concerning the mechanisms underlying water intake, water intake at low or high ambient temperature was blocked when urine excretion was inhibited by ADH administration in cows. This indicates that water retention caused by ceasing urination might block thirst (Kamal et al. 1959).

The drinking behaviour of the high-producing cows increased compared to the low-producing dairy cows, for which two factors were involved. The first one is the higher milk yield, and the other is hot environment since the high milk yielding cows had faster dehydration rates by increased sweating and respiratory water loss as compared to low yielders (Berman et al. 1985) during summer. The higher drinking behaviour of the high-producing cows is consistent with the notion that high-producing cows try to overcome effects of the high temperature of summer. Water requirement by animal is highly influenced by demands to maintain homeothermy during heat stress (Beede and Collier 1986).

12.1.2 Rumen Health

The heat-stressed cows are prone to rumen acidosis, and many of the lasting effects of warm weather (laminitis, low milk fats, etc.) are probably related to a low rumen pH during the summer. Therefore, adequate care should be taken when feeding *hot* rations during the summer months. In addition, obviously fibre quality is important all the time, but it is paramount during the summer as it has some buffering capacity and stimulates saliva production. Dietary HCO^{3-} may be a valuable tool to maintain a healthy rumen pH.

12.1.3 Energy

The most limiting nutrient for lactating dairy cows during summer is usually energy intake, and a common approach to increase energy density is to reduce forage and increase concentrate content of the ration. However, increasing concentrates to greater than 55–60% of the diet dry matter is risky and can result in depressed milk fat content, acidosis, cows going off feed, laminitis and reduced efficiency of nutrient use. High-fibre diets increase heat production, but addition of oil cakes helps in reducing heat stress and improves efficiency of milk production. While heat increment is a consideration for high-fibre diets, total intake has a much greater impact on metabolic heat production by the animal. Growing heifers fed pelleted rations containing 75% alfalfa or 25% alfalfa produced 48.8 and 45.5 MJ/day of heat (Reynolds et al. 1991), but heat production for low and high intake heifers (4.2 and 7.1 kg/day DMI) was 38.2 and 56.1 MJ/day. Therefore, intake has a substantial effect on heat production and must be considered in formulating an effective nutritional and environmental management programme under tropical and subtropical conditions. Intake normally declines for high-fibre diets, and West et al. (1999) demonstrated that the DMI decline for diets with a range of NDF concentration from 27 to 35% was less severe with increasing NDF during hot weather. The total DMI was less during hot weather and suggests that the less severe decline in hot weather was due to lower intake and not higher NDF content.

Added dietary fat is an excellent way to increase energy content of the diet, especially during summer when feed intake is depressed. Fat is high in energy (about 2.25 times as much as carbohydrate), does not add starch to the diet (minimising rumen acidosis) and may reduce heat load in summer. Added dietary fat often boosts milk fat test a point or two. Addition of rumen-protected fat is a good option to increase energy of the ration. One rule of thumb when high fat addition is required is that 1/3 comes from natural feed ingredients, 1/3 comes from oilseeds and 1/3 comes from rumen bypass fats. Despite the fact that heat-stressed cows limitedly oxidise body reserves for energy, feeding dietary fat is probably an effective strategy of providing extra energy. Compared to starch and fibre, fat has a much lower heat increment (Van Soest 1982); thus, it can provide energy without a negative thermal side effect and prove to be a strategy to deal with heat stress. Maximising rumen production of glucose precursors (i.e. propionate) is also an effective strategy to maintain milk production of cows under tropical conditions. Rumen microbes suffer under acidic conditions. The cells of the cow's body also have trouble when they encounter too much acid. Acids change enzyme activities and affect the structure of molecules.

12.1.4 Dietary DCAD

Bovines utilise potassium (K+) as their primary osmotic regulator of water secretion from their sweat glands, and as a consequence, K+ requirements of cows are increased (1.4–1.6% of DM) during the summer. In addition, dietary levels of sodium (Na+) and magnesium (Mg+) should be increased as they compete with K+ for intestinal absorption (West 2002). The loss of potassium increases blood acidity. It has been found that blood pH can be increased by increasing the dietary cation–anion difference (DCAD). This is the difference between the amounts of positively charged cations (especially sodium and potassium) in the diet and the negatively charged anions (especially chloride and sulphur) in the diet. Raising DCAD increases the ability of the cow's blood to buffer acids, and this raises blood pH (decreasing acidity).

Researchers have found positive milk production responses when they have raised DCAD to 35–45 meq/100 g DM. Generally, 1.6–1.8% dietary potassium, 0.34–0.45 kg of added buffer and 0.40% sodium are required to significantly increase DCAD. Chloride levels also need to be controlled (0.40%). Therefore, potassium carbonate should be used rather than potassium chloride.

Feeding diets that have a high dietary cation–anion difference (DCAD) improved DMI and milk yield (Tucker et al. 1988; West et al. 1991). Addition of DCAD in the diet of cattle during heat stress conditions improves DMI.

This suggests that the DCAD equation is more significant than the individual element concentrations. Having a negative DCAD during the dry period and a positive DCAD during lactation is a good strategy to maintain health and maximise production (Block 1994). Keeping the DCAD at a healthy lactating level (~ +20 to +30 meq/100 g DM) remains a good strategy during the warm summer months (Wildman et al. 2007).

12.1.5 Feed Intake

Feed consumption depression is the most important reaction to exposure to elevated temperature in tropical and subtropical conditions (Marai et al. 1994, 2002). Under heat stress conditions, as the quantity of consumed nutrients declines, dry matter (DM) intake including crude protein declines, and a negative nitrogen balance may occur (West 1999). DM digestibility and protein/energy ratio were also found to decrease in heat stress conditions (Moss 1993). Animals in a highly productive state (high producers) have feed intakes and metabolic rates that may be two to four times higher than at maintenance (NRC 1989). Heat stress in such high-producing lactating dairy cows results in reductions in roughage intake and rumination (Collier et al. 1982). The reduction in appetite under heat stress is a result of elevated body temperature and may be related to gut fill (Silanikove 1992). Decreased roughage intake contributes to decreased VFA production and may lead to alterations in the ratio of acetate and propionate. In addition, rumen pH is depressed during heat stress (Collier et al. 1982). However, ruminants well adapted to hot environments are able to maintain their feed intake under heat stress at near maintenance or during moderate growth (Brosh et al. 1988). In chamber experiments, heat-stressed cows changed their feeding pattern and ate when temperatures were cooler (Schneider et al. 1988). In temperature stress experiments with lactating cows, the major decrease in milk production at high ambient temperatures is a result of reduced feed intake (Wayman et al. 1962). Using rumen-fistulated lactating cows, Wayman et al. (1962) demonstrated that the drop in milk production due to heat stress could be reduced by placing feed

rejected due to thermal stress directly into the rumen. A decrease in the efficiency of energy utilisation for milk production when cows were under heat stress was reported. In lactating Murrah buffaloes, digestibility coefficient values for each of DM and crude protein were significantly lower in summer (43.0 and 50.50 ± 0.7, respectively) than in winter (68.31 and 66.83 ± 0.05, respectively) (Verma et al. 2000). Digestion and metabolism of non-pregnant female buffaloes declined when exposed for 2–3 h to solar radiation at air temperature of $42 °C$ (Zhengkang et al. 1994). Nitrogen retention decreased significantly under heat stress conditions ($35 °C$) when compared with the comfort conditions ($10 °C$) in Jersey, Brown Swiss and Holstein heifers. The percentage decline ranged between 25.4 and 49.0 (Kamal et al. 1962). In Holstein breed, the nitrogen balance was positive in calves and negative in cows (Kamal et al. 1970).

13 Effect of Heat Stress on Milk Production in Buffaloes

It has been observed that there are large seasonal variations in breeding and calving in buffaloes in most of the buffalo milk-producing countries of Asia (Ganguli 1981). In India and Pakistan, 80% of the buffaloes calve during June and December causing a decline in milk production in the summer months. However, others have suggested that heat stress and shortage of green fodder in summer can decrease milk production. A dark body, lesser density of sweat glands and thick epidermis make it difficult for the buffaloes to flourish in extreme hot sunny and dry conditions. Buffaloes have developed survival mechanisms to seek water for immersion in these conditions. The buffaloes' milk production and reproductive efficiency are strongly affected, when exposed to extreme summer or winter (Sastry 1983).

13.1 Milk Yield and Its Constituents

13.1.1 Milk Yield

Differences in the physiological responses of cattle due to the form and duration of heat

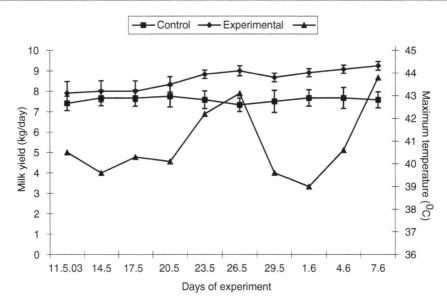

Fig. 3 Milk yield of buffaloes of control group (kept under showers) and experimental group (kept in a water tank) during hot–dry season

stress have been reported, and reflection of such responses has also been noted in productive (and reproductive) performance traits. Season of calving affected significantly milk yield in Indian (Roy Chaudhury and Deskmuykh 1975) and Egyptian buffaloes (Mourad 1978; Mohamed 2000; Marai et al. 2009). The highest milk yield was recorded during spring and winter (by calving during the mild period) and the lowest in summer (by calving during the hot period), in Egyptian buffaloes (El-Khaschab et al. 1984). Some studies have shown no significant effect of season of calving on milk yield in Egyptian buffaloes (Alim 1967; Marai and Habeeb 2010). The insignificant difference in total milk yield due to season of calving may be an evidence for the availability of adequate managerial conditions all year round. Decline in milk yield as a direct result of high environmental temperatures had been reported by many authors (Thatcher 1974; Johnson 1976; Marai et al. 2009). Between 20°C (18.2 kg) and 35°C (16.7 kg), the reduction in milk yield was estimated to be 9%. Particularly, the rise in temperature averages by 1.6, 3.2 and 8.8°C above normal (21°C); results in the decrease in daily milk yield average by 4.5, 6.8 and 14%, respectively. On the other side, the decline in the daily temperature by 7°C below

normal resulted in an increase in daily milk yield by 6.5% in dairy cattle (Petkov 1971). At 30°C, the high- and low-producing animals showed a mean reduction of 2.0 and 0.65 kg/day, respectively (Vanjonack and Johnson 1975). In a study, two cooling methods, namely, water showers and wallowing, were compared. Milk yield of Murrah Buffaloes was found to be significantly higher in buffaloes which were allowed to wallow during hot–dry (Fig. 3) and during hot–humid (Fig. 4) seasons as compared to the buffaloes kept under showers. The results indicated more beneficial effects of wallowing than water showers during heat stress (Anjali and Singh 2008).

13.1.2 Milk Constituents

Milk constituents are greatly affected by hyperthermia. In lactating Holstein cows transferred from an air temperature of 18–30°C, milk fat, solids-not-fat and milk protein percentages decreased with 39.7, 18.9 and 16.9%, respectively (McDowell et al. 1976). Friesian cows maintained under 38°C showed lower averages of total solids, fat, protein, ash and lactose yields than when the same animals were maintained under thermoneutral environmental temperatures. The reduction percentages were 28, 27, 7, 22.7 and 30, respectively (Habeeb et al. 1989). Similar

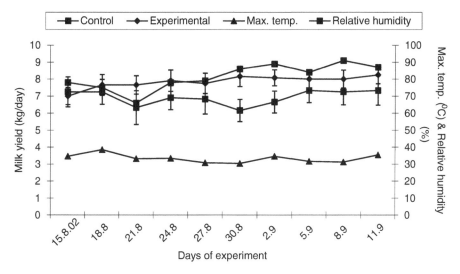

Fig. 4 Milk yield of buffaloes of control group (kept under showers) and experimental group (kept in a water tank) during hot–humid season

reduction values in milk constituents were reported by Habeeb et al. (1993, 1996), Yousef et al. (1996) and Marai et al. (1997a, b). Rodriguez et al. (1985) demonstrated that fat and protein percentages declined between 8 and 37°C and protein to fat ratio decreased at temperatures above 29°C, while chloride content increased above 21°C, in Friesian cows. Averages of phosphorus and magnesium values were also found to be less in summer. Citric acid and calcium contents decreased during early lactation, while potassium decreased in all lactation stages at high temperatures (Kamal et al. 1962). The decrease in milk yield and milk constituents of dairy cattle is a result to the depression in feed consumption which is the most important reaction to heat exposure. In a comparative study, one group of buffaloes were kept under water showers, and the other group of buffaloes were kept in a wallowing pond. The results indicated that in wallowing group of buffaloes, the fat, protein and lactose content of milk was significantly improved as compared to showers group (Aggarwal and Singh 2006).

13.2 Comparison Between Buffaloes and Cattle

The wide distribution of buffaloes in the world indicates that buffaloes are more adaptable than cattle to a large range of the environmental conditions. Buffalo productivity surpassed that of cattle, with males reaching 400 kg in 30 months on a diet of native grasses. Buffaloes are less affected by high humidity as compared to cattle if they are provided with shade or wallows are available. In Southern Brazil, comparison between buffaloes and cattle on subtropical riverine plains has also favoured the buffaloes. Buffaloes' adaptability to the subtropical environment of Egypt was found to be better than for Friesians. The estimated values of adaptability were 89.1 and 82.9% for buffaloes and Friesians, respectively (Marai et al. 2009).

14 Conclusions

Summer heat stress negatively impacts ruminant (especially dairy animals) performance in most areas of the world. The severity of heat stress issues will become more of a problem in the future as global warming progresses, and genetic selection for milk yield continues. Heat stress reduces milk production in cows with high genetic merit for milk production. Strategies to alleviate metabolic and environmental heat loads in early lactation need to be researched and developed. The heat-stressed lactating dairy cow has an extra need for glucose (due to its preferential oxidisation by extra mammary tissue). Therefore, any dietary component that increases propionate production (the primary precursor to hepatic glucose production), without reducing rumen pH, will probably increase milk yield. In addition, reducing systemic insulin sensitivity will increase glucose availability to the mammary and thus also probably increase milk yield. Feeding of rumen-protected fats and proteins are among other endeavours to reduce metabolic heat production and supply the correct profile of nutrients to high-producing cows in early lactation. An adequate supply of nutrients must also include well-balanced mixture of dietary minerals, especially of Na, K, Cl and SO_4^{2-}. These play a pivotal role in the thermal physiology of the cow. Milk production per cow has increased over threefold during last 50 years in response to advances in animal nutrition, in technology and in biotechnology as well as genetic progress for milk production. Using these facts as a basis, it is apparent that genetic selection and other variables enhancing milk production may have resulted in adjustments in factors important to lactation and nutritional physiology of the dairy cow. One such factor, important especially in hot environments, is the thermoregulatory ability and capacity of cows. It is important to quantify thermoregulatory capabilities and the physiological effects of heat stress on high-producing cows in modern dairies. Such empirical data are prerequisites for improving nutrition, fine tuning nutrient supply and adjusting the management of high merit cows so that they can express theirtrue genetic potential for milk production. Determination that adaptation of animals to thermal stress is a homeorhetic process under endocrine control opens new opportunities to use endocrine regulation as means of improving thermal tolerance. Substantial efforts are underway to identify specific genes associated with tolerance and sensitivity to thermal stress. Additional work is needed to reduce energy costs of housing and cooling animals during thermal stress. Accurately identifying heat-stressed cows and understanding the biological mechanism(s) by which thermal stress reduces milk synthesis and reproductive indices is critical for developing novel approaches (i.e. genetic,

managerial and nutritional) to maintain production or minimise the reduction in dairy cow productivity during stressful summer months.

References

Aggarwal A (2004) Effect of environment on hormones, blood metabolites, milk production and composition under two sets of management in cows and buffaloes. PhD thesis submitted to National Dairy Research Institute, Karnal, Haryana

Aggarwal A, Singh M (2006) Effect of water cooling on physiological responses, milk production and composition of Murrah buffaloes during hot-humid season. Indian J Dairy Sci 59:386–389

Aggarwal A, Singh M (2008) Changes in skin and rectal temperature in lactating buffaloes provided with showers and wallowing during hot-dry season. Trop Anim Health Prod 40:223–228

Alim KA (1967) Repeatability of milk yield and length of lactation of the milking buffaloes, in Egypt. Trop Agric Trinidad 44:159–163

Al-Katanani YM, Webb DW, Hansen PJ (1999) Factors affecting seasonal variation in 90-day nonreturn rate to first service in lactating Holstein cows in a hot climate. J Dairy Sci 82:2611–2616

Arave CW, Albright JL (1981) Cattle behavior. J Dairy Sci 64(6):1318–1329

Armstrong DV (1994) Heat stress interaction with shade and cooling. J Dairy Sci 77:2044–2050

Baldwin RL, Smith NE, Taylor J, Sharp M (1980) Manipulating metabolic parameters to improve growth rate and milk secretion. J Anim Sci 51:1416–1428

Barash H, Silanikove N, Shamay A, Ezra E (2001) Interrelationships among ambient temperature, day length, and milk yield in dairy cows under a mediterranean climate. J Dairy Sci 84:2314–2320

Bauman DE, Currie WB (1980) Partitioning of nutrients during pregnancy and lactation: a review of mechanisms involving homeostasis and homeorhesis. J Dairy Sci 63:1514–1529

Baumgard LH, Rhoads RP (2007) The effects of hyperthermia on nutrient partitioning. In: Proceedings of Cornell Nutrition Conference, pp 93–104. Cornell University, New York, USA

Baumgard LH, Moore CE, Bauman DE (2002) Potential application of conjugated linoleic acids in nutrient partitioning. In: Proceedings of the Southwest Nutrition Conference, pp 127–141. Tempe, Arizona

Baumgard LH, Odens LJ, Kay JK, Rhoads RP, VanBaale MJ, Collier RJ (2006) Does negative energy balance (NEBAL) limit milk synthesis in early lactation? In: Proceedings of the Southwest Nutrition Conference, pp 181–187. Tempe, Arizona

Beam SW, Butler WR (1999) Effects of energy balance on follicular development and first ovulation in postpartum dairy cows. J Reprod Fertil 54:411–424

Beede DK, Collier RJ (1986) Potential nutritional strategies for intensively managed cattle during thermal stress. J Anim Sci 62:543–554

Belibasakis NG, Ambatzidis P, Aktsali P, Tsirgogianni D (1995) Effects of degradability of dietary protein on milk production and blood components of dairy cows in hot weather. World Rev Anim Prod 30:21–26

Berman A (2005) Estimates of heat stress relief needs for Holstein dairy cows. J Anim Sci 83:1377–1384

Berman A, Folman Y, Kaim M, Mamen M, Herz Z, Wolfenson D, Arieli A, Graber Y (1985) Upper critical temperatures and forced ventilation effects for high-yielding dairy cows in a subtropical climate. J Dairy Sci 68:1488–1495

Bernabucci U, Lacetera N, Ronchi B, Nardone A (2002) Effects of the hot season on milk protein fractions in Holstein cows. Anim Res 51:25–33

Bianca W (1965) Reviews of the progress of dairy science. Section A, Physiology of cattle in a hot environment. J Dairy Res 32:291–345

Bligh J (1976) Introduction to acclamatory adaptation-including notes on terminology. In: Bligh J, Cloudsley-Thompson JL, Macdonald AG (eds) Environmental physiology of animals. Wiley, New York, pp 219–229

Block E (1994) Manipulation of dietary cation-anion difference on nutritionally related production diseases, productivity, and metabolic responses of dairy cows. J Dairy Sci 77:1437–1450

Brosh A, Chosniak I, Tadmor A, Shkolnik A (1988) Physico-chemical conditions in the rumen of Bedouin goats: effect of drinking, food quality and feeding time J. Agric Sci 11:147–157

Broucek J, Letkovicova M, Kovalcuj K (1991) Estimation of cold stress effect on dairy cows. Int J Biometeorol 35:29–32

Carabaño MJ, Wade KM, Van Vleck LD (1990) Genotype by environment interactions for milk and fat production across regions of the United States. J Dairy Sci 73:173–180

Chillard Y (1991) Physiological constraints to milk production: factors which determine nutrient partitioning, lactation persistency and mobilization of body reserves. In: Speedy A, Ren´e S (eds) Feeding dairy cows in the tropics. FAO animal production and health paper no. 86. FAO, Rome

Colditz PJ, Kellaway RC (1972) The effect of diet and heat stress on feed intake, growth, and nitrogen metabolism in Friesian, F1 Brahman × Friesian, and Brahman heifers. Aust J Agric Res 23:717–725

Cole JA, Hansen PJ (1993) Effects of administration of recombinant bovine somatotropin on the responses of lactating and non-lactating cows to heat stress. J Am Vet Med Assoc 203:113–117

Collier RJ, Beede DK, Thatcher WW, Israel LA, Wilcox CJ (1982) Influences of environment and its modification on dairy animal health and production. J Dairy Sci 65:2213–2227

Collier RJ, Baumgard LH, Lock AL, Bauman DE (2005) Physiological limitations: nutrient partitioning. Chapter 16. In: Wiseman J, Bradley R (eds) Yields of farmed species: constraints and opportunities in the 21st century. Proceedings: 61st Easter School, Nottingham, England. Nottingham University Press, Nottingham, pp 351–377

Collier RJC, Stiening M, Pollard BC, VanBaale MJ, Baumgard LH, Gentryand PC, Coussens PM (2006) Use of gene expression microarrays for evaluating environmental stress tolerance at the cellular level in cattle. J Anim Sci 84:1–13

Coppock CE (1985) Energy nutrition and metabolism of the lactating dairy cow. J Dairy Sci 68:3403–3410

Daly SEJ, Owens RA, Hartmann PE (1993) The short-term synthesis and infant-regulated removal of milk in lactating women. Exp Physiol 78:209–220

Danfaer A, Thysen I, Ostergaard V (1980) The effect of the level of dietary protein on milk production. 1. Milk yield, liveweight gain and health. Beret. Statens Husdyrbrugsfors 492

Drackley JK (1999) Biology of dairy cows during the transition period: the final frontier? J Dairy Sci 82:2259–2273

Edwards JL, Hansen PJ (1997) Differential responses of bovine oocytes and preimplantation embryos to heat shock. Mol Reprod Dev 46:138–145

Edwards JL, Ealy AD, Monterroso VH, Hansen PJ (1997) Ontogeny of temperature-regulated heat shock protein 70 synthesis in preimplantation bovine embryo. Mol Reprod Dev 48:25–33

El-Khaschab S, El-Danasoury MS, Omer S (1984) Studies on some reproductive and productive traits of buffaloes, in Egypt. Minufiya J Agric Res 9:211–237

Finch VA (1985) Comparison of nonevaporative heat transfer in different cattle breeds. Aust J Agric Res 36:497–508

Finch VA (1986) Body temperature in beef cattle: its control and relevance to production in the tropics. J Anim Sci 62:531–542

Flamenbaum I, Wolfenson D, Kunz PL, Maman M, Berman A (1995) Interactions between body condition at calving and cooling of dairy cows during lactation in summer. J Dairy Sci 78(10):2221–2229

Fox DG, Tylutki TP (1998) Accounting for the effects of environment on the nutrient requirements of dairy cattle. J Dairy Sci 81:3085–3089

Fregley MJ (1996) Adaptations: some general characteristics. In: Fregley MJ, Blatteis CM (eds) Handbook of physiology, section 4: Environmental physiology, vol I. Oxford University Press, Oxford, pp 3–15

Fuquay JW (1981) Heat stress as it affects animal production. J Anim Sci 32:164–174

Ganguli NC (1981) Buffalo as a candidate for milk production. Federation Internationale De Laiterie – International Dairy Federation Bulletin 137

Gangwar PC (1985) Importance of photoperiod and wallowing in buffalo production. Ind J Dairy Sci 38:150–155

Goff JP, Horst RL (1997) Physiological changes at parturition and their relationship to metabolic disorders. J Dairy Sci 80:1260–1268

Guerriero V Jr, Raynes DA (1990) Synthesis of heat stress proteins in lymphocytes from livestock. J Anim Sci 68:2779–2783

Habeeb AAM, Abdel-Samee AM, Kamal TH (1989) Effect of heat stress, feed supplementation and cooling technique on milk yield, milk composition and some blood constituents in Friesian cows under Egyptian conditions. In: Proceedings of the 3rd Egyptian- British conference on animal fish and poultry production, Alexandria University, Alexandria, vol 2, pp 629–635

Habeeb AAM, Aboulnaga AJ, Yousef HM (1993) Influence of exposure to high temperature on daily gain, feed efficiency and blood components of growing male Californian rabbits. Egypt J Rabbit Sci 3:73–80

Habeeb AAM, El-Marsy KA, Aboulnaga AI, Kamal TH (1996) The effect of hot summer climate under level of milk yield on blood biochemistry and circulating thyroid and progesterone hormones in Friesian cows. Arab J Nucl Sci Appl 29:161–173

Hansen PJ (1990) Effects of coat color on physiological responses to solar radiation in Holsteins. Vet Rec 127:333–334

Hansen PJ (2007) Exploitation of genetic and physiological determinants of embryonic resistance to elevated temperature to improve embryonic survival in dairy cattle during heat stress. Theriogenology 68S:242–249

Harbin R, Harbough FS, Neeley KL, Find NC (1958) Effect of natural combinations of ambient temperature and relative humidity on the water intake of lactating and unlactating dairy cows. J Dairy Sci 41:1621

Horowitz M (2001) Heat acclimation: phenotypic plasticity and cues to the underlying molecular mechanisms. J Therm Biol 26:357–363

Huber JT, Higginbotham G, Gomez-Alarcon RA, Taylor RB, Chen KH, Chan SC, Wu Z (1994) Heat stress interactions with protein, supplemental fat, and fungal cultures. J Dairy Sci 77:2080–2090

Igono MO, Bjotvedt G, Sanford-Crane HT (1992) Environmental profile and critical temperature effects on milk production of Holstein cows in desert climate. Int J Biometeorol 36:77–87

Ingraham RH, Stanley RW, Wagner WC (1979) Seasonal effects of tropical climate on shade and nonshaded cows as measured by rectal temperature, adrenal cortex hormones, thyroid hormone, and milk production. Am J Vet Res 40:1792–1797

Johnson HD (1976) World climate and milk production. Biometeorology 6:171–175

Johnson HD, Ragsdale AC (1959) Effects of constant environmental temperatures of 50° and 80°F on the growth responses of Holstein, Brown Swiss, and Jersey calves. Columbia: Mo Agric Exp Stn Bull 705

Johnson HD, Vanjonack WJ (1976) Effects of environmental and other stressors on blood hormone patterns in lactating animals. J Dairy Sci 59:1603–1617

Johnson HD, Ragasdale AC, Berry LI, Shanklin D (1963) Temperature-humidity effects including influence of acute heat elimination in feed and water consumption of Holstein cattle. Univ Mo Res Bull 846

Kadzere CT, Murphy MR, Silanikove N, Maltz E (2002) Heat stress in lactating dairy cows: a review. Livest Prod Sci 77:59–91

Kamal TH, Johnson HD (1971) Total body solids as measure of a short-term heat stress in cattle. J Anim Sci 32:306–311

Kamal TH, Johnson HD, Ragsdale RC (1959) Water consumption in dairy cattle as influenced by environmental temperatures and urine excretion. J Dairy Sci 42:926

Kamal TH, Johnson HD, Ragsdale RC (1962) Metabolic reactions during thermal stress (35 to 95°F) in dairy animals acclimated at 50° and 80°F. Mo Agric Exp Stn Res Bull 785:1–114

Kamal TH, Clark JL, Johnson HD (1970) The effect of age on heat tolerance in cattle as determined by the whole body 40K and nitrogen retention. Int J Biometeorol 14:301–308

King VL, Denise SK, Armstrong DV, Torabi M, Wiersma F (1988) Effects of a hot climate on the performance of first lactation Holstein cows grouped by coat color. J Dairy Sci 71:1093–1096

Lacetera NG, Ronchi B, Bernabucci U, Nardone A (1994) Influence of heat stress on some biometric parameters and on body condition score in female Holstein calves. Rivista di Agricoltura Subtropicale e Tropicale 88:80–89

Lee DHK (1965) Climatic stress indices for domestic animals. Int J Biometeorol 9:29–35

Lin JC, Moss BR, Koon JL, Flood CA, Smith III RC, Cummins KA, Coleman DA (1998) Comparison of various fan, sprinkler, and mister systems in reducing heat stress in dairy cows. 14:177–182

Liu X, Robinson GW, Wagner KU, Garrett L, Wynshaw-Boris A, Hennighausen L (1997) Stat5a is mandatory for adult mammary gland development and lactogenesis. Genes and Dev 11:179–186

Lucy MC, Staples CR, Thatcher WW, Erickson PS, Cleale RM, Firkins JL, Clark JH, Murphy MR, Brodie BO (1992) Influence of diet composition, dry matter intake, milk production and energy balance on time of postpartum ovulation and fertility in dairy cows. Anim Prod 54:323–331

Lund LR, Bjorn SF, Sternlicht MD, Nielsen BS, Solberg H, Usher PA, Osterby R, Christensen IJ, Stephens RW, Bugge TH, Dano K, Werb Z (2000) Lactational competence and involution of the mouse mammary gland require plasminogen. Development 127:4481–4492

Maloyan A, Horowitz M (2002) β-Adrenergic signaling and thyroid hormones affect HSP72 expression during heat acclimation. J Appl Physiol 93:107–115

Maltz E, Silanikove N, Shalit U, Berman A (1994) Diurnal fluctuations in plasma ions and water intake of dairy cows as affected by lactation in warm weather. J Dairy Sci 77:2630–2639

Maltz E, Silanikove N (1996) Kidney function and nitrogen balance of high yielding dairy cows at the onset of lactation. J Dairy Sci 79:1621–1666

Manalu W, Johnson HD, Li RZ, Becker BA, Collier RJ (1991) Assessment of thermal status of somatotropin-injected lactating Holstein cows maintained under controlled-laboratory thermoneutral hot and cold environments. J Nutr 121:2006–2019

Marai IFM, Habeeb AAM (2010) Buffalo biological functions as affected by heat stress – a review. Livest Sci 127:89–109

Marai IFM, El-Masry KA, Nasr AS (1994) Heat stress and its amelioration with nutritional, buffering, hormonal and physiological techniques for New Zealand White rabbits maintained under hot summer conditions of Egypt. Options Mediterr 8:475–487

Marai IFM, Habeeb AA, Daader AH, Yousef HM (1995) Effects of Egyptian subtropical summer conditions and the heat stress alleviation technique of water spray and a diaphoretic on the growth and physiological functions of Friesian calves. J Arid Environ 30:219–225

Marai IFM, Daader AM, Abdel-Samee AM, Ibrahim H (1997a) Winter and summer effects and their amelioration on lactating Friesian and Holstein cows maintained under Egyptian conditions. In: Proceedings of the international conference on animal, poultry, rabbits and fish production and health, Cairo

Marai IFM, Daader AM, Abdel-Samee AM, Ibrahim H (1997b) Lactating Friesian and Holstein cows as affected by heat stress and combination of amelioration techniques under Egyptian conditions. In: Proceedings of the international conference on animal, poultry, rabbits and fish production and health, Cairo

Marai IFM, Habeeb AAM, Gad AE (2002) Rabbit's productive, reproductive and physiological traits as affected by heat stress (a review) live. Prod Sci 78:71–90

Marai IFM, Daader AH, Soliman AM, El-Menshawy SMS (2009) Non-genetic factors affecting growth and reproduction traits of buffaloes under dry management housing (in sub-tropical environment) in Egypt. Livest Res Rural Dev 21:3

McDowell RE, Hooven NW, Camoens JK (1976) Effects of climate on performance of Holsteins in first lactation. J Dairy Sci 59:965–973

McGuire MA, Beede DK, DeLorenzo MA, Wilcox CJ, Huntington GB, Reynolds CK, Collier RJ (1989) Effects of thermal stress and level of feed intake on portal plasma flow and net fluxes of metabolites in lactating Holstein cows. J Anim Sci 67:1050–1060

Mishra MS, Sengupta BP, Roy A (1963) Physiological reactions of buffalo cows maintained in two different housing conditions during summer months. Indian J Dairy Sci 16:203

Moe PW (1981) Energy metabolism of dairy cattle. J Dairy Sci 64:1120–1139

Mohamed IAS (2000) The performance of Egyptian buffaloes under desert new reclaimed lands. MSc thesis, Faculty of Agriculture, Zagazig University, Zagazig

Moore CE, Kay JK, VanBaale MJ, Baumgard LH (2005a) Calculating and improving energy balance during times of nutrient limitation. In: Proceedings of southwest nutrition and management conference, pp 173–185. Tempe, Arizona

Moore CE, Kay JK, VanBaale MJ, Collier RJ, Baumgard LH (2005b) Effect of conjugated linoleic acid on heat stressed Brown Swiss and Holstein cattle. J Dairy Sci 88:1732–1740

Moss RJ (1993) Rearing heifers in the subtropics: nutrient requirements and supplementation. Trop Grassl 27:238–249

Mourad KA (1978) Some productive characters of the Egyptian buffalo. MSc thesis, Faculty of Agriculture, Cairo University, Cairo

Mullick DN (1964) A study on the metabolism of food nutrients in cattle and buffaloes under climatic stress. Arid Zone Res 14:137

Murphy MR, Davis CL, McCoy GC (1983) Factors affecting water consumption by Holstein cows in early lactation. J Dairy Sci 66:35–38

Nardone A, Lacetera N, Bernabucci U, Ronchi B (1997) Composition of colostrum from dairy heifers exposed to high air temperatures during late pregnancy and the early postpartum period. J Dairy Sci 80:838–844

Nardone A, Ronchi B, Lacetera N, Ranieri MS, Bernabucci U (2010) Effects of climate changes on animal production and sustainability of livestock systems. Livest Sci 130:57–69

National Research Council (1981) Effect of Environment on Nutrient Requirement of Domestic Animals. National Academy Press Washington, DC

National Research Council (2001a) Nutrient requirements of dairy cattle, 7th rev edn. National Academies Press, Washington, DC

Nessim MG (2004) Heat-induced biological changes as heat tolerance indices related to growth performance in buffaloes. PhD thesis, Faculty of Agriculture, Ain-Shams University, Cairo, Egypt.J. I. DS

Oldham JD (1984) Protein-energy interrelationships in dairy cows. J Dairy Sci 67:1090–1114

Oszterman S, Redbo I (2001) Effect of milking frequency on lying down up behaviour in dairy cows. Appl Anim Behav Sci 70:167–176

Petkov G (1971) Environmental milk production of cows. Veterinaria Shirka 75:23–28

Purwanto BP, Abo Y, Sakamoto R, Furumoto F, Yamamoto S (1990) Diurnal patterns of heat production and heart rate under thermoneutral conditions in Holstein Friesian cows differing in milk production. J Agric Sci 114:139–142

Quarrie LH, Addey CVP, Wilde CJ (1998) Programmed cell death during mammary tissue involution induced by weaning, litter removal, and milk stasis. J Cell Physiol 168:559–569

Ravagnolo O, Misztal I, Hoogenboom G (2000) Genetic component of heat stress in dairy cattle, development of heat index function. J Dairy Sci 83(2120–2125):29

Reynolds CK, Tyrrell HF, Reynolds PJ (1991) Effects of diet forage-to-concentrate ratio and intake on energy metabolism in growing beef heifers: whole body energy and nitrogen balance and visceral heat production. J Nutr 121:994–1003

Rhoads ML, Rhoads RP, Sanders SR, Carroll SH, Weber WJ, Crooker BA, Collier RJ, VanBaale MJ, Baumgard LH (2007) Effects of heat stress on production, lipid metabolism and somatotropin variables in lactating cows. J Dairy Sci 90:230

Rhoads ML, Rhoads RP, VanBaale MJ, Collier RJ, Sanders SR, Weber WJ, Crooker BA, Baumgard LH (2009) Effects of heat stress and plane of nutrition on lactating Holstein cows: I. Production, metabolism and aspects of circulating somatotropin. J Dairy Sci 92:1986–1997

Richards JI (1985) Effect of high daytime temperatures on the intake and utilization of water in lactating friesian cattle. Trop Anim Health Prod 17:209–217

Rivera RJ, Hansen PJ (2001) Development of cultured bovine embryos after exposure to high temperatures in the physiological range. Reproduction 121:107–115

Rivera RJ, Kelley KL, Erdos GW, Hansen PJ (2003) Alterations in ultrastructural morphology of two-cell bovine embryos produced in vitro and in vivo following a physiologically relevant heat shock. Biol Reprod 69:2068–2077

Rodriguez LR, McKonnen G, Wilcox CJ, Martin FG, Krienke WA (1985) Effect of relative humidity and maximum and minimum temperature, pregnancy and stage of lactation on milk composition and yield. J Dairy Sci 68:973–978

Roy Chaudhury PN, Deskmuykh K (1975) Effect of month, season and sequence of calving on milk yield in Italian buffaloes. Indian Vet Med J 46:1059–1068

Sastry NSR (1983) Monograph: buffalo husbandry; constraints to successful buffalo farming and overcoming the same through management. Institute of Animal Management and Breeding, University of Hohenheim, Germany, Discipline – Milk Production, pp 4–6

Schneider PL, Beede DK, Wilcox CJ (1988) Nycterohemeral patterns of acid–base status, mineral concentrations and digestive function of lactating cows in natural or chamber heat stress environments. J Anim Sci 66:112–125

Shamay A, Shapiro F, Barash H, Bruckental I, Silanikove N (2000) Effect of dexamethasone on milk yield and composition in dairy cows. Annals Zootechniques 49:343–352

Shamay A, Shapiro F, Leitner G, Silanikove N (2003) Infusions of casein hydrolyzates into the mammary gland disrupt tight junction integrity and induce involution in cows. J Dairy Sci 86:1250–1258

Sharma AK, Rodriguez LA, Mekonnen G, Wilcox CJ, Bachman KC, Collier RJ (1983) Climatological and genetic effects on milk composition and yield. J Dairy Sci 66:119–126

Shennan DB, McNeillie SA (1994) Milk accumulation down-regulates amino-acid-uptake via system-A and system-L by lactating mammary tissue. Horm Metab Res 26:611

Shwartz G, Rhoads ML, VanBaale MJ, Rhoads RP, Baumgard LH (2009) Effects of a supplemental yeast culture on heat-stressed lactating Holstein cows. J Dairy Sci 92:935–942

Silanikove N (1992) Effects of water scarcity and hot environment on appetite and digestion in ruminants: a review. Livest Prod Sci 30:175–194

Silanikove N (1994) The struggle to maintain hydration and osmoregulation in animals experiencing severe dehydration and rapid rehydration: the story of ruminants. Exp Physiol 79:281–300

Silanikove N (2000) Effects of heat stress on the welfare of extensively managed domestic ruminants. Livest Prod Sci 67:1–18

Silanikove N, Tadmor A (1989) Rumen volume, saliva flow rate, and systemic fluid homeostasis in dehydrated cattle. Am J Physiol 256:809–815

Silanikove N, Shamay A, Shinder D, Moran A (2000) Stress down-regulates milk yield in cows by plasmin induced beta-casein product that blocks K+ channels on the apical membranes. Life Sci 67:2201–2212

Silanikove N, Iscovich J, Leitner G (2005) Therapeutic treatment with casein hydrolyzate eradicate effectively bacterial infection in treated mammary quarters in cows. In: Hogeveen H (ed) Mastitis in dairy production – current knowledge and future solutions. Wageningen Academic Publishers, Wageningen, pp 327–332

Silanikove N, Merin U, Leitner G (2006) Physiological role of indigenous milk enzymes: an overview of an evolving picture. Int Dairy J 16:535–545

Silanikove N, Shapiro F, Shinder D (2009) Acute heat stress brings down milk secretion in dairy cows by up-regulating the activity of the milk-borne negative feedback regulatory system. BMC Physiol 9:13. doi:10.1186/1472-6793-9-13

Sonna LA, Fujita J, Gaffin SL, Lilly CM (2002) Invited review: effects of heat and cold stress on mammalian gene expression. J Appl Physiol 92:1725–1742

Stelwagen K, Van Espen DC, Verkerk GA, McFadden HA, Farr VC (1998) Elevated plasma cortisol reduces permeability of mammary tight junctions in the lactating bovine mammary epithelium. J Endocrinol 159:173–178

Stott GH, Wiersma F, Menefee BE, Radwanski FR (1976) Influence of environment on passive immunity in calves. J Dairy Sci 59:1306–1311

Thatcher WW (1974) Effect of season, climate and temperature on reproduction and lactation. J Dairy Sci 57:350

Tucker WB, Harrison GA, Hemken RW (1988) Influence of dietary cation-anion balance on milk, blood, urine, and rumen fluid in lactating dairy cattle. J Dairy Sci 71:346–354

Tyrrell HF, Moe PW, Flatt WP (1970) Influence of excess protein intake on energy metabolism of the dairy cow. In: Fifth symposium energy metabolism of farm animals, pp 69–71. Vitznau, Switzerland

Umphrey JE, Moss BR, Wilcox CJ, Van Horn HH (2001) Interrelationships in lactating Holsteins of rectal and skin temperatures, milk yield and composition, dry matter intake body weight, and feed efficiency in summer in Alabama. J Dairy Sci 84:2680–2685

Upadhyay RC, Ashutosh AK, Gupta SK, Singh SV, Rani N (2009) Inventory of methane emission from livestock in India. In: Aggarwal PK (ed) Global climate change and Indian agriculture. ICAR, New Delhi, pp 117–122

Van Soest PJ (1982) Nutritional ecology of the ruminant. O & B Books, Inc., Corvallis

Vanjonack WJ, Johnson HD (1975) Effects of moderate heat and yield on plasma thyroxine in cattle. J Dairy Sci 58:507–516

Verma DN, Lal SN, Singh SP, Parkash OM, Parkash O (2000) Effect of season on biological responses and productivity of buffalo. Int J Anim Sci 15:237–244

Vicini JL, Crooker BA, McGuire MA (2002) Energy balance in early lactation dairy cows. In: California animal nutrition conference, pp 1–8

Wayman O, Johnson HD, Merijan CP, Berry IL (1962) Effect of ad libitum on force feeding of two rations of lactating dairy cows subject to temperature stress. J Dairy Sci 45:1472

West JW (1999) Nutritional strategies for managing the heat-stressed dairy cow. Am Soc Anim Sci Am Dairy Sci Assoc 2:21–35

West JW (2002) Physiological effects of heat stress on production and reproduction. In: Proceedings of tri-state dairy nutrition conference, pp 1–9. Ohio State University

West JW (2003) Effects of heat stress on production in dairy cattle. J Dairy Sci 86(6):2131–2144

West JW, Mullinix BG, Johnson JC Jr, Ash KA, Taylor VN (1990) Effects of bovine somatotropin on dry matter intake, milk yield, and body temperature in Holstein and Jersey cows during heat stress. J Dairy Sci 73:2896–2906

West JW, Mullinix BG, Sandifer TG (1991) Effects of bovine somatotropin on physiologic responses of lactating Holstein and Jersey cows during hot, humid weather. J Dairy Sci 74:840–851

West JW, Hill GM, Fernandez JM, Mandebvu P, Mullinix BG (1999) Effects of dietary fiber on intake, milk yield, and digestion by lactating dairy cows during cool or hot, humid weather. J Dairy Sci 82:2455–2465

Wheelock JB, Sanders SR, Shwartz G, Hernandez LL, Baker SH, McFadden JW, Odens LJ, Burgos R, Hartman SR, Johnson RM, Jones BE, Collier RJ, Rhoads RP, VanBaale MJ, Baumgard LH (2006) Effects of heat stress and rbST on production parameters and glucose homeostasis. J Dairy Sci 89:290–291

Wilde CJ, Peaker M (1990) Autocrine control in milk secretion. J Agric Sci (Camb) 114:235–238

Wildman CD, West JW, Bernard JK (2007) Effect of dietary cation-anion difference and dietary crude protein on performance of lactating dairy cows during hot weather. J Dairy Sci 90:1842–1850

Yousef HM, Habeeb AAM Fawzy SA, Zahed SM (1996) Effect of direct solar radiation of hot summer season and using two types of sheds on milk yield and composition and some physiological changes in lactating Friesian cows. In: Proceedings of 7th Scientific Congress, Faculty of Veterinary Medicine, Assiut University, Assiut, pp 63–75

Zimbelman RB, Rhoads RP, Rhoads ML, Duff GC, Baumgard LH, Collier RJ (2009) A re-evaluation of the impact of temperature humidity index (THI) and black globe humidity index (BGHI) on milk production in high producing dairy cows. In: Collier RJ (ed) Proceedings of the southwest nutrition conference, pp 158–169. Tempe, Arizona. Retrieved 2 Feb, from http://cals.arizona.edu/ans/swnmc/Proceedings/2009/14Collier_09.pdf

Heat Stress and Reproduction

Contents

Abstract

Heat stress induces infertility in farm animals and represents a major source of economic loss to the livestock sector. The decrease in animal fertility is caused by elevated body temperature that influences ovarian functions, oestrous expression, oocyte health and embryonic development. Protection from heat stress during dry period is particularly crucial for a high-producing cow since it involves mammary gland involution and subsequent development, rapid fetal growth and induction of lactation. Cows and cycling buffaloes under heat stress have lower plasma inhibin concentrations, reflecting reduced folliculogenesis, since a significant proportion of plasma inhibin comes from small- and medium-sized follicles. Concentrations of plasma FSH are higher during the preovulatory period in summer and are

A. Aggarwal and R. Upadhyay, *Heat Stress and Animal Productivity*,
DOI 10.1007/978-81-322-0879-2_4, © Springer India 2013

associated with lower circulating concentrations of inhibin. The neuroendocrine mechanisms controlling gonadotrophin secretion are more sensitive to heat stress particularly in animals with low concentrations of plasma oestradiol. Environmental temperature and humidity 2 days prior to insemination is critical for conception than at any other phase of the reproductive cycle. A rise in rectal temperature diverts blood from the visceral organs to the peripheral circulation due to redistribution of blood to alleviate heat, which could reduce perfusion of nutrients and hormones to the endometrial and oviductal tissues affecting reproductive functions. In terms of steroid production, the thecal cells are more susceptible than granulosa cells to heat stress and express a delayed effect of heat stress in both medium-sized and preovulatory follicles. A rise in testicular temperature in bulls similar to other mammals with external testes leads to reduced sperm output, decreased sperm motility and an increased proportion of morphologically abnormal spermatozoa in the ejaculate. X and Y spermatozoa are affected differentially by high temperature. The plasma concentrations of insulin, IGF-I and glucose are low in summer months compared to winter months () probably because of low dry matter intake and increased negative energy balance. Insulin is required for the development of follicles and has beneficial effects on oocyte quality. Genetic selection for heat adaptability, both natural and artificial, is likely to modulate the impact of heat stress on reproductive functions, and therefore, genetic selection for thermal tolerance may be a necessity under climate change conditions.

1   Introduction

Heat stress in livestock drives their body temperature above set-point temperature leading to disruptions in reproductive functions. Two homeokinetic mechanisms are involved in body temperature homeostasis compromising reproductive functions. First, the redistribution of blood flow occurs from the body core to the periphery in order to increase sensible heat loss; other homeokinetic homeothermic or thermoregulatory control mechanism leads to reduced voluntary feed intake during heat stress.

A low feed intake that occurs in order to reduce metabolic heat production leads to changes in energy balance and nutrient availability affecting reproductive cyclicity, pregnancy and fetal development. Other reproduction process that may be disrupted in heat-stressed livestock is homeokinetic related to gonadal functions due to impact on hormone synthesis.

Heat stress induces several physiological and biochemical changes in body functions and negatively affects milk production and reproductive efficiency of both male and female animals. High ambient temperature during summer has been observed to negatively impact breeding efficiency and drastically reduce conception rate and increase embryonic loss (Gwazdauskas et al. 1981; Hansen 2005). Berman et al. (1985) has concluded that thermoregulatory ability in the face of heat stress as a result of selection for milk production in high-producing cows magnifies the seasonal depression in fertility due to heat stress.

2   The Effect of Heat Stress on Reproductive Functions

Under intensive and extensive livestock management system, only productive or likely to be productive in near future gets preference. Lactating and pregnant animals get adequate protection and care, and other animals like dry pregnant cows are neglected or provided little protection from heat stress. In fact, during early pregnancy, neglect or limited attention predisposes them to additional stressors due to abrupt physiological changes, and any nutritional insufficiency and environmental changes make them vulnerable. These factors also increase the susceptibility of pregnant cows to heat stress and have a critical influence on fetal and dam's postpartum health and productive performance. Therefore, protection from heat stress during dry period is particularly important for a high-producing cow as it involves mammary gland involution and subsequent development, rapid fetal growth and induction of lactation.

Higher incidence of silent heat and anoestrus is one of the main problems in cows and buffaloes during summer or on exposure to high ambient temperatures. Heat stress compromises oocyte growth in cows by altering progesterone, luteinising hormone and follicle-stimulating hormone secretions during the oestrus cycle (Ronchi et al. 2001) as well as impairing embryo development and increasing embryo mortality (Wolfenson et al. 2000).

Heat stress reduces fertility of dairy cows in summer, and cows poorly express oestrus due to a decline in oestradiol secretion from a dominant follicle developed in a low luteinising hormone environment (De Rensis and Scaramuzzi 2003). *There are two pathways by which heat stress leads to infertility in cows. The first is a direct effect of hyperthermia on the reproductive axis. The second is an indirect effect of heat stress on appetite and dry matter intake, both of which are reduced by heat stress. As a result, animal comes under negative energy balance especially during early lactation and produces lower blood levels of hormones insulin, leptin and IGF-1; however, GH and NEFA increase. These changes in metabolic profile acting via hypothalamic–hypophyseal axis result in decreased GnRH and LH secretion which leads to reduced oestradiol by dominant follicle; therefore, poor oestrus detection and ovulatory failure may result* (Fig. 1). *Heat stress also affects the secretion of thyrotrophic (thyroxine) and adrenocorticotrophic (cortisol) and adrenomedullary hormones (adrenaline). These may potentially impair fertility* (De Rensis and Scaramuzzi 2003). A drop of about 20% in conception rates (Lucy 2002) or decrease in 90-day nonreturn rate to the first service in lactating dairy cows (Al-Katanani et al. 1999) occurs in summer. The thermal stress also increases calving interval; the birth rate is lower, and milk yield per lactation is reduced. Heat stress during pregnancy slows down growth of the fetus, although active mechanisms attenu-

ate excursions in fetal body temperature when cow is thermally stressed. In heat-stressed cows, there is reduced duration and intensity of oestrus (Younas et al. 1993) or remain unchanged (Howell et al. 1994). During summer, motor activity and other behavioural manifestations of oestrus have been reported to decline (Hansen 1997), and the incidence of anoestrus and silent ovulation is increased. These effects lead to a reduced number of mounts in hot season compared to cold season (Pennington et al. 1985) leading to poor detection of oestrus. Holstein cows during summer have been reported to mount on an average 4.5 mounts per oestrus versus 8.6 per oestrus in winter (Nebel et al. 1997). Therefore, in hot climates, there is a reduction in the number of inseminations and an increase in the proportion of inseminations that may not result in pregnancy.

3   The Effect of Heat Stress on the Hypothalamic–Hypophyseal Ovarian Axis

The main factors regulating ovarian activity are gonadotrophin-releasing hormone from the hypothalamus and the gonadotrophins, luteinising hormone (LH) and follicle-stimulating hormone (FSH) from the anterior pituitary gland. The effects of heat stress on these hormones and their concentrations in peripheral blood may be observed to be inconsistent due to variable reasons related to physiology and state of health of the animal. The levels remain unaltered (Gauthier 1986) while others report increase in concentrations (Roman-Ponce et al. 1981), and some report a decline in concentrations (Gilad et al. 1993; Lee 1993) due to heat stress. The LH secretion in heat-stressed cows has a low intensity in LH pulse amplitude (Gilad et al. 1993) and LH pulse frequency (Wise et al. 1988). The effects of heat stress on the preovulatory surge of LH are similarly inconsistent; a reduction of the endogenous LH surge was reported in heifers (Madan and Johnson 1973) but not in cows (Gauthier 1986; Rosemberg et al. 1982). These differences in LH surge during heat stress have been attributed

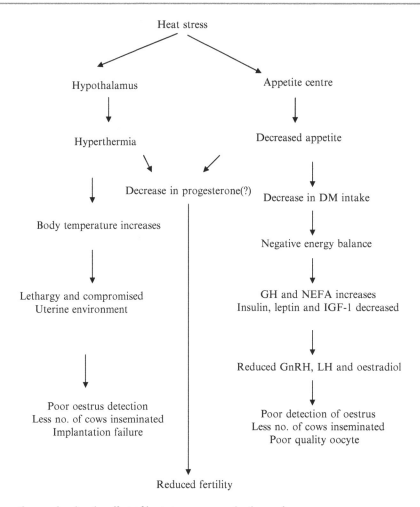

Fig. 1 Two pathways showing the effect of heat stress on reproductive performance

to preovulatory oestradiol levels because the amplitude of tonic LH pulses and GnRH-induced preovulatory plasma LH surges are decreased in cows with low plasma concentrations of oestradiol but not in cows with high plasma concentrations of oestradiol (Gilad et al. 1993). Plasma inhibin concentrations in summer are lower in cows under heat stress (Wolfenson et al. 1995) and in cyclic buffaloes (Palta et al. 1997), reflecting reduced folliculogenesis since a significant proportion of plasma inhibin comes from small- and medium-sized follicles. Concentrations of plasma FSH are higher during the preovulatory period in summer, and this was associated with lower circulating concentrations of inhibin (Ingraham et al. 1974). However, a reduced FSH response

in heat-stressed compared to control cows was observed after administration of a GnRH analogue (Gilad et al. 1993).

Because decreased LH levels have been reported in most studies under heat stress, the dominant follicle develops in an environment of low LH, and this results in reduced oestradiol secretion from the dominant follicle leading to poor expression of oestrus. FSH is increased due to heat stress, and this in turn decreases plasma inhibin production by follicles. FSH increases during summer and appears insufficient to overcome the effect of low LH concentrations and therefore a reduced availability of androgen precursors for oestradiol synthesis (Roth et al. 2000).

Plasma oestradiol concentrations are affected by heat stress in dairy cows (Wolfenson et al. 1995; Wilson et al. 1998), and effects may be consistent with decreased concentrations of LH and reduced dominance of the selected follicle. Similarly, the effect of heat stress on plasma progesterone concentration is unequivocal and controversial. Wilson et al. (1998) found that heat stress had no effect on the plasma progesterone concentrations but that luteolysis was delayed. However, other studies have reported increased progesterone concentrations (Trout et al. 1998; Abilay et al. 1975; Vaught et al. 1977), decreased progesterone concentrations (Ronchi et al. 2001; Rosenberg et al. 1977; Jonsson et al. 1997) or concentration remain unchanged (Roth et al. 2000; Guzeloglu et al. 2001) during heat stress in dairy cows. These differences probably arise because of uncontrolled changes in other factors that affect blood progesterone concentrations. Several factors like the type and magnitude of heat stress (i.e. acute or chronic) and differences in dry matter intake independently affect blood progesterone concentrations. Plasma progesterone concentrations are determined by the differences between the rate of luteal production and the rate of hepatic metabolism, and both of these are also affected by changes in dry matter intake. Low plasma progesterone concentrations during the luteal phase of the preconception oestrous cycle can compromise follicular development leading to abnormal oocyte maturation and early embryonic death (Ahmad et al. 1995). During the conception cycle, low progesterone concentrations lead to the failure of implantation (Mann et al. 1999; Lamming and Royal 2001). In the conception cycle, the effect of progesterone has been probably related to the need for synchronous development of the embryo and delayed or advanced development of the corpus luteum leading to higher rates of implantation failure (Lamming and Royal 2001). The pattern of the postovulatory rise in progesterone has been reported to be associated with fertility (Darwash et al. 1999).

Cows producing high milk production often have high dry matter intake (Staples et al. 1990; Hommeida et al. 2004) and low circulating progesterone concentrations in lactating (Hommeida et al. 2004). Acute feeding reduced circulating progesterone by 25% in pregnant cows (Vasconcelos et al. 2003). Lucy et al. (1998) found that circulating progesterone was lower in cattle genetically selected for high milk production. Sangsritavong et al. (2002) demonstrated that lactating cows have a much greater steroid metabolism than non-lactating cows. As a result, lactating cows may have larger luteal tissue volume on the ovary (Sartori et al. 2002, 2004) yet have lower circulating progesterone and oestradiol concentrations than heifers and dry cows (Wolfenson et al. 2004). Low progesterone secretion compromises fertility in dairy cattle (Mann and Lamming 1999), and an increase in progesterone secretion may facilitate embryonic development. Progesterone provides nourishment for the conceptus via induction of secretion of proteins and other molecules from the endometrium (Garrett et al. 1988). Low peripheral concentrations of progesterone are also associated with increased luteinising hormone (LH) pulses (Ireland and Roche 1982) that can stimulate luteolytic signals in favour of pregnancy failure. Skarzynski and Okuda (1999) reported that blocking the progesterone receptor with a progesterone antagonist (onapristone) increased prostaglandin $F2\alpha$ (PGF2α) production by bovine luteal cells harvested from mid-cycle corpora lutea (CL) (Days 8–12). Also, it was revealed that the bovine corpus luteum (CL) does not undergo apoptosis until progesterone production has declined (Juengel et al. 1993; Rueda et al. 1995).

Increased corticosteroid secretion inhibits GnRH and thus LH secretion (Gilad et al. 1993). The high concentrations of oestradiol can counteract the effect of heat stress, or alternatively, the neuroendocrine mechanism controlling gonadotrophin secretion is more sensitive to heat stress particularly in animals having low concentrations of plasma oestradiol. Heat stress also impairs ovarian functions directly to decrease its sensitivity to gonadotrophin stimulation (Wolfenson et al. 1997). The alteration in the secretory activity of the follicle and the corpus luteum to heat stress may influence fertility in cows and buffaloes.

4   Female Reproduction

Females raised at temperatures between 31 and 33.5°C and 60% of humidity had lower oestradiol in the follicular phase of the cycle and small size of the growing and ovulatory follicles. The length of the luteal phase in heat-stressed cows has been observed to be longer than in females kept in thermoneutral environment. The uterus secreted less PGF2α because of the reduction in oestradiol synthesis and/or because high temperatures can interfere with the release of PGF2α by endometrial cells (Malayer et al. 1990). The uterine endometrium must be 'primed' by oestradiol to produce enough prostaglandin and trigger luteolysis (Silvia et al. 1991). Thermal stress alters the concentrations of FSH and inhibin (Badinga et al. 1994; Roth et al. 2000; Wolfenson et al. 1997; Palta et al. 1997) and corpus luteum function (Wilson et al. 1998), as well as decreases the fluid content of follicles (Badinga et al. 1993). High temperatures reduce the number of granulosa cells and aromatase activity and secretion of androstenedione by theca cells (Wolfenson et al. 1997). Ambient conditions influence oocytes and embryo quality. Oocytes exposed to high temperatures in between the onset of oestrous and insemination produce less number of viable embryos (Putney et al. 1989a). Environmental temperature and humidity 2 days before insemination is critical for conception rates than at any other phase of the cycle, including the period from breeding to 2 days after breeding (Ingraham et al. 1974). A rise in rectal temperature shifts blood flow from the visceral organs to the peripheral circulation and this could reduce perfusion of nutrients, and hormones to the early stages of embryo development (8- to 16-cell stage) are more susceptible to heat stress, but there is also a high risk of embryonic loss at days 13 and 14 of pregnancy (Biggers et al. 1987; Ryan et al. 1993). Also, conceptuses cultured at high temperatures have been observed to reduce the secretion of interferon which impairs implantation and also maintenance of corpus luteum (Putney et al. 1988). As a consequence of these effects, high ambient temperatures on dairy cows have been observed to significantly influence their pregnancy rates. Studies on cows demonstrated that pregnancy rates of cows and heifers were reduced from 80 to 55% (after three inseminations) when the daily maximum temperature increased above 27°C (Orr et al. 1993). Both high and low temperature affect oestrous expression, conception rate and calving per cent. Both adapted and non-adapted breeds experience depression in reproductive traits due to thermal stress (Singh and Mishra 1980).

4.1   Ovarian Follicle

The low summer fertility of about 60% of the world dairy cattle population is associated with high ambient temperatures. However, during the autumn, when air temperatures decrease and cows are no longer exposed to thermal stress, conception rates remain lower than in the winter (Hansen 1997) due to susceptibility of ovarian follicles to heat stress (Badinga et al. 1993; Wolfenson et al. 1995). Normally, it takes about 40–50 days for small antral follicles to develop into large dominant follicles (Lussier et al. 1987). Heat stress inhibits ovarian follicular development leading to diminished reproductive efficiency of domestic animals during summer.

Heat stress affects development of follicle (Badinga et al. 1993; Wilson et al. 1998). The duration of dominance of the preovulatory follicle is increased in summer, and in beef heifers, duration of dominance has been observed to be negatively correlated with fertility (Mihm et al. 1994). More than one dominant follicle may develop, and twinning may occur during summer (Ryan and Boland 1991). Thus, both decrease in follicular steroid secretion and increase twinning rate have been observed under heat stress.

Oestradiol concentration in the follicular fluid and androstenedione production by thecal cells have been observed to be both lower in dominant follicles collected in autumn than in those collected in winter (Wolfenson et al. 1997). Effect of heat stress on steroid production in bovine follicles of medium size has been studied (Roth et al. 2001b), and a delayed effect of heat stress on steroid production and follicular characteristics in both medium-sized and preovulatory follicles has been related to the low fertility of cattle in the autumn.

4.2 Steroid Production

The reduction in reproductive performance under heat stress has been attributed to steroidogenic capacity and its effects on oocyte function (Roth et al. 2001b; Al-Katanani et al. 2002b; Roth and Hansen 2004b). Under heat stress, low oestradiol concentration in the follicular fluid of dominant follicles involves reduced aromatase activity in the granulosa cells (Badinga et al. 1993) and reduced androstenedione production by theca cells (Wolfenson et al. 1997).

In terms of steroid production, the thecal cells are more susceptible than granulosa cells to heat stress and expressed a delayed effect of heat stress in both classes of follicles. The consistent decrease in androstenedione production in both medium-sized and preovulatory follicles in previously heat-stressed cows has been correlated with the decreased concentrations of androstenedione in their follicular fluid. The delayed heat stress response on androstenedione production by thecal cells indicates that the decrease in androstenedione production by dominant follicles is observed in the comfortable season like that in autumn (Wolfenson et al. 1997). Decreased oestradiol concentration in the follicular fluid more likely occurred after exposure to long-term, chronic (summer) heat stress than due to acute heat stress. This response would be consistent with the finding that after chronic summer heat stress, an eight times decrease in androgen production by thecal cells in the autumn was accompanied by a significant decrease in oestradiol concentration in the follicular fluid (Wolfenson et al. 1997).

Low progesterone has been reported to alter ovarian follicular dynamics and persistence of dominant follicles (Sirois and Fortune 1990) and to induce changes in uterine morphology and $PGF_{2\alpha}$ secretion (Shaham-Albalancy et al. 2001). The situation has been aggravated because the achievement of a continuous rise in milk production is associated with a gradual decline in plasma progesterone concentration, consistent with the negative relationship between milk production and progesterone concentration (Lucy and Crooker 1999). This relationship can probably be attributed to accelerated progesterone metabolism in the liver (Parr et al. 1993) and/or to decreased synthesis of the steroid. Decreased progesterone production by the corpus luteum under heat stress conditions could result from:

(a) A suboptimal luteinisation process during hyperthermia
(b) Depressed progesterone synthesis at high temperatures
(c) Heat-induced impairment of the ovulatory follicle, of which the corpus luteum (CL) is subsequently formed

Under chronic summer heat stress conditions, progesterone production is markedly reduced in luteinised theca cells and less so in luteinised granulose cells. These seasonal effects of heat stress are carried over from an impaired follicle to an impaired CL. The decreased progesterone production found in vitro corresponds well with the mild but significant decrease in plasma progesterone during the summer. Alleviation of thermal stress and cooling cows and buffaloes in summer might help in restoring their normal CL function and could contribute to the improvement of fertility. Therefore, the need is to elevate progesterone concentrations as the hormonal strategy to increase the summer conception in dairy cows (Wolfenson et al. 2000).

Information on increase in FSH preceding the emergence of the follicular wave and the role it plays in follicular wave turnover has been well documented (Adams et al. 1992; Ginther et al. 1996). The increase in the number of medium-sized follicles in heat-stressed cows most likely occur as a result of the higher plasma FSH increase that preceded the second follicular wave (Gibbons et al. 1997). A lower plasma FSH surge in heat-stressed cows has been observed (Gilad et al. 1993). Inhibin is an important factor in the regulation of FSH secretion (Findlay 1993; Kaneko et al. 1993). An inverse relationship between plasma FSH and immunoreactive inhibin concentrations was reported throughout the oestrous cycle (Kaneko et al. 1995, 1997). Granulosa cells of large follicles are known to be the main source of plasma inhibin (Findlay 1993). Decreased mRNA for inhibin subunits in the early stages of follicular atresia has been observed (Braw-Tal 1994). The changes in the forms of the inhibin content in the follicular fluid during the growth and regression phases of dominant follicles have been observed (Guilbault et al. 1993). Therefore, heat stress decreases plasma inhibin concentration resulting in changes of granulosa cell functions. A tendency for a decrease in plasma inhibin concentration has been observed in heat-stressed cows (Wolfenson et al. 1995) and in buffalos during summer (Palta et al. 1997). Inhibin and oestradiol secretion by the largest follicle within a wave is thought to mediate the inhibitory effect of the dominant follicle on FSH secretion (Kaneko et al. 1991; Findlay 1993), and a synergistic effect of inhibin and oestradiol anti-serum on FSH secretion during the follicular phase in cows has been reported (Kaneko et al. 1995). The high preovulatory FSH surge in heat-stressed cows is synergistic to low secretion of both oestradiol and inhibin. However, the low plasma oestradiol and the high plasma FSH concentrations in heat-stressed cows during the follicular phase do not conclude about their role in heat. Alterations in follicular dynamics during heat stress in cows may have physiological significance due to their role in the emergence of the second follicular wave during heat stress (Wolfenson et al. 1995). Heat stress has been observed to induce ovulation of an aged dominant follicle. A negative relationship between duration of dominance of the preovulatory follicle and fertility has been shown in spontaneously cyclic cows (Bleach et al. 2004) and in experimentally induced persistent dominant follicles (Mihm et al. 1994; Austin et al. 1999). The growth of more than one large follicle during the first follicular wave (Wolfenson et al. 1995) and during the follicular phase occurs due to a high plasma FSH concentration in heat-stressed cows and not necessarily due to a high pulsatile LH secretion (Wise et al. 1988; Gilad et al. 1993). The finding gets support from investigations of

Adams et al. (1993) who showed that exogenous FSH injection before increased the number of ovulatory follicles. De Castro e Paula et al. (2008) studied the effect of heat stress and follicle class (high oestradiol and low oestradiol) on ovarian follicular concentrations of oestradiol 17 β, progesterone and follicular oxygen concentration. They reported that oestradiol-17β concentration was higher in the high oestradiol-17β class as compared to cows in the low oestradiol-17β group. In addition, progesterone concentration was lower in the high oestradiol-17β class as compared to cows in the low oestradiol-17β group. They did not find effect of heat stress or follicle class × treatment interaction on the concentration of oestradiol-17β or progesterone. Follicular oxygen concentration ranged from 3.9 to 9.2% and was not significantly affected by heat stress.

4.3   The Oocyte

Heat stress affects oocyte development and its functions. Lactating dairy cows are more sensitive to heat stress, and oocyte competence for fertilisation and subsequent development is reduced during heat stress (Zeron et al. 2001; Al-Katanani et al. 2002b; Sartori et al. 2002). High ambient temperatures 10 days prior to oestrus were found to be associated with low fertility (Al-Katanani et al. 1999). Steroid production by cultured granulosa and thecal cells was low when cells were obtained from cows exposed to heat stress 20–26 days previously (Roth et al. 2001a), that is, when follicles were 0.5–1 mm in diameter. The resumption of fertility observed in dairy cows in the autumn could be hastened by removing follicles formed in the summer (Roth et al. 2001b). In goats, heat stress has also been found to reduce plasma concentrations of oestradiol and decline in follicular oestradiol concentration, aromatase activity and LH receptor level and delay in ovulation (Ozawa et al. 2005).

Effects of heat stress on follicular functions involve changes in the follicle function or development and/or the secretion of the pituitary hormones involved for the development of the follicle. Heat stress at the time of ovulation and oocyte maturation may or may not affect the oocyte fertilisation in cows (Putney et al. 1989a) and in mice (Baumgartner and Chrisman 1988; Aroyo et al. 2007; Roth et al. 2008), but the fertilised embryos are more likely to be affected and likely to develop slowly or abnormally. In some studies, the process of oocyte maturation has been observed to be disrupted at elevated temperature (Payton et al. 2004; Roth and Hansen 2005; Wang et al. 2009). Damage to the oocyte during the preovulatory period by heat stress may occur due to the generation of reactive oxygen species, as both in vivo (Roth et al. 2008) and in vitro (Lawrence et al. 2004) effects of heat stress were found to be countered to some extent when antioxidants were administered.

Apoptosis is critical to cell functions, and thermal stress on the maturing oocyte in cattle affects it. A fraction (approx. 15–30%) of oocytes exposed to elevated temperature undergoes apoptosis as determined by TUNEL labelling of the pronucleus (Roth and Hansen 2004a, 2005; Soto and Smith 2009). The effect of elevated culture temperature on oocyte competence for fertilisation and subsequent development was reduced by inhibition of heat-shock-induced apoptosis with a caspase inhibitor (Roth and Hansen 2004a), sphingosine 1-phosphate (Roth and Hansen 2004b, 2005) or a BH4 peptide (Soto and Smith 2009).

4.4 Embryonic Development

The preimplantation embryos are susceptible to maternal heat stress, but the susceptibility declines as development proceeds. In cattle, exposure of lactating cows to heat stress at day 1 after oestrus (2 cell embryo stage) reduced the proportion of embryos that developed to the blastocyst stage at day 8 after oestrus (Ealy et al. 1993). However, heat stress at day 3 (8–16 cells), day 5 (morula) and day 7 (blastocysts) was observed to have no effect on the proportion of embryos that were blastocysts at day 8. A similar pattern of development has also been reported in sheep (Dutt 1964). The adverse effects of heat stress on in vitro cultured embryos of cows have been observed to decline with the advancement of stage or development (Edwards and Hansen 1997; Sakatani et al. 2004). However, sensitivity to elevated temperatures of mouse embryos at the two-cell, four-cell and morula stages of development was observed to be almost similar, and no significant differences were found (Aréchiga and Hansen 1998).

The elevated temperature or heat stress on the preimplantation embryos increased production of reactive oxygen species. Maternal heat stress has been observed to increase reactive oxygen species activity in oviducts and embryos (Ozawa et al. 2002; Matsuzuka et al. 2005a) and reduced glutathione content in recovered embryos (Ozawa et al. 2002; Matsuzuka et al. 2005b). Embryonic development during heat stress was observed to reduce by treatment of female mice with either melatonin (Matsuzuka et al. 2005b) or vitamin E (Sakamoto et al. 2008). Female embryos were better able to survive challenge of elevated temperature than male mice due to the reduced reactive oxygen species production in females (Pe´rez-Crespo et al. 2005). Increased reactive oxygen species production in response to elevated temperature was also reported in cattle (Sakatani et al. 2004, 2008), and treatment with the antioxidant 2-mercaptoethanol was found to alleviate the negative effects of heat stress on embryonic development (Sakatani et al. 2008); however, de Castro e Paula and Hansen (2008) found no effect of treatment on embryonic development.

As indicated earlier also, the heat response is not uniform all throughout the stages of development of embryos, and a decline occurs in response with the stage of development. There could be several reasons for the gain of resistance to elevated temperature as embryo development proceeds. Generation of reactive oxygen species in response to heat shock declines as bovine embryos advance in development (Sakatani et al. 2004), while intracellular concentrations of the cytoplasmic antioxidant glutathione increase (Lim et al. 1996). In addition, there is developmental regulation in the capacity of the embryo to undergo the induced thermotolerance response, whereby exposure to a mild elevation in temperature makes cells more resistant to a subsequent severe temperature elevation or makes them refractory. This thermo-adaptive response may not develop at an early stage or until day 4 in cattle (Paula-Lopes and Hansen 2002a) and the eight-cell stage in mice (Ar´echiga et al. 1995) and varies in other animals like buffalo and goat. Acquisition of the capacity for induced thermotolerance involves synthesis of heat-shock protein 70 (HSP70) which helps stabilisation of intracellular proteins and organelles, and apoptosis is inhibited (Brodsky and Chiosis 2006). High temperature can induce HSPs as early as the two-cell stage in cattle (Edwards and Hansen 1996) and mice (Christians et al. 1997), that is, before dividing fertilised cell acquire thermotolerance. Therefore, other molecular mechanisms are most likely involved in thermotolerance or adaptive response of a fertilised cell in earlier stage of development. Glutathione has been reported to be required for induced thermotolerance in mice (Ar´echiga et al. 1995), and changes in redox status may be an important determinant of development of induced thermotolerance.

Inhibition of apoptosis in bovine embryos with a caspase inhibitor has been observed to increase the magnitude of the reduction in development caused by high temperature (Paula-Lopes and Hansen 2002b). Thus, apoptosis, if limited to the most damaged cells of the embryo, may allow the embryo to continue to develop after thermal challenge. In cattle, induction of apoptosis by high temperature does not occur until the 8–16-cell stage at day 4 after insemination (Paula-Lopes and Hansen 2002a). Most of the effects of thermal challenges that occur at the early stage of development after fertilisation of ovum may not occur due to direct impacts of heat, and some of these effects of elevated temperature on embryonic survival in utero could be indirect results due to the changes in maternal physiology rather than a direct effect on the embryo.

4.5 Fetal Development

In large animals, heat stress affects fetal growth during gestation. Exposure of pregnant ewes to heat stress has been observed to reduce fetal and placental weights. The concentrations of placental hormones in the blood are influenced, and effects on growth are greater during mid-gestation than that occurring during later gestation (Wallace et al. 2005). Some effects of heat stress on placental function represent redistribution of blood to the periphery and reduced perfusion of the placental vascular bed (Alexander et al. 1987). However, reduced perfusion to the placenta is not the only reason of reduced fetal weights because placental blood flow per gram of fetus was similar between heat-stressed and control ewes (Wallace et al. 2005). Perhaps more important is an increase in vascular resistance in the placenta (Galan et al. 2005) caused by alterations in angiogenesis as reflected by aberrant patterns of expression of genes such as vascular endothelial growth factor, its receptors and placental growth factor (Regnault et al. 2002). Heat stress has more effects during mid-gestation than during late gestation because angiogenesis is more extensive during mid-gestation. Glucose transport capacity across the placenta is also reduced by maternal heat stress (Thureen et al. 1992), and this effect involves reduced expression of GLUT8 genes in cotyledonary placenta (Limesand et al. 2004).

Similar effects of maternal heat stress on placental functions and fetal development have also been observed in the cows (Collier et al. 1982). Reduced secretion of placental hormones as a result of heat stress may reduce milk yield in cows (Collier et al. 1982; Wolfenson et al. 1988).

4.6 Uterine Environment

Heat stress reduces blood flow to the uterus, and there is increase in uterine temperature which may affect implantation and embryonic mortality. These effects are likely to be associated with the production of heat-shock proteins by the endometrium during heat stress and reduced production of interferon-tau by the conceptus. Heat stress may affect endometrial prostaglandin secretion leading to premature luteolysis and embryonic loss (Malayer and Hansen 1990). There are distinct breed differences between Brahman and Holstein cows in endometrial responses to culture at high temperature.

5   Male Reproduction

The mammalian species have testes located out-side in groin region, and the testes are suspended in scrotum outside the body cavity to keep intra-testicular temperature slightly lower than core body temperature. This intricate thermoregula-tory system in the testis, involving countercurrent heat exchange from warm blood entering the testis and cool blood draining from the testis through an arteriovenous plexus, that is, pampini-form plexus, helps in maintaining optimum temperature for sperm development. The degree of descend or ascend for cooling is further con-trolled by two muscles, the tunica dartos in the scrotum that regulates scrotal surface area and the cremaster muscle that controls the position of the scrotum relative to the body. Evolution of the scrotum occurred because of the need for low temperatures either for spermatogenesis, sperm storage or to minimise mutations in gamete DNA (Werdelin and Nilsonne 1999; Bedford 2004). Different mammals have different location of tes-tes in the body and system for optimising sperm development. Heat is lost from the testis and scrotum to the environment through the scrotal skin, which is well endowed with sweat glands (Setchell and Breed 2006). *B. indicus* bulls are less sensitive to the effects of high temperatures than *B. taurus* or crossbred bulls, but as they are actually more sensitive to the effects of scrotal insulation (Brito et al. 2003), this would appear to be due to the greater ability of *indicus* animals to keep their testes cool (Brito et al. 2002). *B. indicus* bulls have greater testicular artery length to testicular volume ratios and smaller tes-ticular artery wall thickness and arterial to venous distances, which may be responsible for greater cooling of the arterial blood in the spermatic cord (Brito et al. 2004).

Regardless of the evolutionary reason for the location of the testis and epididymis outside the body, a rise in testicular temperature in mammals with external testes reduces sperm output, dec-reased sperm motility and an increased propor-tion of morphologically abnormal spermatozoa in the ejaculate. Such effects can be observed when a local heat source is applied to the testis,

the scrotum is insulated, the testes are internalised (i.e. cryptorchidism induced) or body temperature is raised because of fever or thermal environment (Setchell 1998). However, low body temperature is not an absolute requirement for spermatogenesis. Birds, which have body temperatures higher than mammals (Prosser and Heath 1991), have internal testes. The cells that are most susceptible to dam-age by high temperature are the spermatocyte and spermatid (Setchell 1998), although B sper-matogonia are also damaged. Oxidative stress is a major cause for thermal damage to spermato-genic cells which leads to apoptosis and DNA strand breaks (Pe´rez-Crespo et al. 2008; Paul et al. 2008, 2009). Effects of cryptorchidism on spermatogenesis enhance in superoxide dismus-tase-1 knockout mice (Ishii et al. 2005).

Heat stress significantly impairs bull fertility during summer. Semen quality decreases when bulls are continually exposed to ambient tempera-tures of 86 °F for 5 weeks or 100 °F for 2 weeks despite no apparent effect on libido (De la Sota et al. 1998). Heat stress decreases sperm concen-tration, lowers sperm motility and increases percentage of morphologically abnormal sperm in an ejaculate. When exposed to long period of heat stress, semen quality does not return to nor-mal for approximately 2 months because of the length of the spermatic cycle, adding to the carry-over effect of heat stress on male reproduction.

High temperatures cause degeneration of mei-otic germ cells in the seminiferous tubules (Lue et al. 1999), influence the structure of spermatozoa DNA (Love and Kenney 1999) and sperm counts and sperm motility are lowered (Setchell 1998). The progressive sperm motility is more sensitive to temperature variation as compared to the number of spermatozoa produced at each ejaculation (Moreira et al. 2001). If heat damaged sperm cells are used to fertilise normal oocytes, premature embryonic death may occur (Burfening and Ulberg 1968).

Paul et al. (2008) observed that in vitro fertilisa-tion with sperm collected from male mice, in which the scrotum was heated to 42°C, resulted in embryos with reduced ability to complete development. In addition, females mated to males exposed to scrotal heating were observed to have conceptuses with smaller fetal and placental weights compared with controls (Jannes et al. 1998; Paul et al. 2008).

5.1   Mechanism of Effects of Heat

The effect of heat stress on the testes and testicular functions varies widely in animals. Different cell types are affected in different ways, and response of one cell type is likely to be different from other cells. The Leydig cells are either not directly affected by heat or are only minimally affected; however, the primary site of action of heat stress may be the Sertoli cell (Setchell 2006). Because of its position in the seminiferous epithelium, it may be influenced. The germ cells depend almost entirely on the Sertoli cells for nutrients, and their development is influenced by the environment components of Sertoli cells. The Sertoli cells could have an influence on cells such as the spermatogonia and preleptotene spermatocytes. Secretion of fluid appears to be reduced under some circumstances, but not in the first 24 h after the testes were made cryptorchid (Setchell 2006). The changes may occur in composition of the secreted fluid without any influence on total volume secreted, and the composition of the secretion is likely to be affected at different temperatures. An effect on the Sertoli cells could also influence chromosome behaviour during the meiotic prophase, and investigations on the effects of heat on the synaptonemal complex may be required to further elucidate responses (Setchell 2006).

Increased metabolism in the testis after heat stress may not be met by a sufficient increase in blood flow and may become hypoxic (Setchell 1998). The response or damage due to hypoxic condition on exposure to heat may not be so much directly, as compared to that caused by the generation of reactive oxygen species and the effect of scavengers for ROS during heating or immediately afterwards. Therefore, changes in enzymes, heat-shock factors and heat-shock proteins need to be understood for their protective or harmful responses on the testis.

Semen characteristics are not likely to be immediately affected by changes in testicular temperature as damaged spermatogenic cells do not enter ejaculates immediately after heat stress. In the bull, where spermatogenesis takes about 61 days, alterations in semen have been observed to occur about 2 weeks after thermal stress and return to normal takes about 8 weeks after the heat stress. Culture of bull spermatozoa at 40°C was observed to have no effect on their fertilising capability and the competence of the resultant embryos to develop to the blastocyst stage (Hendricks et al. 2009). Ejaculated bull and stallion spermatozoa were observed to induce no apoptotic changes when cultured at temperatures characteristic of physiological hyperthermia (Hendricks and Hansen 2009). However, there may be some epigenetic changes in embryonic development associated with damage to the sperm in the reproductive tract. Insemination of rabbit does with sperm exposed to elevated temperature in vitro (Burfening and Ulberg 1968) or in the female reproductive tract (Howarth et al. 1965) was observed to result in reduced preimplantation and post-implantation survival (Burfening and Ulberg 1968). The X and Y spermatozoa are likely to be affected differentially by elevated temperature. The sex ratio of embryos has been observed to skew towards female when female mice were bred to males experiencing scrotal heat treatment on the day of mating (Pe´rez-Crespo et al. 2008). Incubation of sperm at 40°C for 4 h compared to 38.5°C has been observed to reduce the proportion of embryos (Hendricks et al. 2009).

5.2   Hormone Secretion

Limited studies have been carried out in livestock species on hormonal mechanisms involved in depression of male gonadal functions during summer. There are few experiments on the effects of high environmental temperature on the secretion of hormones controlling reproductive functions. Data from both bulls and boars indicate that heat stress causes an initial decline in circulating concentrations of testosterone lasting 2 weeks, but concentrations are restored even in the face of continued heat stress (Rhynes and Ewing 1973; Wettemann and Desjardins 1979). Thyroxine levels show a positive correlation with seminal volume and initial motility while T_3

exhibited a positive correlation with total sperm concentration and percentage of live spermatozoa. Correlation with other seminal and behavioural characteristics was not significant. The sexual interest of buffalo bulls was observed to be low during the summer, and bulls can produce semen throughout the year under appropriate feeding and management conditions (Dixit et al. 1984). Seasonal variations in semen quality, freezability and plasma-luteinising hormone (LH) levels have been studied during summer and spring seasons. In bulls' semen concentration, number of spermatozoa and motile cells per ejaculate were found to be lower in summer than in winter and spring seasons (Mathevon et al. 1998). Semen attributes of buffaloes during different seasons have been studied (Gangwar 1980), and results indicate that summer season affects volume, motility and number of live sperm. Heat stress causes hyperthermia of the scrotum and testes which leads to poorer morphological and functional semen quality. Hansen (1997) reported deterioration of bull fertility caused by heat stress during the summer months. Heat stress has less severe effects on semen quality of zebu bulls than it does on bulls of European breeds or their crosses, and this phenomenon is associated not only with the generally more efficient thermoregulation observed in zebu cattle but also specific adaptations that enhance the local cooling of blood entering the testis (Brito et al. 2004). Plasma LH decreased between summer and spring, but the differences were, however, not significant. Pre-freezing motility did not differ significantly, but post-freezing motility varied significantly ($P < 0.01$) between seasons. Post-freezing motility was observed lowest during summer and highest during winter. Summer spermatozoa may be fragile and are not able to withstand freezing stress affecting reproductive efficiency of buffalo during summer. Therefore, buffalo semen should be collected and frozen during winter and spring for use during hot weather conditions.

Seasonal variations in plasma LH levels were observed to be insignificant in buffaloes (Bahga and Khokar 1991). Hyperthermia can alter luteinising hormone (LH) secretion in females, even in the absence of ovarian steroids

(Schillo et al. 1978), and it is likely that severe heat stress can compromise LH secretion in males similar to that observed in females. However, the major site for disruption of reproductive function appears to be the spermatogenic cell lineage in the testis as indicated earlier.

6   Effect of Heat Stress on Reproduction by Altering Energy Balance

Immediately after calving, a critical phase occurs when dry matter intake is not commensurate with the body needs and does not meet the increased metabolic demands of lactation, and as a result, body is in a state of 'negative energy balance'. During this period, body reserves of fat and protein are mobilised (Bauman and Currie 1980; Butler 2000) and animals have low body condition score (BCS). Both negative energy balance and low BCS are observed to be associated with low fertility in cattle (O'Callaghan and Boland 1999; Butler 2000; Pryce et al. 2001; Pushpakumara et al. 2003). Low body energy or energy deficiency impacts or impairs gonadotrophin secretion. Since an animal reaches this state around parturition, gonadotrophin secretion to support follicular development and ovulation is compromised and reproductive problems (i.e. cystic ovaries) associated with onset of ovarian activity become prevalent (Zulu et al. 2002b). Growth hormone stimulates insulin-like growth factor 1 (IGF-1) production by the liver (Jones and Clemmons 1995), but during negative energy balance in cows, growth hormone receptors are downregulated, and this process is referred to as 'growth hormone resistance' (Donaghy and Baxter 1996). During early lactation in cows under negative balance, the liver becomes refractory to growth hormone due to the decline of growth hormone receptors (Vicini et al. 1991), which in turn results in reduced plasma concentration of IGF-1 (Pell et al. 1993). Follicular growth is stimulated by IGF-1 (Webb et al. 2004), and plasma concentrations are observed to be low in high-producing cows (Rose et al. 2004), and such cows may be observed to take long periods for return to ovarian cyclicity (Taylor et al. 2004). After calving, cows with IGF-I concentrations greater than 50 ng/ml at first service were observed to conceive quickly than those with lower concentrations (Taylor et al. 2004). Heat stress may affect reproductive performance both directly and indirectly through alterations in animal energy balance. In the dairy cow, there is an interaction between dry matter intake, stage of lactation, milk production, energy balance and heat stress that results in reduced LH secretion and a decreased diameter of the dominant follicle in the postpartum period (Ronchi et al. 2001; Jonsson et al. 1997).

Negative energy balance in cows leads to decreased plasma concentrations of insulin, glucose and IGF-I and increased plasma concentrations of GH and nonesterified fatty acid (Butler 2001; Lucy et al. 1992; Jolly et al. 1995). All of these metabolic hormones affect reproduction and other related functions. Metabolic hormones acting on the hypothalamic–hypophyseal–adrenal axis and the gonads (ovary and testes) probably mediate the inhibitory effects of negative energy balance on postpartum fertility.

Heat stress associated with or without negative energy balance is likely to affect reproductive efficiency, and several studies indicate that lactating dairy cows losing greater than 0.5 units BCS within 70 days postpartum had longer calving to first detected oestrus and (or) ovulation interval (Butler 2000; Beam and Butler 1999). Garnsworthy and Webb (1999) reported lowest conception rates in cows that lost more than 1.5 BCS units between calving and insemination. However, Butler (2000) reported that conception rates ranged between 17 and 38% with a decline in BCS by 1 unit or more, between 25 and 53% with a loss in BCS and between 0.5 and 1 unit, and conception was greater than 60% if cows did not lose more than 0.5 units or were observed to gain weight.

In addition to heat stress, another deterrent to high-producing cow fertility is increased blood urea nitrogen concentrations. In terms of effects on fertility, most research has focused on the urea produced as a result of protein metabolism within the rumen. However, elevated urea concentrations are also a consequence of increased skeletal muscle breakdown. The end result of these physiological changes that possibly occur during heat stress is elevated plasma urea nitrogen concentrations in heat stress cows compared to pair-fed cows in thermoneutral conditions. Therefore, elevated plasma urea nitrogen concentrations may be exacerbating the decrease in fertility that is frequently observed during periods of heat stress. Therefore, heat stress alters dry matter intake, and as a consequence, postpartum animals may develop negative energy balance. Decreased concentrations of plasma insulin and IGF-I and eventually even glucose (Richards et al. 1995) are likely to alter folliculogenesis. Their lower plasma concentrations are likely to lead to impaired follicular development, poor oestrus detection and poor quality oocytes affecting reproductive efficiency.

7 Genetic Plasticity Controlling the Magnitude of Heat Stress Effects

Inbreeding represents increased frequency of identical alleles at a gene locus, and the inbreeding per cent is a measure for the genes of an individual that are identical by descent (Wright 1922; Falconer 1981). In general, the reproductive function declines as the level of inbreeding in a population rise, that is, above 6.25% (Hansen 2005). Thompson et al. (2000a, b) observed that calving intervals increase by 12 and 17 days for Jersey and Holsteins cows, respectively, with levels of inbreeding more than 10%. Similarly, inbreeding had pronounced negative effects on fertility particularly at higher levels (10%) of inbreeding (Wall et al. 2005). Animals with an inbreeding coefficient >9% were observed to have fewer transferable embryos following superovulation than animals with a lower inbreeding coefficient (Alvarez et al. 2005). The gene pools of mammals contain allelic variants of specific genes that are responsible for body temperature regulation and cellular response to hyperthermia. Thus, genetic selection, both natural and artificial, is likely to modulate the impact of heat stress on reproductive functions. Thus, genetic selection for thermal tolerance may be a necessity for future livestock production system in tropical climate for sustaining and optimising productivity under global climate change conditions.

Genetic influences on regulation of body temperature have been well studied in cattle. In cattle, estimates of the heritability of rectal temperature range from 0.25 to 0.65°C (Finch 1986). There are distinct breed differences in thermoregulatory ability (Hammond et al. 1996; Hansen 2004; Pereira et al. 2008). One specific gene affecting body temperature regulation of cattle during heat stress, the *slick* gene affecting hair length, has been identified (Olson et al. 2003; Dikmen et al. 2008), and there are undoubtedly many others. The superior thermoregulatory ability of zebu cattle has been ascribed to lower metabolic rate, reduced resistance to heat flow from the body core to the periphery and properties of the hair coat (Hansen 2004). In addition to these, many more metabolic and morphological features differ in Zebu from Taurus cattle.

8 Consequences of Actions of Climate Change on Reproduction for Species Survival and Distribution

As has been clearly indicated, heat stress can have profound effects on most aspects of reproductive functions in male and female livestock, and functions like gamete formation, embryonic development and fetal growth and development may be affected. The potential impact of heat stress can be assessed by examining seasonal trends in reproductive function of different livestock species. A study carried out in Spain indicates that the proportion of inseminated dairy cows that become pregnant during the warm months of the year was 22.1 versus 43.1% of cows inseminated in the cool season (López-Gatius et al. 2004). Indeed, the magnitude of the summer decline in fertility is much less for non-lactating heifers or cows producing low amounts of milk than it is for cows with high milk yield (Badinga et al. 1985; Al-Katanani et al. 1999).

Thus, it is likely that the direct impact of global warming (i.e. consequences for body temperature regulation) on reproduction will be more severe for domestic animals than other mammals. In addition, the existence of allelic variation in genes controlling body temperature regulation and cellular resistance to heat shock indicates that genetic adaptation to increasing global temperature will be possible for many species.

9 Reproduction in Buffaloes

Buffaloes under natural conditions have been observed to breed during specific period and, therefore, are believed to be seasonal breeders. However, this is not entirely true as buffaloes are polyestral and may be observed to breed all year round. The buffalo is also regarded as a difficult breeder mainly because of its inherent susceptibility to heat stress, and prolonged exposure to high solar radiation during summer may cause anoestrus and sub-oestrus. The irregular or anoestrus conditions in buffalo affect and prolong inter-calving periods resulting in economic loss to the farmers. Heat stress on buffalo also affects its feed intake and in turn the nutritional balance and reproductive efficiency. The ideal or optimum climatic conditions for buffalo growth and reproduction as suggested by Payne (1990) are air temperatures of 13–18°C combined with an average relative humidity of 55–65%, a wind velocity of 5–8 km/h and a medium level of sunshine. However, these conditions are likely to vary for different adapted and non-adapted buffalo breeds under different conditions.

Buffaloes have acquired several morphological features which reinforce their ability to thrive well in tropical and hot–humid conditions. The melanin pigmentation of buffalo skin is useful for defence against ultraviolet rays. Hair density in adult buffalo is only one-eighth of that in cattle (Hafez et al. 1955), thus facilitating dissipation of heat by convection and radiation in open conditions. Low number of sweat glands in buffalo compared to cattle results in a lower efficiency of sweating in buffalo than in cattle. Furthermore, the number of sebaceous glands is lower in buffalo than in cattle;

however, sebum secretion shows an opposite trend (Hafez et al. 1955). The sebum provides effective protection to the skin while the buffalo is the mud or water. The skin of buffaloes is also thicker than that of cattle and protects the sparsely covered skin from harmful mechanical and chemical agents dissolved/suspended in water and mud during swimming and wallowing (Badreldin and Ghany 1954; Hafez et al. 1955). The wallowing behaviour, as well as shade seeking, is necessary during the hot season to protect and dissipate body heat. Grooming or rubbing skin against hard surface is important behaviour for body care and for removing irritants form skin surface. Tail swishing is used to remove flies and other irritants. Body areas within reach are licked and scratched, while inaccessible areas are rubbed on available surfaces or groomed by other herd members following solicitation. The buffalo exhibits oestrous throughout the year, but fertility may not be equally distributed as buffalo is more fertile when daylight hours decrease, which coincides in tropical areas with greater forage availability (Zicarelli 1994). The length of the oestrous cycle of buffalo is less than that of cattle, that is, average 21 days, and considerable variations in length of the cycle may be observed in individual animals. In buffaloes, the typical signs of oestrus behaviour (i.e. excitability, standing behaviour) are less pronounced than in cows (Seren et al. 1992; Zicarelli et al. 1992), whereas the courtship behaviour of buffalo bulls is similar to that reported in cattle. Buffalo bulls/teasers are able to detect pre-oestrus buffalo and remain close to them, exhibiting the characteristic 'flehmen' behaviour in response to pheromones in oestrus urine. In buffalo, the fraction of cisternal milk is lower than in cattle (Thomas 2005). As a result, lactating buffalo seems to be sensitive to the minor change in milking routines, which determines a decrease of milk flow and milk yield (Thomas et al. 2005; Saltalamacchia et al. 2007).

9.1 Males

In well-nourished, good body condition buffalo bull, testicular spermatogenic cell divisions are likely to start by about 1 year of age,

and active spermatogenesis can be observed by 15 months of age. However, the ejaculate contains viable spermatozoa only after 24–30 months of age (Perera 1999), indicating that male buffalo matures more slowly than male cattle (Drost 2007) and has a longer time lag between onset of spermatogenesis and the onset or achievement of puberty. Thereafter, even if capable of breeding throughout the year, buffalo bulls are known to show seasonal rise and fall in reproductive functions, which have been recorded in most breeds in different countries (Sengupta et al. 1963; Esposito et al. 1992; Zicarelli 1997). Buffalo bulls may reach sexual maturity at 2–3 years of age. Semen is produced all year round, but it is highly affected by heat stress and low-quality feeds and fodders. The buffalo bull is believed to be most fertile in spring, when the volume of ejaculate and sperm concentration is highest. Sperm vitality is also higher in spring than at other times of the year. The vitality is lowest in summer, and heat stress has a negative effect on libido (Gili et al. 1974).

A morphometric study of the testis helps to evaluate the influence of different factors, such as hormonal fluctuations of the photo-neuroendocrine circuit on reproductive efficiency due to seasonal variation (Yasuo et al. 2006). The morphology and physiological functions of the male gonad are under influence of the pineal and pituitary secretions during the year (Goeritz et al. 2003; Pant et al. 2003). Size and weight of testis are excellent indicators of sperm-producing capacity and spermatogenic functions (Fields et al. 1979; Finch 1986; Yarney et al. 1990). The highest scrotal circumference has been observed in the month of October (Mikelsen et al. 1981). The reduction in testicular measurements (testes weight and length) by exposure to heat stress is due to degeneration in the germinal epithelium and to a partial atrophy in the seminiferous tubules (Chou et al. 1974). This is reflected in the adverse effects on the average number of testicular cells, especially the secondary spermatocytes and spermatids of types B, C and D, the ratio of Sertoli cells to other cells and the diameter of the seminiferous tubules (El-Sherry et al. 1980). However, Habeeb et al. (2007) reported that farm animal's testes size does not undergo marked seasonal changes. Histological examination of testes suggests that only spermatogenic elements disappear and that interstitial material remains unchanged or increases in number and volume as a function of exposure to heat stress (Gomes et al. 1971).

High temperature or heat stress usually affects negatively the processes of spermatogenesis and metamorphosis of sperms leading to semen degeneration (Coser et al. 1979). Accordingly, impairment of spermatozoa fertility and their ability to produce viable embryos may be observed with heat load increase or stress (Wildeus and Hammond 1993). Sahni and Roy (1967) reported that maximum and minimum temperature range for optimum spermatogenesis was 29.4 and 15.6°C, respectively. Seasonal differences, with minimal spermatogenesis occurring during summer months that is usually referred to as 'summer sterility' (Akpokodje et al. 1985), are attributed to reduction of steroidogenic function of the testes and to the decrease of the blood flow through the testes (Setchell 1970).

Despite the thermoregulatory mechanism of the testes, sexual desire, that is, libido, is negatively influenced by high environmental temperature. Other functions that may be adversely effected are ejaculated volume, live sperm percentage, sperm concentration, viability and motility (Gamcik et al. 1979) leading to reduced conception and fertility. The reaction time is generally shorter during summer than during the other seasons. The shortest time (9.4±0.6 min) was observed during summer and the longest (15.9±1.5 min) in autumn. Values of 10.5±0.7 and 14.7±1.1 min were recorded in spring and winter season, respectively (El-Saidy 1988). However, El-Sherbiny (1987) reported that it was significantly longer in summer season than in the other seasons of the year. The optimum climatic conditions for sexual activity were found to be either during autumn or spring season (Hafez 1968; Zeidan 1989), and the lowest sexual activity was recorded in summer season (Zeidan 1989).

9.2 Physical Characteristics of Semen

The quantity and quality of semen vary with seasons, although the degrees of response to seasonal effects vary according to species, breed and locality. Season affects ejaculate volume; however, information on influences of high environmental temperature on semen–ejaculate volume is conflicting. The studies of Zeidan (1989) and Marai et al. (1996) showed that semen–ejaculate volume decreased, while studies of Fawzy (1982) showed remarkable increase with heat elevation. The highest mean value (3.7±0.1 ml) was recorded in winter and the lowest (2.4±0.1 ml) in spring. Mean values of 2.9±0.2 and 2.6±0.1 ml were recorded in summer and autumn, respectively. The effects of season on motility of spermatozoa are inconclusive and conflicting. The initial motility of spermatozoa has been observed to decrease in hot climate conditions (Ax et al. 1989; Zeidan 1989). Other studies indicated that motility of spermatozoa either increased (Oloufa et al. 1959; El-Azab 1980) or did not show any change due to season of the year or elevation of temperature (Zeidan 1989; Silvia et al. 1991).

Zeidan (1989) observed that live sperm were 87.5±0.5, 76.2±0.8, 74.5±1.5 and 85.8±0.6% during spring, summer, autumn and winter, respectively. Everett et al. (1978) recorded the highest total sperm output during summer season. Spring and summer seasons were characterised by the highest percentage of ejaculates with white milky (47.4%) and opaque creamy (28.5%) appearance and by the lowest percentage with greyish soapy (14%) and translucent watery (10%) appearance. In contrary, autumn showed the highest percentage of bad semen appearance. In other words, the highest percentage (47.4%) of ejaculates showed a white milky appearance, whereas (28.5%) was opaque creamy, 14% greyish soapy and 10% translucent white watery. Generally, semen appearance varied from opaque creamy to translucent watery. The difference in the results may be due to type and duration of heat exposure, intensity of environmental heat and differences in species, breed and age of the experimental animals.

Murrah buffalo bulls subjected to surface cooling had good quality (creamy and light creamy) semen more frequently with high sperm concentration, mass activity, motility % and live sperm production in cooled than uncooled cows but had no effect on semen volume and pH (Mandal et al. 2002).

9.3 Females

Buffaloes are found only in certain regions in the world, principally Asia, some Mediterranean countries, some countries in Eastern Europe and in many countries in Latin America. These regions are widely different in the geographical conditions, no other domestic animal can thrive and be similarly useful and economical. Most of the buffaloes in India and Pakistan and the countries of Southeast Asia, where the vast majority of the world's population of ~170 million buffaloes (FAO 2003) reside, are in the care of small holders. Buffaloes in these countries are kept primarily for draught purposes and as a source of milk. By contrast, most cattle are raised in herds and on farms. Cattle have been selected specifically as dairy animals or beef animals largely, through the use of AI and more recently embryo transfer. Globally, there has been minimal genetic selection for fertility in buffalo, which are generally regarded as poor breeders (Bhattacharya 1974). Poor breeding is attributed to late maturity, poor oestrus expression, prolonged calving intervals and seasonal reproductive patterns. Wild or feral female buffalo reaches sexual maturity at 2–3 years of age. Domesticated buffalo that is cared for and fed properly is likely to reach puberty early. Puberty is highly affected by management factors and practices. Body size is more important than age of animal, and a Murrah heifer should attain a two-third of mature body weight around 325 kg at first insemination or mating and 450–500 kg at her first calving. The age of puberty in buffalo is 36–42 months in India. It is comparatively late compared to other countries like Italy, where the age at first calving is between 28 and 32 months on average (Borghese 2005).

The tubular genitalia of the buffalo are generally more muscular and firmer, and the uterine horns are more coiled as compared to those of the cow. The body of the uterus is much shorter (1–2 cm) than that of the cow (2–4 cm). The cervix of the water buffalo is smaller than that of the cow (length 3–10 cm, diameter 1.5–6.0 cm), and its canal is more tortuous, which probably accounts for less dilation of the external os during oestrus. The average number of cervical folds in water buffalo is three (Sarabia 2004). The inactive ovaries of the mature water buffalo (El-Wishy 2007) are smaller ($3.0 \times 1.4 \times 1.0$ cm; 2.9–6.1 g) versus ($3.7 \times 2.5 \times 1.5$ cm; 5–15 g) in the cow (Roberts 1986). There are differences due to variations in breed, environmental conditions, season and management practices.

9.4 The Reproductive Cycle of a Buffalo

9.4.1 Puberty

The onset of puberty occurs late in buffaloes. First oestrus occurs at around 15–18 and 21–24 months in river buffalo and swamp buffalo, respectively; it varies considerably with the status of nutrition and body condition score (Jainudeen 1986). The age at first oestrus in some cases may be higher, and in heifers, expression of behavioural oestrus may be low.

9.4.2 Oestrous Cycle

Water buffalo is seasonally polyestrous with an average cycle length of 20 days (range 18–24 days) and an average duration of oestrus of 18 h (range 5–36 h). The oestrous behaviour in water buffalo is much more subtle, and homosexual behaviour is rare. Secondary oestrus signs such as swollen vulva, reddening of the vulvar mucosa, mucous vaginal discharge and frequent urination are also observed but may not be true indicators of oestrus. Ovulation occurs after about 30 h of the onset of oestrus. The diameter of an ovulatory follicle is ~10 mm. The diameter of the mature corpus luteum (CL) ranges from 10 to 15 mm versus 12.5 to 25.0 mm in the bovine. The ovulation papilla, or crown, of the CL does not protrude much beyond the surface of the ovary, making it more difficult to identify by palpation per rectum. The CL of pregnancy is invariably

Table 1 Signs of oestrus in buffaloes during different seasons

Symptoms	Medium monsoon	Peak winter	Low summer
Bellowing	–	+	–
Mucus discharge	+	+	+/–
Swollen vulva	+/–	+	–
Activity	+	+	+
Frequent urination	+	+	+
Uterine tone	+	+	+
Open cervix	+	+	+
Nature of cervical	Periodic thick	Frequent thin	Patchy thick
Discharge	Yellowish	Glassy copius	Opaque very less

Source: Modified from Janakiraman (1978)
+ present, – absent, +/– present in some cases only

located ipsilateral to the gravid horn (Baruselli et al. 1997). As in cattle, follicular growth occurs in waves in buffaloes, and twin ovulations are rarely observed. Two-wave cycles in buffalo are common (63.3%) followed by three-wave cycles (33.3%) and a single wave cycle (3.3%). The number of waves influences the length of the luteal phase and the oestrous cycle.

The oestrus cycle of buffalo varies in different breeds. The total duration of oestrus is usually 24 h but varies from 12 to 72 h. The most reliable sign of oestrus in buffaloes is frequent urination. The behavioural signs of oestrus are much less pronounced in buffalo than in cattle (Table 1). Many buffaloes show oestrus only at night time and for short period. A lactating buffalo may slightly decrease in milk yield on the day of heat. The buffalo in heat may be more restless and difficult to milk (Bhikane and Kawitkar 2000).

- Age at puberty: 36–42 months
- Length of oestrus cycle: 20 ± 2 days
- Duration of heat: 12–24 h
- Time of ovulation: 10–14 h after end of oestrus
- Period of maximum fertility: last 8 h of oestrus
- Gestation period: 310 days
- Period of involution of uterus: 25–35 days

9.5 Reproductive Performance in Buffalo

The reproductive efficiency in buffaloes is determined by many different processes, which result from interaction among genetic and environmental factors. The processes involved, individually or together, include age of puberty or maturity, pattern of oestrus cycle and oestrus behaviour, length of breeding, ovulation rate, lactational anoestrus period, postpartum anoestrus, inter-calving period and reproductive life span. A combination of these traits is used to measure breeding efficiency or breeding performance in farm animals (Agarwal 2003). Reproductive efficiency in buffalo is reported to be alarmingly low, causing severe economic losses to milk producers. However, the reproductive efficiency of some buffaloes reared under farm conditions is better.

In Northern India, conception rate is highest during the cool season (October to January) and lowest during the hot–dry season (May to July). In Southern India, the period from October to April is more favourable for conception in buffaloes. Maximum conception rates have been observed during winter in buffaloes (Rao et al. 1973; Rao and Kodagali 1983). GnRH analogue improves the conception rate in repeat breeder Nili–Ravi buffaloes when administered as a single dose at the time of insemination (Ahmad et al. 2002).

Studies have established a negative influence of ambient temperature and duration of sunshine on incidence of oestrus, and ambient humidity has a favourable effect. The period of the year from October–March with moderate RH and low ambient temperature is most congenial for resumption of postpartum oestrus. However, high humidity (hot–humid season) coupled with moderate temperature has a negative effect on postpartum oestrus (Reddy 1985). It has been observed that a maximum % (75.02%) of open animals come into heat in October and minimum (43.18%) in April. Conception rate is the highest in October (33.9%) and lowest in the month of June (18.09%).

9.6   Problems Observed with Buffalo Reproductive Performance

1. The first postpartum heat varies greatly with season, breed and individual. It has been observed to appear within less than 60 days in some cases and over 230 in others. Average postpartum oestrus in the Murrah breed of India has been reported to be 100 days. The first postpartum oestrus is not always fertile.
2. At birth, buffaloes have fewer primordial cells in the ovary than cattle have.
3. Compared to cows, buffaloes suffer from higher atresia of follicles – 20,000 versus 100,000 (Borghese 2005).
4. Buffaloes have a high proportion of silent oestrus and short duration oestrus. This is one of the most important problems in buffalo reproductive efficiency. It is even more problematic during the hot and humid months when it is compounded by thermal stress. Short and silent oestrus is the main reason for undetected heat in buffaloes.
5. A large number of buffaloes also suffer from postpartum anoestrus, a complete absence of oestrus cycle and no signs of heat. This is one of the most common causes of buffalo infertility.

9.7   Factors Causing Poor Reproductive Performance

1. Climate affects both production and reproduction of farm animals. Buffaloes are susceptible to extreme conditions of heat and cold, and their performance improves during the cool months. In India, 70–80% of buffaloes conceive between July and February, and a lower number of services are needed during the July to February than in the March to June season (Agarwal 2003). In India, buffaloes are photoperiodic and sexually activated by decreased

daylight. In Italy, the usual calving season is from September to December.

2. Buffaloes have poor thermal tolerance on account of an underdeveloped thermoregulatory system and are unable to get rid of excess body temperature. If their housing is not designed to take care of this special species-specific requirement for adequate shade and ventilation, it adversely affects production and reproduction (Ramesh et al. 2002).
3. Nutrition plays a major role in the reproductive performance of buffalo, as with other farm animals. However, there is a strong possibility that the consequences of poor nutrition are often interpreted as seasonality of breeding in buffalo. Under feeding, overfeeding or unbalanced feeding, as well as deficiencies in minerals, vitamins or trace elements are likely to negatively impact fertility in buffalo. A poor body condition score at calving affects fertility, characterised by prolonged postpartum intervals, reduced conception rates and more services per conception. A very low protein diet can cause cessation of oestrus (Agarwal 2003).
4. One of the reasons buffalo suffers from long postpartum anoestrus is because their natural behaviour of wallowing in dirty water pools, and unhygienic shed conditions, causes buffalo to suffer from a high incidence of endometritis. The loose broad uterine ligaments and rolling in water cause torsion of uterus cases in buffalo. Buffalo also suffers from uterine prolapse and retention of the placenta. All these lead to uterine infections, delayed involution of the uterus and endometritis.

9.8   Approaches for Improving Reproductive Efficiency

1. Providing the right kind of housing for buffalo to suit their natural behavioural requirements is important for their optimum performance. Free stall as well as tied systems works well for buffalo. However, it is important that the housing provides sufficient shelter from both heat and extreme cold. During summer, they

need to be protected from extreme heat, while in winter they need to be protected from extreme cold as well (Ramesh et al. 2002).

2. Showers or foggers with fans or wallowing tanks should be made available to buffalo during the hottest part of the day. Thermal ameliorative measures such as sprinkling and cooling are known to increase comfort levels and feed intake in buffalo (Thomas et al. 2005; Aggarwal and Singh 2008).

3. Balanced feeding with mineral supplements, plenty of green fodder and concentrate as per each animal's specific need, is necessary to bring buffalo into normal reproductive cycles.

4. Regular testing of all buffaloes and bulls for infectious reproductive diseases like brucellosis and regular culling of infected animals are crucial for good reproductive health in the herd. Attending cases of difficult birth and retained placenta in time and maintaining good hygiene during parturition are also crucial to prevent reproductive disorders such as endometritis.

5. Wall charts, breeding wheels, herd monitors and individual buffalo records are important oestrus detection aids. The key to successful use of these inexpensive management aids is to accurately record every heat, beginning with the first heat after calving, and to make daily use of the information to identify those buffaloes that are due to return to oestrus (Ramesh et al. 2002).

---

10 Management and Production Systems for Improving Fertility of Cows and Buffaloes in Summer Months

Research has resulted in dramatic improvements in dairy cow management in hot environments. Two primary strategies are:

• To minimise heat gain by reducing solar heat load

• Maximise heat loss by reducing air temperature around the animal or increasing evaporative heat loss directly from the animals

The use of cooling systems in hot weather has a beneficial effect, but these alone do not restore normal fertility in buffaloes. Additionally, summer infertility can be alleviated by the provision of high-quality forage and feed to overcome negative energy balance and by the use of hormonal treatments to induce normal cyclicity.

Following are several strategies to potentially help reduce the negative impacts of heat stress on reproduction in lactating dairy cows.

10.1 Identifying the Hot Spots

Locating where heat stress is occurring on the dairy farm by identifying hot spots is key to implementing the proper cooling or management strategy to eliminate these hot spots. Temperature devices have been used to monitor core body temperatures in cows by attaching a temperature monitor to a blank intravaginal drug release (CIDR®, Pfizer Animal Health, New York, NY) device for practical on-farm use. The device is inserted into the cow's vagina for measuring core body temperature every minute for up to 6 days. This allows monitoring of the cow's body temperature and identification of where the cow is experiencing heat stress.

10.2 Hormonal Therapy

An alternative approach for improving summer fertility is the use of reproductive hormones to modulate follicular development and improve fertility. In heat-stressed cows, the administration of GnRH has been observed to induce follicular development and a healthy preovulatory follicle (Guzeloglu et al. 2001). In summer, the administration of GnRH to lactating dairy cows at oestrus increased the conception rate from 18 to 29%. However, luteal support from a single administration of hCG (3,000 IU) on day 5 or 6 after insemination did not improve summer fertility (Schmitt et al. 1996a, b). Similar results have been reported following exogenous administration of progesterone with the CIDR intravaginal delivery device (Wolfenson et al. 1994). The effect of timed artificial insemination on fertility in summer has been investigated, and the results suggest that exogenous hormone administration can help overcome the effects of heat stress and reduce summer infertility in cattle (Wolfenson et al. 1994; De La Sota et al. 1998).

The use of fixed time insemination (AI) has the distinct advantage of not requiring the detection of oestrus and effective synchronisation methods for fixed time AI have been developed. They are based on administration of GnRH or hCG to induce ovulation, followed by a luteolytic dose of prostaglandin $F_{2\alpha}$ 6–7 days later and a second treatment with GnRH or hCG 24–60 h after the luteolytic treatment to induce a fertile ovulation (Schmitt et al. 1996b; Pursley et al. 1995, 1997). During summer, these programmes did not increase the number of cows pregnant to the fixed time insemination, but they did increase the number of cows pregnant by 120 days postpartum and reduced the numbers of days open (Almier et al. 2002; Cartmill et al. 1999). These results suggest that the principal benefit of these treatments is to induce cyclicity and the development of normal corpora lutea leading to good fertility. These approaches lead to an increase in the number of pregnant cows by increasing the number of inseminations at oestrus.

10.3   Improving Summer Fertility

It takes approximately 40–50 days for antral follicles to develop into large dominant follicles and ovulate (Roth et al. 2001b). If heat stress occurs during this time period, both the follicle and oocyte inside the follicle become damaged. Once ovulation occurs, the damaged oocyte has reduced chances of fertilising and developing into a viable embryo. Cooling dry cows may reduce heat stress effects on the antral follicle destined to ovulate 40–50 days later, which coincides with the start of most breeding periods and possibly increases first service conception rates. Heat alleviation or cooling dry cows with feed line sprinklers, fans and shades has been observed to be beneficial for reducing services per conception, reproductive culls and

days open, as well as increasing milk yield with a significant return on investment (Avendano-Reyes et al. 2006; Urdaz et al. 2006). In addition to thermal stress alleviation or proper cooling, changing management practices may also help reduce the severity of heat stress on animals during summer or hot–humid conditions. There should be at least 38–45 square feet of shade per mature dairy cow to reduce solar radiation. Spray and fan systems should be used in the holding pen, over feeding areas, over the feeding areas in some free stall barns and under shades on dry lot dairies in arid climates. Exit lane cooling is an inexpensive way to cool cows as they leave the parlour. Providing enough access to water during heat stress is critical. Water needs increase 1.2–2 times during heat stress conditions. Lactating cattle require 130–175 l/day of water/day. Since milk is approximately 90% water, water intake is vital for milk production and to maintain thermal homeostasis. Various other attempts have been made to overcome the effects of heat stress on fertility, including the use of shade, fans, air-conditioning and sprinkler systems to cool animals during summer (Igono et al. 1987; Bucklin et al. 1991; Huber 1996; Hansen 1997; Aggarwal 2009). The most widely used methods are cooling systems that mist the cows with water from overhead sprays and cool the air. The use of the animal cooling systems has produced some improvement of fertility, but they are unable to match the level of normal winter fertility (Armstrong 1994; Huber 1996).

In a study on Tharparkar cows, the average age at first conception was lower in cows kept under comfort conditions (with shelter), that is, 23.87 months as against 25.88 months in animals under stress condition (Razdan 1965). It was found that physiological maturity was attained by the animals in both normal and heat-stressed cows simultaneously, but either the ova did not survive or the inseminations were not successful due to excessive heat loads on the animal.

Hot–humid environments have been associated with low milk production and poor reproductive efficiency (Hafez 1968) because of increased metabolic heat production associated with lactation (Brody 1945). Cooling the cows by mister system results in improvement in conception rate in cows and buffaloes (Aggarwal 2009). Conception rate was observed to be higher for cooled cows than control cows (83.3 vs. 66.6%) when THI ranged between 79.44 and 87.72. Therefore, evaporative cooling increases conception and pregnancy rates during heat stress. In control group cows, during hot–humid season, the number of inseminations required ranged from 2 to 9 with an average of 4.8 inseminations in comparison to 2–4 (average 3.1 inseminations). Conception rate to all services was higher for cooled cows than control cows (83.3 vs. 66.6%).

The change in vaccination schedule to avoid peak summer days may help in avoiding stress imposed due to vaccine reactions. Possibly, during severe heat waves, it would prove beneficial to delay vaccinations at dry-off if the dry pen does not contain adequate cooling.

Embryo transfer may become a more effective strategy to increase pregnancy rates as compared to AI in lactating cows during periods of heat stress, and the magnitude of the increased temperature does not seem to influence overall success following transfer (Hansen and Arechiga 1999). As embryos advance in their development, the effects of elevated temperatures become less significant because embryos become more resistant to the deleterious effects of elevated temperatures (Ealy et al. 1992, 1995; Ealy and Hansen 1994; Edwards and Hansen 1997; Rivera and Hansen 2001). As a result, pregnancy rates following embryo transfer during heat stress are higher than pregnancy rates to AI (Putney et al. 1989b; Ambrose et al. 1999; Al-Katanani et al. 2002a) although not in the absence of heat stress. One potential constraint for embryo transfer in lactating cows is the short duration of oestrus and lack of intense mounting activity seen in dairy cows (Dransfield et al. 1998). This phenomenon is exacerbated by heat stress (Nebel et al. 1997) and will limit the number of embryos transferred in lactating cows in a programme that is dependent upon oestrus detection. The first report of a timed embryo transfer (TET) protocol, where ovulation was synchronised using an Ovsynch protocol, was by Ambrose et al. (1999) who

evaluated the efficiency of TET using either fresh or frozen–thawed in vitro produced (IVP) embryos and TAI under heat stress conditions. Pregnancy rates in cows that received a fresh IVP embryo were higher compared to cows in the TAI group.

Embryo transfer can significantly improve pregnancy rates during the summer months. Embryo transfers can bypass the period (i.e. before day 7) in which the embryo is more susceptible to heat stress and to bypass the harmful effects of heat stress on oocyte quality that limit embryonic development (Al-Katanami et al. 2002; Zeron et al. 2001). A study (Al-Katanami et al. 2002) used timed embryo transfer to study the effect of heat stress on fertility in lactating dairy cows and showed that timed embryo transfer improved pregnancy rates under heat stress conditions but only when fresh embryos were transferred (Al-Katanami et al. 2002). Nevertheless, embryo transfer is not a widely adopted technique. Improvements need to be made in the in vitro embryo production techniques, embryo freezing and timed embryo transfer, and lowering the cost of commercially available embryos before this becomes a feasible solution.

11   Conclusions

Heat stress has a wide range of effects on the reproductive axis. Some of these effects directly affect individual reproductive organs such as the hypothalamus, the anterior pituitary gland, the uterus, the follicle and its oocyte and the embryo itself, while the other effects of heat stress are indirect and probably mediated by changes in the metabolic axis in response to low feed intake or reduced dry matter intake. There is no single mechanism by which heat stress can reduce postpartum fertility in dairy cows, and this problem is due to the accumulation of the effects associated with several biophysical factors. Thus, it is likely that the direct impact of global warming (i.e. consequences for body temperature regulation) on mammalian reproduction will be severe for domestic animals. Buffaloes are suited to the hot and humid climates although they exhibit signs of great distress when exposed to direct solar radiation or when working in the sun during hot weather. Elevation of ambient temperature affects male reproductive functions deleteriously. Such phenomenon leads to testicular degeneration and reduces percentages of normal and fertile spermatozoa in the ejaculate of males. The ability of the male to mate and fertilise is also affected. Thermal stress adversely affects the conceptus and dams subsequent postpartum performance traits, as well as different stages of gestation. Thermal stress alleviation by cooling the cows and buffaloes may be a profitable and effective way to improve both milk production and reproduction during the summer months. Even generally milder climates experience heat stress or heat waves that dramatically reduce fertility. Dry cows are also susceptible to heat stress and should be provided cooling to improve subsequent fertility after calving. Managemental strategies to reduce negative effects of heat stress on fertility by artificial cooling, ration adjustments and reproductive protocol changes will improve efficiency and profitability.

References

Abilay A, Johnson HD, Madan ML (1975) Influence of environmental heath on peripheral plasma progesterone and cortisol during the bovine oestrus cycle. J Dairy Sci 58:1836–1840

Adams GP, Matteri RL, Kastelic JP, Ko JCH, Ginther OJ (1992) Association between surges of follicle-stimulating hormone and the emergence of follicular waves in heifers. J Reprod Fertil 94:177–188

Adams GP, Kot K, Smith CA, Ginther OJ (1993) Selection of dominant follicle and suppression of follicular growth in heifers. Anim Reprod Sci 30:259–271

Agarwal KP (2003) Augmentation of reproduction in buffaloes. 4th Asian Buffalo Congress lead papers, 121

Aggarwal A (2009) Effect of mist cooling on milk urea and pregnancy rate in lactating cows. Indian Vet J 86:838–839

Aggarwal A, Singh M (2008) Skin and rectal temperature changes in lactating buffaloes provided with showers and wallowing during hot-dry season. Trop Anim Health Prod 40:223–228

Ahmad N, Schrick FN, Butcher RL, Inskeep EK (1995) Effect of persistent follicles on early embryonic losses in beef cows. Biol Reprod 52:1129–1135

Ahmad G, Saeed MA, Bashir IN (2002) Use of GnRH to improve conception rate in repeat breeder buffaloes during low breeding season. Pakistan Vet J 22:42–44

Akpokodje JU, Dede TI, Odili PI (1985) Seasonal variations in seminal characteristics of West African Dwarf sheep in the humid tropics. Trop Vet 3:61–65

Alexander G, Hales JRS, Stevens D, Donnelly JB (1987) Effects of acute and prolonged exposure to heat on regional blood flows in pregnant sheep. J Dev Physiol 9:1–15

Al-Katanami YM, Paula-Lopes FF, Hansen PJ (2002) Effect of season and exposure to heat stress on oocyte quality of Holstein cows. J Dairy Sci 58:171–182

Al-Katanani YM, Webb DW, Hansen PJ (1999) Factors affecting seasonal variation in 90-day non return rate to first service in lactating Holstein cows in a hot climate. J Dairy Sci 82:2611–2616

Al-Katanani YM, Drost M, Monson RL, Rutledge JJ, Krininger-III CE, Block J, Thatcher WW, Hansen PJ (2002a) Pregnancy rates following timed embryo transfer with fresh or vitrified in vitro produced embryos in lactating dairy cows under heat stress conditions. Theriogenology 58:171–182

Al-Katanani YM, Paula-Lopes FF, Hansen PJ (2002b) Effect of season and exposure to heat stress on oocyte competence in Holstein cows. J Dairy Sci 85:390–396

Almier M, De RG, Grasso F, Napolitana F, Bordi A (2002) Effect of climate on the response of three oestrus synchronisation techniques in lactating dairy cows. Anim Reprod Sci 71:157–168

Alvarez RH, da Silva MV, de Carvalho JB, Binelli M (2005) Effects of inbreeding on ovarian responses and embryo production from superovulated Mantiqueira breed cows. Theriogenology 64:1669–1676

Ambrose JD, Drost M, Monson RL, Rutledge JJ, Leibfried-Rutledge ML, Thatcher MJ, Kassa T, Binelli M, Hansen PJ, Chenoweth PJ, Thatcher WW (1999) Efficacy of timed embryo transfer with fresh and frozen in vitro produced embryos to increase pregnancy rates in heat-stressed dairy cattle. J Dairy Sci 82:2369–2376

Aréchiga CF, Hansen PJ (1998) Response of preimplantation murine embryos to heat shock as modified by developmental stage and glutathione status. In Vitro Cell Dev Biol Anim 34:655–659

Aréchiga CF, Ealy AD, Hansen PJ (1995) Evidence that glutathione is involved in thermotolerance of preimplantation murine embryos. Biol Reprod 52:1296–1301

Armstrong DV (1994) Heat stress interaction with shade and cooling. J Dairy Sci 77:2044–2050

Aroyo A, Yavin S, Arav A, Roth Z (2007) Maternal hyperthermia disrupts developmental competence of follicle-enclosed oocytes: in vivo and ex vivo studies in mice. Theriogenology 67:1013–1021

Austin EJ, Mihm M, Ryan MP, Williams DH, Roch JF (1999) Effect of duration of dominance of the ovulatory follicle on onset of oestrus and fertility in heifers. J Anim Sci 77:2219–2226

Avendano-Reyes L, Alvarez-Valenzuela FD, Correa-Calderon A, Saucedo-Quintero JS, Robinson PH, Fadel JG (2006) Effect of cooling Holstein cows during the dry period on postpartum performance under heat stress conditions. Livest Sci 105:198–206

Ax RL, Gilbert GR, Shook GE (1989) Sperm in poor quality semen from bulls during heat stress have a lower affinity for binding hydrogen-3 heparin. J Dairy Sci 70:195–200

Badinga L, Collier RJ, Thatcher WW, Wilcox CJ (1985) Effects of climatic and management factors on conception rate of dairy cattle in subtropical environment. J Dairy Sci 68:78–85

Badinga L, Thatcher WW, Diaz T, Drost M, Wolfenson D (1993) Effect of environmental heat stress on follicular steroidogenesis and development in lactating Holstein cows. Theriogenology 39:797–810

Badinga L, Thatcher WW, Wilcox CJ, Morris G, Entwistle K, Wolfenson D (1994) Effect of season on follicular dynamics and plasma concentrations of oestradiol-17b, progesterone and luteinizing hormone in lactating Holstein cows. Theriogenology 42:1263–1274

Badreldin AL, Ghany MA (1954) Species and breed differences in the thermal reaction mechanism. J Agric Sci 44:160–164

Bahga CS, Khokar BS (1991) Effect of different seasons on concentration of plasma luteinizing hormone and seminal quality vis-à-vis freezability of buffalo bulls (*Bubalus bubalis*). Int J Biometeorol 35(4):222–224. doi:10.1007/BF01047289

Baruselli PS, Mucciolo RG, Vistin JA, Viana WG, Arruda RP, Maduriera EH, Oliveira CA, Molero-Filho JR (1997) Ovarian follicular dynamics during the oestrous cycle in buffalo (*Bubalus bubalis*). Theriogenology 47:1531–1547

Bauman DE, Currie WB (1980) Partitioning of nutrients during pregnancy and lactation: a review of mechanisms involving homeostasis and homeorhesis. J Dairy Sci 63(15):141–529

Baumgartner AP, Chrisman CL (1988) Analysis of post-implantation mouse embryos after maternal heat stress during meiotic maturation. J Reprod Fertil 84:469–474

Beam SW, Butler WR (1999) Effects of energy balance on follicular development and first ovulation in postpartum dairy cows. J Reprod Fertil 54:411–424

Bedford JM (2004) Enigmas of mammalian gamete form and function. Biol Rev 79:429–460. doi:10.1017/S146479310300633X

Berman A, Folman Y, Kaim M, Mamen M, Herz Z, Wolfenson D, Arieli A, Graber Y (1985) Upper critical temperatures and forced ventilation effects for high-yielding dairy cows in a subtropical environment. J Dairy Sci 68:1488–1495

Bhattacharya P (1974) Reproduction. In: Cockrill WR (ed) The husbandry and health of the domestic buffalo. Food and Agricultural Organization of the United Nations, Rome, pp 105–166

Bhikane AV, Kawitkar SB (2000) Hand book for veterinary clinician. Venkatesh Books, Udgir

Biggers BG, Geisert RD, Wettemann RP, Buchanan DS (1987) Effect of heat stress on early embryonic development in the beef cow. J Anim Sci 64:1512–1518

Bleach EC, Glencross RG, Knight PG (2004) Association between ovarian follicle development and pregnancy rates in dairy cows undergoing spontaneous oestrous cycles. Reproduction 127(5):621–629

Borghese A (2005) Buffalo production and research. FAO Ed REU Tech Ser 67:1–315

Braw-Tal R (1994) Expression of mRNA for follistatin and inhibin–activin subunits during follicular growth and atresia. J Mol Endocrinol 13:253–264

Brito LFC, Silva AEDF, Rodrigues LH, Vieira FV, Deragon LAG, Kastelic JP (2002) Effect of age and genetic group on characteristics of the scrotum, testes and testicular vascular cones, and on sperm production and semen quality in AI bulls in Brazil. Theriogeneology 58:1175–1186

Brito LFC, Silva AEDF, Marbosa RT, Unanian MM, Kastelic JP (2003) Effects of scrotal insulation on sperm production, semen quality and testicular echo-texture in *Bos indicus* and *Bos indicus x Bos Taurus* bulls. Anim Reprod Sci 79:1–15

Brito LF, Silva AE, Barbosa RT, Kastelic JP (2004) Testicular thermoregulation in *Bos indicus*, crossbred and *Bos taurus* bulls: relationship with scrotal, testicular vascular cone and testicular morphology, and effects on semen quality and sperm production. Theriogenology 61:511–528

Brody S (1945) Bioenergetics and growth: with special reference to the efficiency complex in domestic animals. Reinhold Publishing Corporation/Waverly Press, Baltimore

Brodsky JL, Chiosis G (2006) Hsp70 molecular chaperones: emerging roles in human disease and identification of small molecule modulators. Curr Top Med Chem 6:1215–1225

Bucklin RA, Turner LW, Beede DK, Bray DR, Hemken RW (1991) Methods to relieve heat stress for dairy cows in hot humid climates. Appl Eng Agric 7:241–247

Burfening PJ, Ulberg LC (1968) Embryonic survival subsequent to culture of rabbit spermatozoa at 38 and 40 8C. J Reprod Fertil 15:87–92

Butler WR (2000) Nutritional interactions with reproductive performance in dairy cattle. Anim Reprod Sci 60–61:449–457

Butler WR (2001) Nutritional effects on resumption of ovarian cyclicity and conception rate in postpartum dairy cows. In Diskin MG (ed) Fertility in the high-producing dairy cow, vol 26. BSAS Edinburgh, Occasional Publication, pp 133–145

Cartmill JA, Hensley BA, El-Zarkouny SZ, Rozell TG, Smith JF, Stevenson JS (1999) An alternative AI-breeding protocol during summer heat stress. J Dairy Sci 82:48 (Abstract)

Chou IP, Chuan L, Chen-Chao C (1974) Effect of heating on rabbit spermatogenesis. Chin Med J 6:365–367

Christians E, Michel E, Adenot P, Mezger V, Rallu M, Morange M, Renard JP (1997) Evidence for the involvement of mouse heat shock factor 1 in the atypical expression of the HSP70.1 heat shock gene during mouse zygotic genome activation. Mol Cell Biol 17:778–788

Collier RJ, Doelger SG, Head HH, Thatcher WW, Wilcox CJ (1982) Effects of heat stress during pregnancy on maternal hormone concentrations, calf birth weight and postpartum milk yield of Holstein cows. J Anim Sci 54:309–319

Coser AME, Godinho HP, Fonseea VO (1979) Effect of high temperatures on spermatogenesis in Brazilian Wooless rams under experimental conditions. Arquivos ad Escola de Veterinaria da Uitiversidad Federal de Minas Gerais 31:147–154

Darwash AO, Lamming GE, Woolliams JA (1999) The potential for identifying heritable endocrine parameters associated with fertility in post-partum dairy cows. Anim Reprod Sci 68(1999):333–347

de Castro e Paula LA, Hansen PJ (2008) Modification of actions of heat shock on development and apoptosis of cultured preimplantation bovine embryos by oxygen concentration and dithiothreitol. Mol Reprod Dev 75:1338–1350

de Castro e Paula LA, Andrzejewski J, Julian D, Spicer LJ, Hansen PJ (2008) Oxygen and steroid concentrations in preovulatory follicles of lactating dairy cows exposed to acute heat stress. Theriogenology 69:805–813

De la Sota RL, Burke JM, Risco CA, Moreira F, DeLorenzo MA, Thatcher WW (1998) Evaluation of timed insemination during summer heat stress in lactating dairy cattle. Theriogenology 49:761–770

De Rensis F, Scaramuzzi RJ (2003) Heat stress and seasonal effects on reproduction in the dairy cow—a review. Theriogenology 60:1139–1151

Dikmen S, Alava E, Pontes E, Fear JM, Dikmen BY, Olson TA, Hansen PJ (2008) Differences in thermoregulatory ability between slick-haired and wild-type lactating Holstein cows in response to acute heat stress. J Dairy Sci 91:3395–3402

Dixit NK, Agarwal SP, Agarwal VK, Dwaraknath PK (1984) Seasonal variations in serum levels of thyroid hormones and their relation with seminal quality and libido in buffalo bulls. Theriogenology 22:497–507

Donaghy AJ, Baxter RC (1996) Insulin-like growth factor bioactivity and its modification in growth hormone resistant states. Baillieres Clin Endocrinol Metabol 10:421–446

Dransfield MB, Nebel RL, Pearson RE, Warnick LD (1998) Timing of insemination for dairy cows identified in estrus by a radiotelemetric estrus detection system. J Dairy Sci 81:1874–1882

Drost M (2007) Bubaline versus bovine reproduction. Theriogenology 68:447–449

Dutt RH (1964) Detrimental effects of high ambient temperature on fertility and early embryo survival in sheep. Int J Biometeorol 8:47–56

Ealy AD, Hansen PJ (1994) Induced thermotolerance during early development of murine and bovine embryos. J Cell Physiol 160:463–468

Ealy AD, Drost M, Barros CM, Hansen PJ (1992) Thermoprotection of preimplantation bovine embryos from heat shock by glutathione and taurine. Cell Biol Int Rep 16:125–131

Ealy AD, Drost M, Hansen PJ (1993) Developmental changes in embryonic resistance to adverse effects of maternal heat stress in cows. J Dairy Sci 76:2899–2905

Ealy AD, Howell JL, Monterroso VH, Arechiga CF, Hansen PJ (1995) Developmental changes in sensitivity of bovine embryos to heat shock and use of antioxidants as hermoprotectants. J Anim Sci 73:1401–1407

Edwards JL, Hansen PJ (1996) Elevated temperature increases heat shock protein 70 synthesis in bovine two-cell embryos and compromises function of maturing oocytes. Biol Reprod 55:340–346

Edwards JL, Hansen PJ (1997) Differential responses of bovine oocytes and preimplantation embryos to heat shock. Mol Reprod Dev 46:138–145

El-Azab AI (1980) The interaction of season and nutrition on semen quality in buffalo bulls. PhD thesis, Faculty of Veterinary Medicine, Cairo University, Cairo

El-Saidy EI (1988) Studies on reproduction in male goats. MSc thesis, Faculty of Agriculture, Mansoutra University

El-Sherbiny AM (1987) Seasonal variation in seminal characteristics of rabbits. MSc thesis, Faculty of Agriculture, Ain-Shams University, Cairo

El-Sherry MI, El-Naggar MA, Nassar SM (1980) Experimental study of summer stress in rabbits 2. The quantitative pathogenesis spermatogenic cell cycle in rabbits. Assiut Vet Med J 7:17–31

El-Wishy AB (2007) The postpartum buffalo II. Acyclicity and anestrus. Anim Reprod Sci 97:216–236

Esposito L, Campanile G, Di Palo R, Boni R, Di Meo C, Zicarelli L (1992) Seasonal reproductive failure in buffaloes bred in Italy. In: Proceedings of the twelfth international Congress, The Hague, Anim Reprod (ICAR) 4, pp 2045–2047

Everett RW, Bean B, Foote RH (1978) Sources of variation of semen output. J Dairy Sci 61:90–95

Falconer DS (1981) Introduction to quantitative genetics, 2nd edn. Longman House, Harlow, Essex

Fawzy SAH (1982) Effect of feeding elephant grass to bulls on their performance and semen quality. M.Sc. Thesis, Faculty of Agriculture, Tanta University, Tanta, Egypt

Fields MJ, Burns WC, Warmick AC (1979) Age, season and breed effects on testicular volume and semen traits in young beef bulls. J Anim Sci 48:1299–1304

Finch VA (1986) Body temperature in beef cattle: its control and relevance to production in the tropics. J Anim Sci 62:531–542

Findlay JK (1993) An update on the roles of inhibin, activin, and follistatin as local regulators of folliculogenesis. Biol Reprod 48:15–23

Food and Agricultural Organization (FAO) FAOSTAT (2003) Agricultural data; 2003, http://apps.fao.org

Galan HL, Anthony RV, Rigano S, Parker TA, de Vrijer B, Ferrazzi E, Wilkening RB, Regnault TR (2005) Fetal hypertension and abnormal Doppler velocimetry in an ovine model of intrauterine growth restriction. Am J Obstet Gynecol 192:272–279

Gamcik P, Mesaros P, Schvare F (1979) The effect of season on some semen characters in Slovakian Merino rams. Zivocisna Vvrobo 24:625–630

Gangwar PC (1980) Climate and reproduction in buffaloes. A review. Indian J Dairy Sci 33:419–426

Garnsworthy PC, Webb R (1999) The influence of nutrition on fertility in dairy cows. In: Garnsworthy, P.C., Wiseman, J. (Eds.), Recent Advances in Animal Nutrition. Nottingham University Press, Nottingham, UK, pp 39–57

Garrett JE, Geisert RD, Zavy MT, Morgan GL (1988) Evidence for maternal regulation of early conceptus growth and development in beef cattle. Reprod Fertil 84:437–446

Gauthier D (1986) The influence of season and shade on estrous behaviour, timing of preovulatory LH surge and the pattern of progesterone secretion in FFPN and Creole heifers in a tropical climate. Reprod Nutr Dev 26:767–775

Gibbons JR, Wiltbank MC, Ginther OJ (1997) Functional interrelationships between follicles greater than 4 mm and the follicle-stimulating hormone surge in heifers. Biol Reprod 57:1066–1073

Gilad E, Meidan R, Berman A, Graber Y, Wolfenson D (1993) Effect of heat stress on tonic and GnRH-induced gonadotrophin secretion in relation to concentration of oestradiol in plasma of cyclic cows. J Reprod Fertil 99:315–321

Gili RS, Gangwar PC, Takkar OP (1974) Seminal attributes in buffalo bulls as affected by different seasons. Indian J Anim Sci 44:415–418

Ginther OJ, Wiltbank MC, Fricke PM, Gibbons JR, Kotm K (1996) Selection of the dominant follicle in cattle. Biol Reprod 55:1187–1194

Goeritz F, Quest M, Wagener A (2003) Seasonal timing of sperm production in roe deer: interrelationship among changes in ejaculate parameters, morphology and function of testis and accessory glands. Theriogenology 59:1487–1502

Gomes WR, Buttler WR, Johnson AD (1971) Effects of elevated ambient temperature on testis and blood levels and in vitro biosynthesis of testosterone in the ram. J Anim Sci 33:804–807

Guilbault LA, Rouillier P, Matton P, Glencross RG, Beard AJ, Knight PG (1993) Relationships between the level of atresia and inhibin contents (alpha-subunit and alpha–beta dimer) in morphologically dominant follicles during their growing and regressing phases of development in cattle. Biol Reprod 48:268–276

Guzeloglu A, Ambrose JD, Kassa T, Diaz T, Thatcher MJ, Thatcher WW (2001) Long term follicular dynamics and biochemical characteristics of dominant follicles in dairy cows subjected to acute heat stress. Anim Reprod Sci 66:15–34

Gwazdauskas FC, Thatcher WW, Kiddy CA, Paape MJ, Wilcox CJ (1981) Hormonal patterns during heat-stress following PGF2α-tham salt induced luteal regression in heifers. Theriogenology 16:271–285

Habeeb AAM, Fatma FIT, Osman SF (2007) Detection of heat adaptability using heat chock proteins and some hormones in Egyptian buffalo calves. Egypt J Appl Sci 22:28–53

Hafez ESE (1968) Adaptation of domestic animals. Lea and Febiger, Philadelphia

Hafez ESE, Badreldin AL, Shafie MM (1955) Skin structure of Egyptian buffaloes and cattle with particular reference to sweat glands. J Agric Sci 46:19–30

Hammond AC, Olson TA, Chase CC Jr, Bowers EJ, Randel RD, Murphy CN, Vogt DW, Tewolde A (1996) Heat tolerance in two tropically adapted Bos Taurus breeds, Senepol and Romosinuano, compared with Brahman, Angus, and Hereford cattle in Florida. J Anim Sci 74:295–303

Hansen PJ (1997) Effects of environment on bovine reproduction. In: Youngquist RS (ed) Current therapy in large animal theriogenology. WB Saunders, Philadelphia, pp 403–415

Hansen PJ (2004) Physiological and cellular adaptations of zebu cattle to thermal stress. Anim Reprod Sci 82–83:349–360

Hansen PJ (2005) Managing the heat-stressed cow to improve reproduction. In: Proceedings of 7th Western Dairy Management conference, Reno, NV, pp 63–70

Hansen PJ, Arechiga CF (1999) Strategies for managing reproduction in the heat-stressed dairy cow. J Anim Sci 77:36–50

Hendricks KEM, Hansen PJ (2009) Can programmed cell death be induced in post-ejaculatory bull and stallion spermatozoa? Theriogenology 71:1138–1146. doi:10. 1016

Hendricks KEM, Martins L, Hansen PJ (2009) Consequences for the bovine embryo of being derived from a spermatozoan subjected to post-ejaculatory aging and heat shock: development to the blastocyst stage and sex ratio. J Reprod Dev 55:69–74

Hommeida A, Nakao T, Kubota H (2004) Luteal function and conception in lactating cows and some factors influencing luteal function after first insemination. Theriogenology 62:217–225

Howarth B Jr, Alliston CW, Ulberg LC (1965) Importance of uterine environment on rabbit sperm prior to fertilization. J Anim Sci 24:1027–1032

Howell JL, Fuquay JW, Smith AE (1994) Corpus luteum growth and function in lactating Holstein cows during spring and summer. J Dairy Sci 77:735–739

Huber JT (1996) Amelioration of heat stress in dairy cattle. In: Philips CJC (ed) Progress in dairy science. CAB International, Oxon, pp 211–243

Igono MO, Johnson HD, Stevens BJ, Krause GF, Shanklin MD (1987) Physiological, productive, and economic benefits of shade, spray, and fan system versus shade for Holstein cows during summer heat. J Dairy Sci 70:1069–1079

Ingraham RH, Gillette DD, Wagner WD (1974) Relationship of temperature and humidity to conception rate of Holstein cows in subtropical climate. J Dairy Sci 57:476–482

Ireland JJ, Roche JF (1982) Effect of progesterone on basal LH and episodic LH and FSH secretion in heifers. J Reprod Fertil 64:295–302

Ishii T, Matsuki S, Iuchi Y, Okada F, Toyosaki S, Tomita Y, Ikeda Y, Fujii J (2005) Accelerated impairment of spermatogenic cells in SOD1-knockout mice under heat stress. Free Radic Res 39:697–705. doi:10. 1080/10715760500130517

Jainudeen MR (1986) Reproduction in the water buffalo. In: Morrow DA (ed) Current therapy in theriogenology, vol 2. WB Saunders Company, Philadelphia, pp 443–449

Janakiraman K (1978) Control and optimizing reproductive cycle in buffaloes. In: Proceedings of FAORSIDA seminar on buffalo reproduction and artificial insemination, India

Jannes P, Spiessens C, van der Auwera I, D'Hooghe T, Verhoeven G, Vanderschueren D (1998) Male subfertility induced by acute scrotal heating affects embryo quality in normal female mice. Hum Reprod 13:372–375

Jolly PD, McDougall S, Fitzpatrick LA, Macmillan KL, Entwhitsle K (1995) Physiological effects of under nutrition on postpartum anoestrous in cows. J Reprod Fertil Suppl 49:477–492

Jones JI, Clemmons DR (1995) Insulin-like growth factors and their binding proteins: biological actions. Endocr Rev 16:3–34

Jonsson NN, McGowan MR, McGuigan K, Davison TM, Hussain AM, Kafi M (1997) Relationship among calving season, heat load, energy balance and postpartum ovulation of dairy cows in a subtropical environment. Anim Reprod Sci 47:315–326

Juengel JL, Garverick HA, Johnson AL, Youngquist RS, Smith MF (1993) Apoptosis during luteal regression in cattle. Endocrinology 132:249–254

Kaneko H, Terada T, Taya K, Watanabe G, Sasamoto S, Hasegawa Y, Igarashi M (1991) Ovarian follicular dynamics and concentration of oestradiol 17, progesterone, luteinizing hormone and follicle stimulating hormone during the preovulatory phase of the oestrous cycle in the cow Reproduction. Fertil Dev 3:529–535

Kaneko H, Nakanishi Y, Taya K, Kishi H, Watanabe G, Sasamotom S, Hasegawa Y (1993) Evidence that inhibin is an important factor in the regulation of FSH secretion during mid-luteal phase in cows. Endocrinology 136:35–42

Kaneko H, Nakanishi Y, Akagi S, Arai K, Taya K, Watanabe G, Sasamoto S, Hasegawa Y (1995) Immunoneutralization of inhibin and estradiol during the follicular phase of the estrous cycle in cows. Biol Reprod 53:931–939

Kaneko H, Taya K, Watanabe G, Noguchi J, Kikuchi K, Shimada AY, Hasegawa Y (1997) Inhibin is involved in the suppression of FSH secretion in the growth phase of the dominant follicle during the early luteal phase in cows. Domest Anim Endocrinol 14:263–271

Lamming GE, Royal MD (2001) Ovarian hormone patterns and subfertility in dairy cows. In: Diskin MG (ed) Fertility in the high-producing dairy cow, vol 26. BSAS Edinburgh, Occasional Publication, pp 105–118

Lawrence JL, Payton RR, Godkin JD, Saxton AM, Schrick FN, Edwards JL (2004) Retinol improves development of bovine oocytes compromised by heat stress during maturation. J Dairy Sci 87:2449–2454

Lee N (1993) Environmental stress effect on bovine reproduction. Vet Clin North Am 9:263–273

Lim JM, Liou SS, Hansel W (1996) Intracytoplasmic glutathione concentration and the role of b-mercaptoethanol in preimplantation development of bovine embryos. Theriogenology 46:429–439. doi:10.1016/ 0093-691X(96)00165-3

Limesand SW, Regnault TR, Hay WW Jr (2004) Characterization of glucose transporter 8 (GLUT8) in the ovine placenta of normal and growth restricted fetuses. Placenta 25:70–77. doi:10.1016/j.placenta. 2003.08.012

López-Gatius F, Santolaria P, Yániz JL, Garbayo JM, Hunter RH (2004) Timing of early foetal loss for single and twin pregnancies in dairy cattle. Reprod Domest Anim 39:429–433

Love CC, Kenney RM (1999) Scrotal heat stress altered sperm chromatin structure associated with a decrease in protamine disulfide bonding in the stallion. Biol Reprod 60:615–620

Lucy MC (2002) Reproductive loss in farm animals during heat stress. In: Proceedings 15th conference on biometeorology and aerobiology, pp 50–53. University of Missouri, Columbia

Lucy MC, Crooker BA (1999) Physiological and genetic differences between low and high index dairy cows. In: Proceedings of the occasional meeting of British Society of Animal Science, Galway, pp 27–28

Lucy MC, Savio JD, Badinga L, de la Sota RL, Thatcher WW (1992) Factors that affect ovarian follicular dynamics in cattle. J Anim Sci 70:3615–3626

Lucy MC, Weber WJ, Baumgard LH, Seguin BS, Koenigsfeld AT, Hansen LB (1998) Reproductive endocrinology of lactating dairy cows selected for increased milk production. J Dairy Sci 81:246 (Abstract)

Lue TH, Hikim APS, Swerdoff RS, Im P, Taing KS, Bui T, Leung A, Wang C (1999) Single exposure to heat stress induces stage-specific germ cell apoptosis in rats: role of intratesticular testosterone on stage specificity. Endocrinology 140:1709–1717

Lussier JG, Matton P, Dufour JJ (1987) Growth rates of follicles in the ovary of the cow. J Reprod Fertil 81:301–307

Madan ML, Johnson HD (1973) Environmental heat effects on bovine luteinizing Hormone. J Dairy Sci 56:575–580

Malayer JR, Hansen PJ (1990) Differences between Brahman and Holstein cows in heat-shock induced alterations of protein secretion by oviducts and uterine endometrium. J Anim Sci 68:266–280

Malayer JR, Hansen PJ, Gross TS, Thatcher WW (1990) Regulation of heat shock-induced alterations in the release of prostaglandins by the uterine endometrium of cows. Theriogenology 34:219–230

Mandal DK, Nagpaul PK, Gupta AK (2002) Effects of body surface cooling during hot dry and hot humid seasons on seminal attributes of Murrah buffalo bulls. Indian J Anim Sci 72:192–194

Mann GE, Lamming GE (1999) The influence of progesterone during early pregnancy in cattle. Reprod Domest Anim 34:269–274

Mann GE, Lamming GE, Robinson RS, Wathes DC (1999) The regulation of interferon-tau production and uterine hormone receptors during early pregnancy. J Reprod Fertil 54:317–328

Marai IFM, Ayyat MS, Gabr HA, Abdel-Monem UM (1996) Effects of heat stress and its amelioration on reproduction performance of New Zealand White adult female and male rabbits, under Egyptian Conditions, Toulouse. World Rabbit Congress, vol 2, pp 197–202

Mathevon M, Buhr MM, Dekkers JCM (1998) Environmental, management, and genetic factors affecting semen production in Holstein bulls. J Dairy Sci 81:3321–3330

Matsuzuka T, Ozawa M, Nakamura A, Ushitani A, Hirabayashi M, Kanai Y (2005a) Effects of heat stress on the redox status in the oviduct and early embryonic development in mice. J Reprod Dev 51:281–287. doi:10.1262/jrd.16089

Matsuzuka T, Sakamoto N, Ozawa M, Ushitani A, Hirabayashi M, Kanai Y (2005b) Alleviation of maternal hyperthermia-induced early embryonic death by administration of melatonin to mice. J Pineal Res 39:217–223. doi:10.1111/j.1600-079X.2005.00260.x

Mihm M, Bagnisi A, Boland MP, Roche JF (1994) Association between the duration of dominance of the ovulatory follicle and pregnancy rate in beef heifers. J Reprod Fertil 102:123–130

Mikelsen WD, Paisley LG, Dahmen JJ (1981) The effect of semen on the scrotal circumference and sperm motility and morphology in rams. Theriogenology 16:45–51

Moreira F, Orlandi C, Risco CA, Mattos R, Lopes F, Thatcher WW (2001) Effects of pre-synchronization and bovine somatotropin on pregnancy rates to a timed artificial insemination protocol in lactating dairy cows. J Dairy Sci 84:1646–1659

Nebel RL, Jobst SM, Dransfield MBG, Pandolfi SM, Bailey TL (1997) Use of radio frequency data communication system, HeatWatch®, to describe behavioral estrus in dairy cattle. J Dairy Sci 80:179

O'Callaghan DO, Boland MP (1999) Nutritional effects on ovulation, embryo development and the establishment of pregnancy in ruminants. Anim Sci 68:299–314

Oloufa MM, Sayed AA, Badreldin AL (1959) Seasonal variations in reaction time in Egyptian buffalo bulls and physico-chemical characteristics of their semen. Indian J Dairy Sci 12:10–17

Olson TA, Lucena C, Chase CC Jr, Hammond AC (2003) Evidence of a major gene influencing hair length and heat tolerance in *Bos taurus* cattle. J Anim Sci 81:80–90

Orr WN, Cowan RT, Davison TM (1993) Factors affecting pregnancy rate in Holstein-Friesian cattle mated during summer in a tropical upland environment. Aust Vet J 70:256–278

Ozawa M, Hirabayashi M, Kanai Y (2002) Developmental competence and oxidative state of mouse zygotes heat-stressed maternally or in vitro. Reproduction 124:683–689. doi:10.1530/rep. 0.1240683

Ozawa M, Tabayashi D, Latief TA, Shimizu T, Oshima I, Kanai Y (2005) Alterations in follicular dynamics and steroidogenic abilities induced by heat stress during follicular recruitment in goats. Reproduction 129:621–630. doi:10.1530/rep. 1.00456

Palta P, Mondal S, Prakash BS, Madan ML (1997) Peripheral inhibin levels in relation to climatic variations and stage of estrus cycle in buffalo (*Bubalus bubalis*). Theriogenology 47:989–995

Pant HC, Sharma RK, Patel SH (2003) Testicular development and its relationship to semen production in Murrah buffalo bulls. Theriogenology 60:27–34

Parr RA, Davis IF, Miles MA, Squires TJ (1993) Liver blood flow and metabolic clearance rate of progesterone in sheep. Res Vet Sci 55:311–316

Paul C, Murray AA, Spears N, Saunders PT (2008) A single, mild, transient scrotal heat stress causes DNA damage, subfertility and impairs formation of blastocysts in mice. Reproduction 136:73–84

Paul C, Teng S, Saunders PT (2009) A single, mild, transient scrotal heat stress causes hypoxia and oxidative stress in mouse testes, which induces germ cell death. Biol Reprod 80:913–919

Paula-Lopes FF, Hansen PJ (2002a) Heat-shock induced apoptosis in bovine preimplantation embryos is a developmentally-regulated phenomenon. Biol Reprod 66:1169–1177

Paula-Lopes FF, Hansen PJ (2002b) Apoptosis is an adaptive response in bovine preimplantation embryos that facilitates survival after heat shock. Biochem Biophys Res Commun 295:37–42. doi:10.1016/ S0006-291X (02)00619-8

Payne WJA (1990) Cattle and buffalo meat production in the tropic, Intermediate tropical agriculture series. Longman Science and Technology, Harlow

Payton RR, Romar R, Coy P, Saxton AM, Lawrence JL, Edwards JL (2004) Susceptibility of bovine germinal vesicle-stage oocytes from antral follicles to direct effects of heat stress in vitro. Biol Reprod 71:1303–1308. doi:10.1095/biolreprod.104.029892

Pe´rez-Crespo M, Ramı´rez MA, Ferna´ndez-Gonza´lez R, Rizos D, Lonergan P, Pintado B, Gutie´rrez-Adan A (2005) Differential sensitivity of male and female mouse embryos to oxidative induced heat-stress is mediated by glucose- 6-phosphate dehydrogenase gene expression. Mol Reprod Dev 72:502–510. doi:10.1002/mrd.20366

Pe´rez-Crespo M, Pintado B, Gutie´rrez-Adan A (2008) Scrotal heat stress effects on sperm viability, sperm DNA integrity, and the offspring sex ratio in mice. Mol Reprod Dev 75:40–47. doi:10.1002/ mrd.20759

Pell JM, Saunders JC, Gilmour RS (1993) Differential regulation of transcription initiation from insulin-like growth factor-I (IGF-I) leader exons and of tissue IGF-I expression in response to changed growth hormone and nutritional status in sheep. Endocrinology 132:1797–1807

Pennington JA, Albright JL, Diekman MA, Callahan CJ (1985) Sexual activity of Holstein cows: seasonal effects. J Dairy Sci 68:3023–3030

Pereira AM, Baccari F Jr, Titto EA, Almeida JA (2008) Effect of thermal stress on physiological parameters, feed intake and plasma thyroid hormones concentration in Alentejana, Mertolenga, Frisian and Limousine cattle breeds. Int J Biometeorol 52:199–208. doi:10.1007/ s00484-007-0111-x

Perera OBM (1999) Reproduction in water buffalo: comparative aspects and implications for management. J Reprod Fertil Suppl 54:157–168

Prosser CL, Heath JE (1991) Temperature. In: Prosser CL (ed) Comparative animal physiology, environmental and metabolic animal physiology, 4th edn. Wiley, New York, pp 109–166

Pryce JE, Coffey MP, Simm G (2001) The relationship between body condition score and reproductive performance. J Dairy Sci 84:1508–1515

Pursley JR, Mee MO, Wiltbank MC (1995) Synchronization of ovulation in dairy cows using PGF2alpha and GnRH. Theriogenology 44:915–923

Pursley JR, Wiltbank MC, Stevenson JS, Ottobre JS, Gaverick HA, Anderson LL (1997) Pregnancy rates per artificial insemination for cows and heifers inseminated at synchronized ovulation or synchronized estrus. J Dairy Sci 80:295–300

Pushpakumara PG, Gardner NH, Reynolds CK, Beever DE, Wathes DC (2003) Relationships between transition period diet, metabolic parameters and fertility in lactating dairy cows. Theriogenology 60:1165–1185

Putney DJ, Malayer JR, Gross TS, Thatcher WW, Hansen PJ, Drost M (1988) Heat stress-induced alterations in the synthesis and secretion of proteins and prostaglandins by cultured bovine conceptuses and uterine endometrium. Biol Reprod 39:717–728

Putney DJ, Drost M, Thatcher WW (1989a) Influence of summer heat stress on pregnancy rates of lactating dairy cattle following embryo transfer or artificial insemination. Theriogenology 31:765–778

Putney DJ, Mullins S, Thatcher WW, Drost M, Gross TS (1989b) Embryonic development in superovulated dairy cattle exposed to elevated ambient temperatures between the onset of estrus and insemination. Anim Reprod Sci 19:37–51

Ramesh V, Thanga TV, Varadhrajan A (2002) Improvement of reproductive performances of buffaloes. Pashudhan 17:1–4

Rao NM, Kodagali SB (1983) Onset of oestrus signs and optimum time of insemination in Surti buffaloes. Indian J Anim Sci 53:553–555

Rao BR, Patel UG, Tahman SS (1973) Studies on reproductive behavior of surti buffaloes. III. Fertility status. Indian Vet J 50:664–667

Razdan MN (1965) Studies on physiological reactions of dairy cattle under different environmental conditions. PhD thesis, University of Calcutta, Calcutta

Reddy AO (1985) Studies on certain patterns in Murrah buffaloes under loose housing system. PhD thesis, Kurukshetra University, Kurukshetra

Regnault TR, Orbus RJ, de Vrijer B, Davidsen ML, Galan HL, Wilkening RB, Anthony RV (2002) Placental expression of VEGF, PlGF and their receptors in a model of placental insufficiency-intrauterine growth restriction (PI-IUGR). Placenta 23:132–144. doi:10. 1053/plac.2001.0757

Rhynes WE, Ewing LL (1973) Testicular endocrine function in Hereford bulls exposed to high ambient temperature. Endocrinology 92:509–515. doi:10.1210/endo-92-2-509

Richards MW, Spicer LJ, Wettemann RP (1995) Influence of diet and ambient temperature on bovine serum insulin like growth factor I and thyroxine: relationship with non-esterified fatty acids, glucose, insulin, luteinizing hormone and progesterone. Anim Reprod Sci 36:267–279

Rivera RM, Hansen PJ (2001) Development of cultured bovine embryos after exposure to increased temperatures in the physiological range. Reproduction 121:107–115

Roberts SJ (ed) (1986) Veterinary obstetrics and genital diseases. (tph.er5i.ogenology). Author, Woodstock

Roman-Ponce H, Thatcher WW, Wilcox CJ (1981) Hormonal interrelationships and physiological responses of lactating dairy cows to shade management system in a tropical environment. Theriogenology 16:139–154

Ronchi B, Stradaioli G, Verini Supplizi A, Bernabucci U, Lacetera N, Accorsi PA, Nardone A, Seren E (2001) Influence of heat stress and feed restriction on plasma progesterone, estradiol-17(LH, FSH, prolactin and cortisol in Holstein heifers. Livest Prod Sci 68:231–241

Rose MT, Weekes TE, Rowlinson P (2004) Individual variation in the milk yield response to bovine somatotropin in dairy cows. J Dairy Sci 87:2024–2031

Rosemberg M, Folman Y, Herz Z, Flamenbaum I, Berman A, Kaim M (1982) Effect of climatic condition on peripheral concentrations of LH, progesterone and oestradio-17beta in high milk-yielding cows. J Reprod Fertil 66:139–146

Rosenberg M, Herz Z, Davidson M, Folman J (1977) Seasonal variations in post-partum plasma progesterone levels and conceptions in primiparous and multiparous dairy cows. J Reprod Fertil 51:363–367

Roth Z, Hansen PJ (2004a) Involvement of apoptosis in disruption of developmental competence of bovine oocytes by heat shock during maturation. Biol Reprod 71:1898–1906

Roth Z, Hansen PJ (2004b) Sphingosine 1-phosphate protects bovine oocytes from heat shock during maturation. Biol Reprod 71:2072–2078. doi:10.1095/biolreprod.104.031989

Roth Z, Hansen PJ (2005) Disruption of nuclear maturation and rearrangement of cytoskeletal elements in bovine oocytes exposed to heat shock during maturation. Reproduction 129:235–244. doi:10.1530/rep. 1.00394

Roth Z, Meidan R, Braw-Tal R, Wolfenson D (2000) Immediate and delayed effects of heat stress on follicular development and its association with plasma

FSH and inhibin concentration in cows. J Reprod Fertil 120:83–90. doi:10.1530/reprod/120.1.83

Roth Z, Arav A, Bor A, Zeron Y, Braw-Tal R, Wolfenson D (2001a) Improvement of quality of oocytes collected in the autumn by enhanced removal of impaired follicles from previously heat-stressed cows. Reproduction 122:737–744

Roth Z, Meidan R, Shaham-Albalancy A, Braw-Tal R, Wolfenson D (2001b) Delayed effect of heat stress on steroid production in medium-sized and preovulatory bovine follicles. Reproduction 121:745–751

Roth Z, Aroyo A, Yavin S, Arav A (2008) The antioxidant epigallocatechin gallate (EGCG) moderates the deleterious effects of maternal hyperthermia on follicle-enclosed oocytes in mice. Theriogenology 70:887–897

Rueda BR, Tilly KI, Hansen TR, Hoyer PB, Tilly JL (1995) Expression of superoxide dismutase, catalase and glutathione peroxidase in the bovine corpus luteum: evidence supporting a role for oxidative stress in luteolysis. Endocrine 3:227–232

Ryan DP, Boland MP (1991) Frequency of twin births among Holstein X Friesian cows in a warm dry climate. Theriogenology 36:1–10

Ryan DP, Prochard JF, Kopel E, Godke RA (1993) Comparing early embryo mortality in dairy cows during hot and cold season of the year. Theriogenology 39:719–737

Sahni KL, Roy A (1967) A note on summer sterility in Romney Marsh rams under tropical conditions. Indian J Vet Sci 37:335–338

Sakamoto N, Ozawa M, Yokotani-Tomita K, Morimoto A, Matsuzuka T, Ijiri D, Hirabayashi M, Ushitani A, Kanai Y (2008) DL-a-Tocopherol acetate mitigates maternal hyperthermia-induced pre-implantation embryonic death accompanied by a reduction of physiological oxidative stress in mice. Reproduction 135:489–496

Sakatani M, Kobayashi S, Takahashi M (2004) Effects of heat shock on in vitro development and intracellular oxidative state of bovine preimplantation embryos. Mol Reprod Dev 67:77–82. doi:10.1002/mrd.20014

Sakatani M, Yamanaka K, Kobayashi S, Takahashi M (2008) Heat shock-derived reactive oxygen species induce embryonic mortality in *in vitro* early stage bovine embryos. J Reprod Dev 54:496–501

Saltalamacchia F, Tripaldi C, Castellano A, Napolitano F, Musto M, De Rosa G (2007) Human and animal behaviour in dairy buffalo at milking. Anim Welf 16:139–142

Sangsritavong S, Combs DK, Sartori R, Armentano LE, Wiltbank MC (2002) High feed intake increases liver blood flow and metabolism of progesterone and estradiol-17beta in dairy cattle. J Dairy Sci 85:2831–2842

Sarabia AS (2004) Basic anatomy and physiology of buffalo reproduction. Pre-congress training on the use of reproductive biotechniques in water buffalo. In: Proceedings of the 7th world buffalo Congress, Manila, pp 1–17

Sartori R, Sartor-Bergfelt R, Mertens SA, Guenther JN, Parrish JJ, Wiltbank MC (2002) Fertilization and early embryonic development in heifers and lactating cows in summer and lactating and dry cows in winter. J Dairy Sci 85:2803–2812

Sartori R, Haughian JM, Shaver RD, Rosa GJ, Wiltbank MC (2004) Comparison of ovarian function and circulating steroids in estrous cycles of Holstein heifers and lactating cows. J Dairy Sci 87:905–920

Schillo KK, Alliston CW, Malven PV (1978) Plasma concentrations of luteinizing hormone and prolactin in the ovariectomized ewe during induced hyperthermia. Biol Reprod 19:306–313. doi:10.1095/biolreprod19.2.306

Schmitt EJ, Barros CM, Fields PA, Fields MJ, Diaz T, Kluge JM, Thatcher WW (1996a) A cellular and endocrine characterization of the original and induced corpus luteum after administration of a gonadotropin-releasing hormone agonist or human chorionic gonadotropin on day five of the estrous cycle. J Anim Sci 74:1915–1929

Schmitt EJ, Diaz T, Drost M, Thatcher WW (1996b) Use of a gonadotropin-releasing hormone agonist or human chorionic gonadotropin for timed insemination in cattle. J Anim Sci 74:1084–1091

Sengupta BP, Misra MS, Roy A (1963) Climatic environment and reproductive behaviour of buffaloes. I. Effect of different seasons on various seminal attributes. Indian J Dairy Sci 16:150–165

Seren E, Parmeggiani A, Zicarelli L, Montemurro N, Pacelli C, Terzano GM (1992) Periestrous endocrine changes in Italian buffaloes. In: Proceedings of international symposium on prospects of buffalo production in the Mediterranean and the Middle East, Cairo, pp 393–396

Setchell PB (1970) Testicular blood supply, lymphatic drainage and secretion of fluid. In: Johnson AD, Gomes WR, Van Demark NL (eds) The testis: I. Development, anatomy and physiology. Academic, New York

Setchell BP (1998) The Parkes lecture. Heat and the testis. J Reprod Fertil 114:179–194. doi:10.1530/jrf.0.1140179

Setchell BP (2006) The effects of heat on the testes of mammals. Anim Reprod 3:81–91

Setchell BP, Breed WG (2006) Anatomy, vasculature and innervation of the male reproductive tract. In: Neill JD (ed) Knobil and Neill's physiology of reproduction. Elsevier, San Diego, pp 771–825

Shaham-Albalancy A, Folman Y, Kaim M, Rosenberg M, Wolfenson D (2001) Delayed effect of low progesterone on bovine uterine prostaglandin F2α secretion in the subsequent estrous cycle. Reproduction 122:643–648

Silvia WJ, Lewis GS, McCracken JA, Thatcher WW, Wilson L Jr (1991) Hormonal regulation of uterine secretion of prostaglandin F2α during luteolysis in ruminants. Biol Reprod 45:655–663

Singh AS, Mishra M (1980) Physiological responses and economic traits of Holstein, Jersey, Crossbred and Hariana cows in hot and humid environment. Indian J Dairy Sci 33:174–178

Sirois J, Fortune JE (1990) Lengthening the bovine estrous cycle with low levels of exogenous progesterone: a model for studying ovarian follicular dominance. Endocrinology 127:916–925

Skarzynski DJ, Okuda K (1999) Sensitivity of bovine corpora lutea to prostaglandin F2 is dependent on progesterone, oxytocin, and prostaglandins. Biol Reprod 60:1292–1298

Soto P, Smith LC (2009) BH4 peptide derived from BclxL and Bax-inhibitor peptide suppresses apoptotic mitochondrial changes in heat stressed bovine oocytes. Mol Reprod Dev 76:637–646

Staples CR, Thatcher WW, Clark JH (1990) Relationship between ovarian activity and energy status during the early postpartum period of high producing dairy cows. J Dairy Sci 73:938–947

Taylor VJ, Cheng Z, Pushpakumara PG, Beever DE, Wathes DC (2004) Relationships between the plasma concentrations of insulin-like growth factor-I in dairy cows and their fertility and milk yield. Vet Rec 155:583–588

Thomas CS (2005) Machine milking buffaloes. Bubalus Bubalis 11:25–29

Thomas CS, Nordstrom J, Svennersten-Sjaunja K, Wiktorsson H (2005) Maintenance and milking behaviours of Murrah buffaloes during two feeding regimes. Appl Anim Behav Sci 91:261–276

Thompson JR, Everett RW, Hammerschmidt NL (2000a) Effects of inbreeding on production and survival in Holsteins. J Dairy Sci 83:1856–1864

Thompson JR, Everett RW, Wolfe CW (2000b) Effects of inbreeding on production and survival in Jerseys. J Dairy Sci 83:2131–2138

Thureen PJ, Trembler KA, Meschia G, Makowski EL, Wilkening RB (1992) Placental glucose transport in heat-induced fetal growth retardation. Am J Physiol 263:578–585

Trout P, McDowell LR, Hansen PJ (1998) Characteristics of the oestrous cycle and antioxidant status of lactating Holstein cows exposed to stress. J Dairy Sci 81: 1244–1250

Urdaz JH, Overton MW, Moore DA, Santos JEP (2006) Effects of adding shade and fans to a feedbunk sprinkler system for preparturient cows on health and performance. J Dairy Sci 89:2000–2006

Vasconcelos JL, Sangsritavong S, Tsai SJ, Wiltbank MC (2003) Acute reduction in serum progesterone concentrations after feed intake in dairy cows. Theriogenology 60:795–807

Vaught LW, Monty DW, Foote WC (1977) Effect of summer heat stress on serum LH and progesterone values in Holstein-Friesian cows in Arizona. Am J Vet Res 38:1027–1032

Vicini JL, Buonomo FC, Veenhuizen JJ, Miller MA, Clemmons DR, Collier RJ (1991) Nutrient balance and stage of lactation affect responses of insulin, insulin-like growth factors I and II, and insulin-like growth factor-binding protein 2 to somatotropin administration in dairy cows. J Nutr 121:1656–1664

Wall E, Brotherstone S, Kearney JF, Woolliams JA, Coffey MP (2005) Impact of nonadditive genetic effects in the estimation of breeding values for fertility and correlated traits. J Dairy Sci 88:376–385

Wallace JM, Regnault TR, Limesand SW, Hay WW Jr, Anthony RV (2005) Investigating the causes of low birth weight in contrasting ovine paradigms. J Physiol 565:19–26

Wang JZ, Sui HS, Miao DQ, Liu N, Zhou P, Ge L, Tan J (2009) Effects of heat stress during in vitro maturation on cytoplasmic versus nuclear components of mouse oocytes. Reproduction 137:181–189

Webb R, Garnsworthy PC, Gong JG, Armstrong DG (2004) Control of follicular growth: local interactions and nutritional influences. J Anim Sci 82:63–74

Werdelin L, Nilsonne A (1999) The evolution of the scrotum and testicular descent in mammals: a phylogenetic view. J Theor Biol 196:61–72. doi:10.1006/jtbi.1998.0821

Wettemann RP, Desjardins C (1979) Testicular function in boars exposed to elevated ambient temperature. Biol Reprod 20:235–241. doi:10.1095/biolreprod 20.2.235

Wildeus S, Hammond AC (1993) Testicular, semen and blood parameters in adapted and non adapted *Bos taurus* bulls in the semi-arid tropics. Theriogenology 40:345–355

Wilson SJ, Marion RS, Spain JN, Spiers DE, Keisler DH, Lucy MC (1998) Effects of controlled heat stress on ovarian function of dairy cattle. Cow J Dairy Sci 81:2139–2144

Wise ME, Armstrong DV, Huber JT, Hunter R, Wiersma F (1988) Hormonal alterations in the lactating dairy cow in response to thermal stress. J Dairy Sci 71:2480–2485

Wolfenson D, Flamenbaum I, Berman A (1988) Dry period heat stress relief effects on prepartum progesterone, calf birth weight, and milk production. J Dairy Sci 71:809–818

Wolfenson D, Kaim M, Rosemberg M (1994) Conception rate of cows supplemented with progesterone postinsemination in the summer. J Anim Sci 72:280

Wolfenson D, Thatcher WW, Badinga L, Savio JD, Meidan R, Lew BJ, Braw-Tal R, Berman A (1995) Effect of heat-stress on follicular development during the estrous cycle in lactating dairy cattle. Biol Reprod 52:1106–1113

Wolfenson D, Lew BJ, Thatcher WW, Graber Y, Meidan R (1997) Seasonal and acute heat stress effects on steroid production by dominant follicles in cows. Anim Reprod Sci 47:9–19

Wolfenson D, Roth Z, Meidan R (2000) Impaired reproduction in heat-stressed cattle: basic and applied aspects. Anim Reprod Sci 60–61:535–547

Wolfenson D, Inbar G, Roth Z, Kaim M, Bloch A, Braw-Tal R (2004) Follicular dynamics and concentrations of steroids and gonadotropins in lactating cows and nulliparous heifers. Theriogenology 15:1042–1055

Wright S (1922) Coefficients of inbreeding and relationship. Am Nat 56:330–338

Yarney TA, Sanford LM, Palmer WM (1990) Pubertal development of ram lambs. Body weight and testicular size measurements as indices of postpubertal reproductive function. J Anim Sci 70:139–147

Yasuo S, Nakao N, Ohkura S (2006) Long-day suppressed expression of type 2 deiodinase gene in the mediobasal hypothalamus of the Saanen goat, a short-day breeder: implication for seasonal window of thyroid hormone action on reproductive neuroendocrine axis. Endocrinology 147:432–440

Younas M, Fuquay JW, Smith AE, Moore AB (1993) Estrous and endocrine responses of lactating Holsteins to forced ventilation during summer. J Dairy Sci 76:430–436

Zeidan AEB (1989) Physiological studies on Friesian cattle. MSc thesis, Faculty of Agriculture, Zagazig University, Zagazig

Zeron YO, Cheretny A, Kedar O, Borochov A, Sklan D, Arav A (2001) Seasonal changes in bovine fertility: relation to developmental competence of oocytes, membrane properties and fatty acid composition of follicles. Reproduction 121:447–454

Zicarelli L (1994) Management in different environmental conditions. Buffalo J 2:17–38

Zicarelli L (1997) Reproductive seasonality in buffalo. In: Proceedings of 3rd course on biotechnology of reproduction in buffaloes, Caserta, pp 29–52

Zicarelli L, Campanile G, Seren E, Borghese A, Parmeggiani A, Barile VL (1992) Periestrous endocrine changes in Italian buffaloes with silent oestrus or ovarian disorders. In: Proceedings of international symposium on prospects of buffalo production in the Mediterranean and the Middle East, Cairo, pp 397–400

Zulu VC, Nakao T, Sawamukai Y (2002a) Insulin-like growth factor-I as a possible hormonal mediator of nutritional regulation of reproduction in cattle. J Vet Med Sci 64:657–665

Zulu VC, Sawamukai Y, Nakada K, Kida K, Moriyoshi M (2002b) Relationship among insulin like growth factor-I, blood metabolites and postpartum ovarian function in dairy cows. J Vet Med Sci 64:879–885

Heat Stress and Immune Function

Contents

Abstract

Heat stress suppresses different components of the immune system and thereby enhances susceptibility of an animal to various diseases. Some of the responses of heat stress can be prevented or overcome through alternative management practices and some nutritional strategies. Thermal stress imposes significant economic burden on the productivity of cows and buffaloes. Homeothermic processes within a reasonable limit bring back the immune system to a baseline response level after an immune challenge due to heat stress. Haemato-poietic system reacts by decreasing erythro-cytes, haemoglobin and packed cell volume during heat stress. The responses of the immune system are of innate and adaptive type. Livestock species of significant economic importance like cattle and buffaloes have a large population of lymphocytes in their blood, and T-lymphocytes recognise antigens through membrane receptors and are responsible for the regulation of the immune response. The immune system of cows contains a large pro-portion of $\gamma\delta$ T-lymphocytes, and the number varies with age and is considerably higher in young animals than in adults, where they con-stitute 5–10% of the total peripheral blood lymphocytes. The higher concentration of catecholamine exhibits a negative impact on immunity of heat-exposed cells through IL-1α and IL-1β. Glucocorticoids and catechola-mines, through their effects on Th1 and Th2 cytokine secretion, may cause suppression of

A. Aggarwal and R. Upadhyay, *Heat Stress and Animal Productivity*,
DOI 10.1007/978-81-322-0879-2_5, © Springer India 2013

cellular immunity and cause a shift towards Th2-mediated humoral immunity. Acute and chronic stressors affect the immune responses, and these responses may vary. Chronic stress most often leads to suppression or dysfunction of innate and adaptive immune responses however, acute responses may be variable, that is, immunosuppressive or immunomodulator (Dhabhar and McEwen 2006). The use of various nutritional strategies to enhance the immune system of livestock throughout various stages of production has been investigated extensively, and ameliorative measures have been suggested. Some of the nutritional supplements suggested may be used to enhance immune functions. Vitamin E and Zn have received most attention as immune-stimulatory nutrients. Micro-minerals modulate immune responses primarily through their critical roles in enzyme activity or efficiency complex, and a deficiency or an excess of minerals can alter the activities of the immune system.

1   Introduction

Livestock experience variety of environmental, managemental and nutritional stressors during the entire production cycle that may adversely affect their overall productivity and health status because of neuroendocrine disruption and stress-induced immunosuppression. All living organisms have evolved with mechanisms to cope with environmental stresses. However, there are numerous challenges within an animal's environment that may evoke a stress response. The response of the immune system is one of the mechanisms to defend and cope against environmental stresses. Stressors suppress different components of the immune system and enhance susceptibility of an organism to various diseases. Heat stress imposes significant economic burdens on the livestock production system in tropical climatic conditions and can be prevented or overcome through alternative management practices and various nutritional strategies.

With the global warming, in regions with seasonally high ambient temperature, heat stress is becoming one of the critical factors affecting health of livestock and production system. The heat stress due to temperature rise in Southeast Asia is likely to negatively impact not only an animal's growth performance but also the immune competence and disease resistance (Upadhyay 2011). On exposure to high ambient temperature, both primary and secondary lymphoid organs lose their weight, profiles of circulating leucocytes are negatively affected, $CD4^+$ T-cells and CDS^+ T-cells are decreased in blood and antibody response has also been found to decrease (Trout and Mashaly 1994).

Most livestock species have physiological and behavioural mechanisms that enable them to cope with environmental stimuli and stresses that strain and evoke a stress response. Homeothermic processes within a reasonable limit bring back the immune system to a baseline response level after an immune challenge, and hypothalamic–hypophyseal axis plays an important role. White blood cells, red blood cells, haemoglobin, haematocrit and globulin as indicators of body immunity are adversely affected by exposure to heat stress. The white blood cells (leucocytes) increase by 21–26% in Friesian cattle (Abdel-Samee 1987), under heat stress conditions due to thyromolymphatic involution. Among leucocytes and various soluble factors, neutrophils, macrophages, natural killer (NK) cells and soluble factors such as complement and lysozyme mediate the innate immune response, whereas lymphocytes, macrophages and soluble components such as immunoglobulins compose the specific immune response (Sordillo et al. 1997; Tizard 2000). The red blood cells decrease significantly (12–20%) in cattle under heat stress conditions (Salem 1980; Habeeb 1987) due to destruction of erythrocytes (Shafferi et al. 1981) and haemodilution effect. Haemoglobin concentration decreases during heat stress (Yousef 1990; Marai et al. 1995) due to depression of haematopoiesis and to haemodilution (Shebaita and Kamal 1973). Haematocrit percentage decreases in heat-stressed animals (Marai et al. 1997a, b) due to red cell destruction and/or to haemodilution (Shebaita and Kamal 1973).

2   Innate and Adaptive Immunity

The responses of the immune system are of innate and adaptive type. Innate immunity is the predominant defence during early stages of infections or a challenge and is activated by antigens, but the response is not amplified by repeated exposure to the same antigen (Tizard 2000). Innate immunity includes physical barriers such as the skin, mucosal secretions, tears, urine, acid in stomach as well as complement and antigen-nonspecific cellular components, and is considered to be the first line of defence against pathogens, namely, bacterial, viral, protozoal or fungal. Beneficial microorganisms in the intestine and respiratory tract which compete against invading pathogens are also an important part of innate immunity. The innate system provides the opportunity and time required by the acquired immune system to develop an antibody response against a specific pathogen, usually several days to several weeks. Innate immunity, although always present to some degree, is regulated and may be strengthened or weakened by many factors, like wounds, dehydration, nutritional status, genetics and stress. The phagocytic cells are the major players of innate immunity, and they also serve as the connection between innate and adaptive immunity (Tizard 2000).

Adaptive immunity is an antigen-specific immune response that evokes over time and is more complex response. Specialised cells like macrophages and dendritic cells initiate adaptive immune responses by presenting antigen to naïve lymphocytes to initiate a cell-mediated or humoral response. More specifically, these antigen presenting cells present antigen to activate naïve T-cells carrying receptors for a specific antigen, thus initiate T-cell immunity. Activated CD4+ T-cells commit early to a pathway of differentiation that results in the formation of two distinct subsets called T helper 1 (Th1) or T helper 2 (Th2) cells (Mosmann and Coffman 1989). Differentiation of either of these subset is established during priming of naïve CD4+ T-cells, and this differentiation process in the early phase of the immune process may be influenced by factors like

cytokines, receptors on the cell surface, antigen dose, nature of the antigen and direct cell to cell interaction with the antigen presenting cell (Constant and Bottomly 1997; Kidd 2003). The nature of the innate immune response to an antigen or pathogen dictates whether the subsequent adaptive CD4+ T-cell response will be mainly Th1 or Th2 (Janeway and Medzhitov 2002). These two distinct subsets of T helper cells are believed to be responsible for different functions of host defence and are distinguished by the spectrum of cytokines secreted by them. The key cytokines are interferon-γ (IFN-γ) and IL-4, which are central to stimulatory and inhibitory roles of a Th subset (Coffman 2006). Thus, IFN-γ and IL-4 inhibit differentiated Th1 or Th2 cells by blocking the differentiation of these subsets from naïve precursors. IFN-γ has been found to inhibit Th2, whereas IL-4 and IL-10 inhibit Th1 (Coffman 2006). The IL-12 favours Th1 and has no effect on Th2 (Constant and Bottomly 1997). None of the cytokines specific to one particular subset are exclusive products of Th cells because other leucocytes can contribute to Th1- or Th2-type responses (Mosmann and Sad 1996).

An enhanced immune reactivity may occur due to immunological memory or repeated host encounters with the same antigen (Janeway et al. 2001; Sordillo and Streicher 2002). Both innate and acquired immune responses have the ability to recognise conserved components of pathogens called pathogen-associated molecular patterns (PAMPs) such as lipopolysaccharide (LPS), peptidoglycan and bacterial DNA (Hornef et al. 2002). The host cell recognition of pathogen-associated molecules relies on a number of membrane receptors, that is, the toll-like receptors (TLRs), which provide cellular signalling during the initiation of the immune response (Medzhitov et al. 1997; Janeway et al. 2001).

Livestock species like cattle and buffaloes have a large population of lymphocytes in their blood and account more than 50% of leucocytes population in a healthy animal. Lymphocytes consist of two different subsets, namely, T- and B-cells, which differ in functions and molecule secretions. T-lymphocytes recognise antigens through membrane receptors and are responsible for the regulation of the immune response (Tizard 2000; Janeway et al. 2001). T-cells are the predominant blood lymphocyte subpopulation in ruminants (Tizard 2000). The T-lymphocytes have been subdivided into two main classes, $\alpha\beta$ and $\gamma\delta$ T-cells, depending on the expression of antigenic markers on the cell surface and cytokine production. $\alpha\beta$ T-cells can be further subdivided into both T helper (CD4[+]) and T-cytotoxic/suppressor (CD8[+]) lymphocytes.

In response to recognition of antigen-MHC-II complexes and co-stimulatory molecules on antigen presenting cell (APC) CD4[+] T-cells are activated and activated CD4[+] cells secrete certain cytokines that either facilitate a cell-mediated (Th1) or a humoral (Th2) immune response (Janeway et al. 2001; Sordillo and Streicher 2002). The Th1 immune response is characterised by increased secretion of IL-2 and INF-γ which in turn enhance cellular responses against intracellular pathogens and viruses, whereas Th2 immune response is characterised by higher production of IL-4, IL-5 and IL-10 supporting humoral immunity (Kehri et al. 1999). However, in contrast, CD8[+] cytotoxic cells have the capacity to kill specific target cells such as tumour cells or virus-infected cells in combination with the MHC-I associated molecules. The CD8[+] suppressor cells may produce different sets of cytokines such as IL-10 and transforming growth factor beta (TGF-β) that suppress the immune response (Janeway et al. 2001). The ruminant immune system contains a large proportion of $\gamma\delta$ T-lymphocytes, and the number varies with age and is considerably higher in young animals than in adults, where it constitutes 5–10% of the total peripheral blood lymphocytes (Hein and Mackay 1991). The $\gamma\delta$ T-cells have a wide range of functions, like secretion of cytokines like interferon gamma (IFN-γ), and cytotoxic activity in response to intracellular infections. They play an important role in the early response to infections, prior to antigen-specific responses are evident setting a Th1 immune response (Bluestone et al. 1995; Baldwin et al. 2002; Ismaili et al. 2002; Pollock and Welsh 2002).

3 Heat Stress and Cell-Mediated Immunity

Heat stress affects peripheral blood mononuclear cells (PBMC) of cattle and buffaloes, and the responses are variable and related to physiological state of animals. Lymphocyte proliferation of buffaloes in vitro and IL-1α and β levels are affected by heat exposure. The levels of catecholamines had significant ($P<0.01$) negative effect on lymphocyte proliferation index (LPI) which indicated that the higher concentration of catecholamines exhibits a negative impact on immunity of heat-exposed cells through IL-1α and IL-1β (Devaraj and Upadhyay 2007). The effects of heat stress on cell-mediated immunity of bovines have not been evaluated in depth, and conflicting results have been reported. Soper et al. (1978) reported that hot weather increases proliferation of PBMC in Holstein–Friesian cows in temperate climatic conditions. However, Elvinger et al. (1991) demonstrated that proliferation of bovine lymphocytes was reduced when cells were incubated for 60 h at 42 °C after stimulation with PHA, PWM or Con A. Kamwanja et al. (1994) reported that in vitro exposure of bovine lymphocytes to 45 °C for 3 h decreased the number of viable cells and reduced responsiveness of PBMC to mitogens. PBMC from dairy cattle experiencing temperature–humidity index (THI) values >72 exhibit reduced proliferation in vitro in response to mitogenic stimulation compared with PBMC from cattle experiencing THI values <72 (Lacetera et al. 2005). Incubation of cattle PBMC at high temperature (42 °C) for 7 h reduced proliferation compared with incubation at 38.5 °C (Elvinger et al. 1991). The precise mechanism underlying reduced cellular immune functions during heat stress in cattle are unclear particularly regarding role of cytokines (Lacetera et al. 2005). The impaired bovine lymphocyte functions in hot environments might be due to reduced cellular immunity which in turn influence the T helper Th1/Th2 balance in favour of the secretion of Th2 cytokines affecting lymphocyte proliferation (Lacetera et al. 2005).

4 Role of Glucocorticoids and Catecholamines in Regulation of Stress Response

The hypothalamic–sympathetic and the hypothalamic–pituitary–adrenal (HPA) systems provide central and peripheral control of stress responses. The hypothalamic–sympathetic system, beginning with neurons in the paraventricular nucleus (PVN) of the hypothalamus, causes release of catecholamines from the brain and the adrenal medulla (Swanson and Sawchenko 1980; Fulford and Harbuz 2005; Levine 2005). The activation of the HPA axis leads to production and secretion of corticotrophin-releasing factor (CRF) or corticotrophin-releasing hormone, primarily from the PVN of the hypothalamus via the median eminence and into the hypothalamic–hypophyseal portal system (Swanson and Sawchenko 1980; Swanson et al. 1980). Endocrine cells in the anterior pituitary respond to CRF by synthesising and secreting pro-opiomelanocortin or its products, namely, β-endorphin, adrenocorticoptrophic hormone (ACTH) and melanocyte-stimulating hormone (MSH). Pituitary ACTH travels through the blood and reaches to the adrenal cortex, where cells of the zona fasciculata secrete glucocorticoids (Fulford and Harbuz 2005). The cortisol is the primary glucocorticoid in cattle and swine (Minton 1994). The glucocorticoids provide negative feedback to the PVN to inhibit CRF release and catecholamine synthesis (Minton 1994; Fulford and Harbuz 2005).

Glucocorticoids effectively help inhibit the neuroendocrine stress response. Corticotrophin-releasing factor from the PVN activates norepinephrine neurons and neuron tracts in the locus coeruleus (LC). The LC also contains CRF neurons that activate catecholamine neurons. Under normal conditions devoid of any stress, CRF secretion within the LC is restrained by basal levels of glucocorticoids (Valentino and Van Bockstaele 2005). Glucocorticoids may cause a shift from a Th1 immune-driven response to a Th2 response (Wiegers et al. 2005). A potential mechanism by which glucocorticoids affect the Th1/Th2 balance may be through the inhibition of the production of and responsiveness to IL-12 (DeKruyff et al. 1998; Elenkov et al. 2000). Catecholamines also inhibit IL-12 and enhance IL-10 production (Elenkov et al. 1996). Thus, glucocorticoids and catecholamines, through their effects on Th1 and Th2 cytokine secretion, may cause suppression of cellular immunity and cause a shift towards Th2-mediated humoral immunity (Elenkov 2002).

Physiologic concentrations of cortisol illicit no response or are ineffective in suppressing IL-10 or IL-12p70 in human whole blood cultures (Visser et al. 1998). In contrast, human dendritic cells treated with cortisol produced less IFN-γ and greater IL-10 and IL-5 (de Jong et al. 1999). Cortisol was observed to suppress Th1 and Th2 cytokines in pig splenocytes (Skjolaas et al. 2002). This indicates that IFN-γ is less sensitive to cortisol suppression than IL-10 and that IL-2 may be resistant to cortisol (Skjolaas et al. 2002; Skjolaas and Minton 2002). The combined immunologic effects of glucocorticoid hormones and catecholamines prevent overstimulation of innate immunity and the TH1 cytokines, while simultaneously priming the humoral immune response through stimulation of the TH2 cells. Therefore, the immune response in an animal is an outcome of the overall effect of stress hormones on the production of the TH1 and TH2 cytokines (Elenkov and Chrousos 2002).

4.1 Role of Cytokines

The glucocorticoids affect the Th1/Th2 balance most likely by inhibiting the production of cytokines and specifically IL-12 during stress (DeKruyff et al. 1998; Elenkov et al. 2000). Catecholamine release during stress may also inhibit IL-12 and enhance IL-10 production (Elenkov et al. 1996). Thus, glucocorticoids and catecholamines concomitantly, through their effects on Th1 and Th2 cytokine secretion, may induce suppression of cellular immunity and a shift towards Th2-mediated humoral immunity (Elenkov 2002). The key cytokines interferon-γ (IFN-γ) and IL-4 are central to immune response and either stimulate or/and inhibit roles of a Th subset (Coffman 2006). Thus, IFN-γ and IL-4 do not directly inhibit differentiated Th1 or Th2 cells, but they inhibit by blocking the differentiation of these subsets from naïve precursors. The IFN-γ has been shown to inhibit Th2, whereas IL-4 and IL-10 inhibit Th1 (Coffman 2006), and the IL-12 favours Th1 and has no effect on Th2 (Constant and Bottomly 1997). However, it may be specified that none of the cytokines specific to one particular subset are exclusive products of Th cells because other leucocytes can also contribute to Th1- or Th2-type responses (Mosmann and Sad 1996).

Cytokines released upon activation of the immune system in response to stress stimulate the HPA axis and increase peripheral levels of glucocorticoids and catecholamines, which in turn suppress the synthesis and release of cytokines. Glucocorticoids inhibit a large number of cytokines, including IL-4, IL-5, IL-6, IL-12, IFN-γ and tumour necrosis factor-α (Wiegers and Reul 1998; Richards et al. 2001; Sapolsky et al. 2001). However, not all cytokines are suppressed by glucocorticoids and IL-10 secretion is increased by glucocorticoids (Blotta et al. 1997; Richards et al. 2001), whereas IL-1, IL-4 and IL-6 act synergistically with glucocorticoids in humans (Wiegers et al. 2005). The glucocorticoids enhance IL-4 production (Wu et al. 1991; Blotta et al. 1997). Also, IL-4 is enhanced by IL-12, but glucocorticoids inhibit IL-12 (Wu et al. 1998; Elenkov et al. 2000). The inhibition of cytokines by glucocorticoids is regarded as a protective mechanism that prevents overshooting of the immune defences. In general, glucocorticoids inhibit proinflammatory cytokine synthesis or induce the cytokines that have immunosuppressive potential (Wiegers et al. 2005).

4.2   Effect of Stress Hormones

Stress hormones released in response to activation of the HPA axis (CRF, ACTH and cortisol) have been observed to have an effect on the immune system. The incubation of cattle and porcine immune cells with cortisol has been shown to suppress lymphocyte proliferation, IL-2

production and neutrophil function (Blecha and Baker 1986; Salak et al. 1993). A high dose of dexamethasone has been found to profoundly increase the number of circulating neutrophils in heifers but inhibited neutrophil cell surface marker expression (Weber et al. 2001). The number of apoptotic cells increased, whereas the number of proliferating cells decreased in calves receiving dexamethasone injections twice daily for 4 days (Norrman et al. 2003), thus leading to an increase in the ratio of apoptotic cells to proliferating cells. In these same calves, T-cells were observed to increase, but B-cells decreased.

Activation of the HPA axis by administering exogenous ACTH, cortisol or by blocking cortisol synthesis has been used to investigate the effects of cortisol on immune function. In pigs, administration of an intravenous bolus of ACTH caused an increased natural killer (NK) activity and IL-2-stimulated NK activity (McGlone et al. 1991), whereas an ACTH injection suppressed neutrophil cellular function in Japanese Black steers (Ishizaki and Kariya 1999). A pharmacologically induced, threefold increase in plasma cortisol concentration via cortisol injection was observed to have no effect on NK cytotoxicity, but an infusion of 400 μg of cortisol resulted in reduced NK activity at 1 h post injection, but not at 2 h (Salak-Johnson et al. 1996). Blocking cortisol synthesis by feeding metyrapone to pigs resulted in low plasma cortisol concentrations and reduced NK cytotoxicity (Salak-Johnson et al. 1996). Central injection of CRF has been observed to decrease NK cell activity in rodents (Irwin et al. 1990), marginal reduction in pigs was observed, but neutrophil chemotaxis was significantly suppressed (Salak-Johnson et al. 1997) by central CRF injection. Administration of central CRF resulted in reduced concanavalin-A (Con A)-induced proliferation in pigs (Johnson et al. 1994) but had no effect on phytohaemagglutinin (PHA)-induced proliferation (Salak-Johnson et al. 1997).

Prior to the administration of LPS (0.5 μg/kg body weight), temperamental bulls had higher cortisol and epinephrine concentrations as compared to calm or intermediate bulls. Cortisol concentrations increased following LPS administration but were not affected by temperament. Epinephrine concentrations peaked 1 h after LPS administration in calm bulls. Temperamental bulls did not exhibit an epinephrine response to LPS challenge. These data demonstrate that the temperament of calves can modulate the physiological, behavioural and endocrine responses of prepubertal Brahman bulls to endotoxin challenge. Specifically, temperament differentially affected the rectal temperature, sickness behaviour and epinephrine, but not cortisol, responses to LPS challenge (Burdick et al. 2011). The study indicated that aggressive behavioural or temperamental animals are able to cope with challenges to immune system much better than nonaggressive or calm animals.

4.3   Colostral Ig

Favourable environmental conditions are vital in promoting calf health, minimising risk of diseases and mortality, subsequently encouraging growth rates. Moderate heat stress does not modify significantly the protective value of colostrum, as evaluated by determining the concentration of Ig fractions. The total Ig concentrations in summer colostrum do not differ or were higher than those recorded during other seasons (Kruse 1970; Shearer et al. 1992). Heat stress (THI >86 from 0900 to 2000 h and 76 from 2100 to 0800 h) significantly reduced IgG and IgA in colostrum of primiparous cows (Nardone et al. 1997). A reduced passage of IgG from the bloodstream to the udder result into an impairment of the immune reactivity of the mammary gland plasmacytes to synthesise IgA during heat stress in cows. Extreme heat can negatively influence a cow's ability to produce high quality colostrum and can also negatively affect a calf's ability to absorb IgG from colostrum (Stott 1980).

Summer heat stress is likely to mask the immunosuppression taking place in periparturient dairy cows (Mallard et al. 1998). Heat stress reduces thermogenic and immunosuppressant hormones like growth hormone and glucocorticoids and may be related to the higher reactivity of the immune system observed in cows exposed to heat stress (Webster 1983).

5   Acute and Chronic Stressors

The stress response considered an all-or-nothing biologic activity associated with the fight or flight behaviour of Selye is not immunosuppressive. Indeed, stress may elicit bidirectional effects on immune response, acute stress may be immunomodulator or enhancer, whereas chronic stress may be immunosuppressive (Carroll and Forsberg 2007). Therefore, the immune response of an animal to stress depends on the type of stress encountered (i.e. acute vs. chronic). In acute stress, hormones associated with priming the immune system are released in a manner that prepares for potentially countering invading pathogens and subsequent infection. However, under prolonged or chronic stress, the effect of stress hormones on the immune system are no longer initial or preparatory initiate events that are of suppressive type, first at the cellular level and then, eventually, across the entire immune system.

Acute and chronic stressors affect the immune responses that may vary. Chronic stress often leads to suppression of the immune system, but acute responses may or may not cause immunosuppression. In pigs, acute heat exposure and transport stress had no effects on various immune measures (McGlone et al. 1993; Hicks et al. 1998). But acute transportation stress reduces chemiluminescence response of alveolar macrophages and increases the ratio of CD4+ to CD8+ cells in cattle (Ishizaki et al. 2005). Acute cold stress in pigs has been observed to cause an increase in NK cytotoxicity (Hicks et al. 1998), whereas NK cytotoxicity had both positive and negative response subjected to acute restraint stress. Specifically, NK was increased during the early phase (0–1 h) and decreased during the late (3–4 h) phase of the stressor (Wrona et al. 2001). Chronic stress may have differential effects on the immune system. Chronic heat stress had no effect on concanavalin-A or PHA-induced lymphocyte proliferation (Bonnette et al. 1990; Morrow-Tesch et al. 1994).

A higher incidence of a variety of intramammary infections during summer occurs (Waage et al. 1998; Cook et al. 2002), and a higher morbidity rate of *Corynebacterium pseudotuberculosis* infection in Israeli dairy cattle has been observed during summer months (Yeruham et al. 2003). The studies performed to assess the relationships between heat stress and immune cell function in bovines are inconclusive (Soper et al. 1978; Kelley et al. 1982a, b; Elvinger et al. 1991; Kamwanja et al. 1994; Lacetera et al. 2002). In particular, with regard to lymphocyte function in cows exposed to hot environments, Soper et al. (1978) reported an improvement, Elvinger et al. (1991) and Kamwanja et al. (1994) described an impairment, while Kelley et al. (1982b) and Lacetera et al. (2002) reported no effect of heat exposure. Extreme events (heat waves) were observed to be associated with depressed cellular immunity, as assessed by measuring DNA synthesis in PBMC stimulated with mitogens, an enhanced humoral response, as assessed by measuring antibody secretion in PBMC stimulated with Pokeweed mitogen (PWM) and higher concentrations of plasma cortisol (Lacetera et al. 2005). Studies have demonstrated that the effects of summer or winter season on disease resistance or immunoresponsiveness of domestic animals depend on variables like species and breed, duration of the exposure, severity of stress and the type of immune response considered (Kelley 1982).

Summer THI conditions associated with moderate heat stress in dairy cows did not impair cell-mediated immunity, colostral Ig content and the passive immunisation of calves (Lacetera et al. 2002). Hahn et al. (2002) reported that heat waves can push vulnerable animals beyond their survival threshold limits and that heat wave events for cattle can cause large economic losses as a result of death and decline in performance. Summer conditions characterised by the occurrence of extreme events (heat waves) can be responsible also for a profound shift from cellular to humoral immune responses, which may lead to modifications of resistance to diseases and thus to further economic losses. On the other hand, previous studies have demonstrated that the occurrence of certain infections in cattle is higher during hot months (Smith et al. 1985; Waage et al. 1998; Cook et al. 2002; Yeruham et al. 2003). The effects of heat stress on the immune response of dairy cows depend on the specific immune function taken into consideration, and neuroendocrine changes due to heat stress may play a role in the perturbation of immune functions (Lacetera et al. 2005).

6 Mechanism of Action of Heat Stress on Immunity

Cytokines provide the link between the innate and adaptive immune systems and help maintain T-cell homeostasis during infection (Bot et al. 2004). The hallmark cytokine of Th2 immunity is IL-4. If IL-4 is overexpressed, it negatively interferes with the immune defence mechanisms, thus decreasing the recruitment, expansion or activity of major effector cells such as the Th1 cells (Bot et al. 2004). During a viral infection, a strong bias towards Th2 responses may interfere with viral clearance. However, if the opposite scenario occurs (i.e. elevated Th1 and reduced Th2 immunity), normal viral clearance can occur (Bot et al. 2004). It is possible that certain stressors may disrupt this balance by interfering directly or indirectly with the mechanistic immune processes. A balanced Th1/Th2 response may be favoured in some cases of disease challenge to achieve a compromise between defence mechanisms and immune homeostasis (Salak-Johnson and McGlone 2007).

The mechanism whereby heat stress affects immune function may be mediated through changes in the prolactin signalling pathway. Suppressors of cytokine signalling (SOCS) proteins and cytokine-inducible SH2-containing proteins (CIS) compose a family of intracellular proteins (Yoshimura et al. 2007) that are stimulated by PRL, act through feedback to inhibit cytokine signalling (Wall et al. 2005) and regulate the responses of immune cells to cytokines (Yoshimura et al. 2007). In particular, SOCS-1 and SOCS-3 have been shown to bind to cytokine receptors or to receptor-associated Janus-associated kinases (JAK) to inhibit activation of signal transducers and activators of transcription (STAT) members, and ultimately interferon signalling (Sasaki et al. 1999). The SOCS proteins also regulate tumour necrosis factor-α (TNF-α)-mediated cellular apoptosis by inhibiting phosphatidylinositol 3-kinase and p38 mitogen-activated protein kinase pathways (Kinjyo et al. 2002). With changes in PRL concentration in plasma, it is plausible that effects of heat stress on immune function may be mediated through the PRL-signalling pathway and lymphocyte cytokine production.

The intracellular proteins SOCS-1 and SOCS-3 inhibit signal transduction of type I and type II cytokine receptors (such as PRL-R), possibly through action at the level of receptors and JAKs; however, CIS and SOCS-2 act mainly by competition with STAT factors for recruitment to activated receptor complexes (Dalpke et al. 2008). Lymphocytes from cows which were provided with cooling had lower expression of SOCS-1 and SOCS-3 mRNA. The protein SOCS-1 is an essential negative regulator for T-cell activation by dendritic cells and for maintaining immunological tolerance by restricting CD8⁺ T-cell proliferation. Thus, low expression of SOCS-1 may enhance JAK2 activity, thereby promoting cell proliferation (Yoshimura et al. 2007). The lymphocytes of cooled cows proliferate more than those of heat-stressed cows and expressed less SOCS-1 mRNA. However, there was no effect of heat stress on mRNA expression of SOCS-2 or CIS.

In addition, HSPA5 mRNA, which encodes a member of the HSP70 protein family, was upregulated in lymphocytes from heat-stressed cows at +20 day relative to calving compared to cooled cows; however, expression was not affected during the dry period in which heat stress was imposed. Likewise, no effect of summer heat stress was reported in the expression of HSP70 in lymphocytes isolated from grazing beef cattle (Eitam et al. 2009). In vitro data suggest that HSP70 require temperatures above 42 °C for activation (Williams et al. 1993). The metabolic stress of lactation might also play a role in lymphocyte HSPA5 mRNA expression. In fact, increased expression of HSP70 has been shown to be protective of cells to metabolic stress (Williams et al. 1993). Large farm animals exposed to heat stress conditions have differential induction of serum HSP70 levels compared with lymphocytes treated in vitro. Furthermore, both hot–humid and dry-heat stressors effectively induce HSP70 in lymphocytes (Mishra et al. 2011). Nonetheless, hot–humid condition caused lesser increase in HSP70 concentration than hot–dry. Thermal stress causes increased HSP70 synthesis both in isolated lymphocytes as well as in intact animals. Increase in serum HSP70 level to either dry–hot or hot–humid stress substantially differs with isolated in vitro treatments of stress and should be considered while evaluating different effects of stressors. However, increase in HSP70 as a thermo-adaptive response does not serve immune protection at physiological level (Mishra et al. 2011).

In terms of cytokine production, lymphocytes isolated from heat-stressed cows also produced less TNF-α compared to the cows provided with cooling. It has been reported that SOCS-1 inhibits TNF-α secretion (Kinjyo et al. 2002). In contrast, CIS can be induced by TNF-α (Starr et al. 1997). Indeed, CIS mRNA expression in lymphocytes from cooled cows was numerically greater (120% vs. 83%; SEM = 24%; $P = 0.30$) than in those of heat-stressed cows. IFN-γ and IL-4 secretion from the lymphocytes did not differ between treatments. Lower IFN-γ and IL-4 production from the heat-stressed lymphocytes were expected, as SOCS-1 inhibits both IFN-γ and IL-4 (Dalpke et al. 2008). However, IL-6 secretion did not differ between lymphocytes from heat-stressed and cooled cows. PRL-R and SOCS mRNA are differentially expressed between lymphocytes isolated from both group of cows. The greater PRL-R and lower SOCS-1 and SOCS-3 mRNA expressions in lymphocytes from cooled cows were associated with greater lymphocyte proliferation and may be a mechanism whereby cooling improves immune cell function. In addition, the increased TNF-α cytokine production of lymphocytes of cool cows provides evidence of cell communication to mount an appropriate regulated immune response (Amaral et al. 2010).

7 Nutrition and Immunity

Attempts are being made to decrease or eliminate the use of antimicrobials in livestock production system due to a growing concern that existing production practices may lead to an increase in the number of antibiotic-resistant human pathogens

(Carroll and Forsberg 2007). Therefore, nutritional supplements are being used to support or enhance immune function, to improve the overall well-being and performance of the livestock and provide non-antimicrobial alternative management practices.

All physiologic processes of the body, including the immune system, are influenced by availability of nutrients or by their deficiency. The use of various nutritional strategies to enhance the immune system of livestock throughout various stages of production has received considerable attention. Not only vitamin and minerals but also supplementation of lipid, protein and amino acid forms and concentrations within livestock diets has been evaluated extensively for their potential impact on the immune system. The immunomodulatory properties of various vitamins (A, B6 and B12, C, D and E), minerals (Zn, Cu, Fe, magnesium and selenium), animal products (spray-dried plasma, fish oil and fish meal), yeast products and plants within domestic livestock diets have been documented. Vitamin E and Zn have received most attention as immunomodulatory micronutrients (Calder and Kew 2002). Vitamins are important dietary components due to the diverse functions in efficiency complex and their deficiencies negatively affect efficiency complex that predispose or lead to a wide range of diseases (Horst 1986; NRC 2001). Specific nutritional requirements for different vitamins have also been associated with changes in the immune response and disease resistance. Therefore, vitamin supplementation to bovines helps prevent diseases like mastitis, retained fetal membranes and metritis. Insufficient contents of trace elements in ruminant diets have been related with low disease resistance (Spears 2000). Several micronutrients such as cobalt (Co), copper (Cu), selenium (Se) and zinc (Zn) have been observed to influence efficiency complex and components of the immune system (Paterson and MacPherson 1990; Reddy and Frey 1990). Relative bioavailability of some important minerals and vitamins has been given in Table 1.

7.1 Vitamins

Vitamin deficiencies have been observed to adversely impact immune response leading to

Table 1 Relative bioavailability of selected mineral and vitamin sources

Element	Source	Bioavailability
Selenium	Sodium selenite	High
Copper	Copper oxide	Low
	Copper carbonate	Intermediate
	Copper sulphate	High
Zinc	Zinc oxide	High
	Zinc sulphate	High
Iron	Iron oxide	Unavailable but may interfere with absorption of other minerals
Vitamin E	RRR-α-tocopherol	High
	All-rac-α-tocopheryl	Low

Source: Modified from McDowell (2003)

immune system disorders and diseases. The proper development and function of the immune system can be linked to an adequate level of one particular vitamin or another. Vitamins exert essential roles in haematopoiesis, maintenance and function of lymphocytes, NK cells and neutrophils and even antibody production in the body. Vitamins act as antioxidants and also play important role in inactivating harmful reactive oxidative species (ROS) such as oxygen ions, free radicals and peroxides produced through normal cellular activity, such as oxygen metabolism, that can destroy cellular membranes, cellular proteins and nucleic acids. Lymphocytes, for example, are highly active cells that, as part of their normal cellular activity, generate ROS continuously. Immune cells are particularly susceptible to oxidative damage for two reasons: (1) one mechanism by which cells of the immune system provide protection is by phagocytising and killing pathogens through an oxidative bactericidal mechanism termed the 'respiratory burst', which generates large amounts of ROS, and (2) immune cells have a high percentage of polyunsaturated fatty acids in their plasma membranes, which makes them more sensitive to oxidative stress (Grimble 2001). As a result, immune cells are dependent particularly on high levels of antioxidants to protect them from ROS-mediated cell and membrane damage. Although the body produces a large number of endogenous antioxidants to defend against ROS, under conditions of high oxidative stress, their ability to eliminate ROS can be compromised or exceeded. In such cases, dietary supplementation of antioxidants such as vitamin E, vitamin C, carotenoids, Zn and Se can be beneficial to eliminate damaging ROS and in maintaining normal cellular functions and health (Chew and Park 2004; Chandra and Aggarwal 2009; Maurya 2011; Aggarwal et al. 2012).

7.1.1 Vitamin A

Vitamin A and its retinoids have significant effects on various components of immunity, including lymphopoiesis, apoptosis, cytokine expression and antibody production. Vitamin A has also been observed associated with the inhibition of Type 1 lymphocyte cytokine production (Frankenburg et al. 1998) and the proliferation of lymphocytes (Semba 1994).

Most of the fat-soluble vitamins, such as retinol, α-tocopherol and β-carotene plays role as antioxidant in the body, decrease at the time of parturition and have been reported to be associated with severe health problems in high-producing cows (Chandra and Aggarwal 2009; Maurya 2011). Periparturient cow undergoes intense mammary growth, copious synthesis and secretion of carbohydrates, fats, proteins as well as marked accumulation of colostrum and milk. Since colostrum is rich in vitamins A and E, cow requires increased supply of these vitamins prior to parturition. Providing vitamin A orally to dry cow or by intramuscular injection at drying off prevents deficiency of vitamin A. Retinol levels above 0.8 μg/ml are necessary for optimum immune function. β-carotene supplementation enhances the per cent intracellular kill by blood phagocytes. Dairy cows supplemented with 300 and 600 mg BC/day have higher killing ability against bacteria in the peripartum period (Michal et al. 1994). However, supplementation with 500 mg BC/d/h results in reduction in somatic cell count (SCC) by 300,000 cells/ml milk and more number of animals were observed to have SCC below 200,000 (Hasselmann et al. 2000). The results on supplementation with vitamin A and B-carotene on mastitis are inconsistent. Positive effects have been reported when cows were given diets that approximately met NRC

(2001) requirements for vitamin A and supplemented with 300 mg/day of β-carotene (Dahlquist and Chew 1985). Supplementation of vitamins and minerals is required during the dry period and early lactation. Plasma concentration of β-carotene in dairy cows should be >3 mg/L for optimum udder health. Elevated serum retinol during the prepartum period has been shown to be associated with a decreased risk of clinical mastitis during the first 30 days postpartum.

7.1.2 Vitamin C

Vitamin C, also termed as ascorbic acid, is a water-soluble antioxidant vitamin that is important in the formation of collagen, tendons, ligaments, muscle and blood vessels. Vitamin C also plays an important role in the synthesis of the norepinephrine, a neurotransmitter and hormone produced in the adrenal medulla. Norepinephrine is critical to brain function and the body's response to stress. The requirement of vitamin C in the body is very less, to protect proteins, lipids, carbohydrates, DNA and RNA from damage caused by free radicals and reactive oxygen species, which are generated during normal metabolic processes and immune cell activity, especially in periparturient cows. An equally important property of vitamin C may be its ability to regenerate other antioxidants, such as vitamin E (Carr and Frei 1999). The beneficial role of vitamin C has also been observed in cattle due to its interaction with other antioxidants, such as vitamin E (Cusack et al. 2005).

7.1.3 Vitamin E

Vitamin E is an important and essential fat-soluble antioxidant vitamin for body immunity and reproductive function of animals. As an antioxidant, vitamin E protects cells against free radicals. In addition to its antioxidant properties, vitamin E has a supportive role for maintenance of the immune system, DNA repair and other metabolic processes (Traber and Arai 1999; Aggarwal et al. 2012). Vitamin E is an important constituent of all the cell membranes (plasma, mitochondrial and nuclear) and is the major antioxidant in body tissues. Vitamin E is found throughout all cells of the body and is more concentrated in the immune cells to provide protection against the destructive free radicals used by white blood cells to destroy pathogenic organisms.

In India, animals are mainly reared on straw-based diets which are low in vitamins particularly vitamin E. Supplementation of vitamin E maintains proper antioxidant status in animals and improves the ability to resist infections. Low plasma concentration of α-tocopherol at parturition is considered as a significant risk factor for intramammary infection and mastitis during the first week of lactation. During periparturient period when there is significant decrease in α-tocopherol, the cow's immunity status and neutrophil functions are depressed; as a result, high incidence (30–50%) of clinical mastitis occurs during the first month of postpartum. Vitamin E is reported to enhance the intracellular kill of *E. coli* and *staphylococcus aureus* by bovine blood neutrophils (Hogan et al. 1990). Supplementation of vitamin E at 1,000 IU vitamin E/day from 15, 30 and 60 days prepartum to 30 days postpartum has been observed to decrease incidence of subclinical mastitis. Vitamin E also modulates prostaglandin release from activated macrophages during bacterial infection. The stimulation index for cell-mediated immunity is improved by vitamin E supplementation (1,000 IU/day) in cows during first week of parturition (Chawla and Kaur 2005). Vitamin E and A levels were significantly higher in cows supplemented with α-tocopherol acetate indicating improvement in antioxidant status and immunity of cows (Chandra and Aggarwal 2010; Aggarwal et al. 2012).

The 1989 NRC recommendation for vitamin E was 150 IU/day for dry cows and 300 IU/day for lactating cows. However, the beneficial effect of vitamin E on the cow's defence against mastitis required greater intakes than the NRC recommendation. Dry cows often are fed poorer quality feeds than lactating cows and therefore are likely to be benefitted by 1,000 IU/day of supplemental vitamin E throughout the dry period. Lactating cows should be supplemented with 500 IU/day of vitamin E because they are fed higher quality feeds than dry cows. This recommendation is based on a significant reduction in udder infections, clinical mastitis and somatic cell counts seen when cows were supplemented with these

amounts of vitamin E (Chawla and Kaur 2005; Chandra and Aggarwal 2010). In buffaloes, when 1500 IU/day of vitamin E was supplemented during dry period, it resulted in improved milk yield and also oxidative stability of milk was better (Panda and Kaur 2007). The amount that needs to be supplemented depends on forage quality and losses during storage. Cattle receiving about 50% of their forage from above average quality pasture probably do not need supplemental vitamin E. The various forms of supplemental vitamin E fed have different availabilities to the cow. DL-alpha-tocopherol acetate, a commonly supplemented form, has an activity of 1 IU/mg, while D-alpha-tocopheryl acetate has an activity of 1.36 IU/mg. Therefore, attention should be paid so that it reaches a level of 1,000 IU/day in feed.

7.2 Minerals

Minerals form an important component of efficiency complex which modulate immune responses primarily through their critical roles in enzyme activity, and a deficiency or an excess of minerals can alter different components of immune system. Mineral deficiencies can have detrimental effects on immune functions through alterations in specific aspects of immunity, including antibody responses, cell-mediated immunity and NK cell activity. Minerals also play a special role in efficiency complex to ensure efficient growth, reproduction and immune competence in animals. Uncontrolled oxidation reactions may impair the animal's immune status (Spears 2000). The herds have been observed to increase risks of metritis, mastitis, locomotion problems or diarrhoea in calves when zinc (Zn) or copper (Cu) status is either marginal or deficient (Enjalbert et al. 2006). The low or suboptimal levels of minerals may be restored by dietary supplementation. The most important way to balance oxidative damage and antioxidant defence in dairy cows is to optimise the dietary intake of antioxidant minerals. Dairy cow feeds typically contain a range of different compounds that possess antioxidant activities, many of which are minerals or are mineral dependent. The key trace elements involved in animal feed are Zn, Cu, selenium (Se), iron (Fe) and manganese (Mn) (Surai et al. 2003; Surai and Dvorska 2002).

The mineral-dependent antioxidant enzymes that can be synthesised in the body are able to deal effectively with free radicals but require a continuous supply of feed-derived mineral cofactors. Selenium (Se) is an essential part of a family of enzymes called glutathione peroxidases (GSH-Px) and thioredoxin reductases (Berry et al. 1991). However, Zn, Cu and Mn are integral parts of superoxide dismutase (SOD), and Fe is an essential part of catalase. When these metals are supplied through feed in sufficient quantities, adequate antioxidant enzymes are synthesised in the body. However, deficiency of these elements results in oxidative stress, leading to potential damage of tissues, biological molecules and membranes.

7.2.1 Selenium (Se)

Selenium is an essential component of a range of selenoproteins, including glutathione peroxidase, thioredoxin reductase and iodothyronine deiodinase. Selenium is found in raw feed materials used for animals in varying quantities, and some of them are deficient or may have higher toxic levels. There are two major sources of Se for animals: (1) Se from feed materials, in the form of selenoamino acids, including selenomethionine and selenocysteine; (2) inorganic Se in the form of selenate or selenite. The physiological requirement for Se is low in animals, but if this is not met, the antioxidant system is likely to be compromised and may affect animal health (Spears 2000). Excessive levels of Se (in plants and soils) are associated with toxicity; therefore, adequate measures need to be adopted for preventing toxicity of selenium.

Selenium content of feeds may vary as per the soil content. Vitamin E and selenium supplementation are required or may be necessary to enhance the body defences from challenge of infectious agents. Supplementation of 0.3 ppm Se and 500 IU/day vitamin E during lactation and vitamin E 1,000 IU/day during the dry period has been shown to reduce prevalence and severity of mastitis and reduce somatic cell count in milk (Chawla and Kaur 2005; Chandra and Aggarwal 2009). The vitamin E content of basal diet is highly variable; therefore, NRC (2001) has recommended supplemental vitamin E regardless of the dietary vitamin E content. The requirements that are based on measures of immune function are usually higher than those based on production or reproduction. NRC (2001) has suggested the requirement of vitamin E during last 60 days of gestation: 80 IU/kg DMI for dry cows and 20 IU/kg DMI for lactating cows. Selenium should be provided only if the soil is deficient in selenium in that particular area.

Although Se and vitamin E functions independently, studies have shown that administration of both Se and vitamin E may result synergistically to enhance immune response (Spears 2000). Vitamin E supplementation of lactating dairy cows (Hogan et al. 1990) and young calves fed on a milk substitute (Eicher et al. 1994) also increased blood neutrophil bactericidal activity. Compared with non-vitamin E-supplemented cows, supplementing dairy cows with 3,000 mg vitamin E per day for 4 weeks prepartum and 8 weeks postpartum prevented a decline after parturition in neutrophil superoxide anion production and IL-1 production and major histocompatibility class II antigen expression by blood monocytes (Politis et al. 1995).

Se plays an important role in removing hydrogen peroxide and organic hydroperoxides through its effects as a component of the antioxidant enzyme glutathione peroxidase, and a deficiency in Se can induce a state of oxidative stress in the host (Chaudiere and Tappel 1984). Se deficiencies have been reported to associate with lower resistance to infections, possibly due to decreased antibody production and an impaired lymphocyte proliferative response (Chandra and Chandra 1986; Reffett et al. 1988). However, other studies have identified additional mechanisms by which Se supplementation enhances immune function, including neutrophil killing activity (Hogan et al. 1990) and neutrophil adherence (Maddox et al. 1999). The altered neutrophil adherence could also affect the ability of neutrophils to attack and sequester pathogens (Spears 2000).

Many of the beneficial health effects of Se are mediated by antioxidant selenoenzymes, with selenocysteine at the active sites. Cytosolic glutathione peroxidase (GPX1) is the selenoenzyme most often associated with antioxidant functions. Determination of cellular or plasma GPX1 activity is a diagnostic tool for assessing Se status of dairy cows. Many other bovine antioxidant selenoproteins exist including selenoprotein P, five different isoforms of GPX and three thioredoxin reductase (TrxR) isoenzymes (Grignard et al. 2005; Hara 2001). Both GPX1 and TrxR1 play a critical role in reducing both H_2O_2 and fatty acid hydroperoxides to less reactive water and alcohols, respectively.

7.2.2 Zinc (Zn)

Zn is the second most abundant trace element in mammals and birds and forms a structural component of over 300 enzymes and plays catalytic and regulatory element affecting efficiency complex. An important role played by zinc is in antioxidant defence system as an integral part of the essential enzyme superoxide dismutase (SOD) (Underwood 1999; National Research Council 2001). Zinc has been reported to influence several components of immunity, including cell-mediated immune functions, tissue regeneration, protein synthesis and inflammatory responses (Erickson et al. 2000; Kruse-Jarres 1989). Zn supports humoral and cell-mediated immunity by facilitating proliferative reactions in response to stimulus by different mitogens by way of its action on immune cells as a cofactor for essential enzymes. Zn deficiency has been associated with decreased T-cell function and antibody responses (Kruse-Jarres 1989).

Tomlinson et al. (2002) have summarised results of 12 experiments and reported an overall significant reduction (196,000 vs. 294,000) in SCC when Zn-Met was supplemented (about 200 mg of Zn/d in five experiments and about 380 mg of Zn/d in seven studies). Whitaker et al. (1997) compared the effects of providing supplemental Zn from a mixture of Zn proteinate (250 mg of Zn/day) and inorganic Zn (140 mg/day) or from all inorganic sources (390 mg of Zn/day). Diets contained approximately 50 ppm total

Zn (about 25 ppm supplemental and 25 ppm from basal diet). Source of Zn had no effect on infection rate, new infections, clinical mastitis and SCC. Animal diets should contain about 20 ppm of copper and 50–60 ppm of Zn.

Zinc also plays a role in maintenance of epithelial tissue. Since teat skin is the first line of defence, any deterioration in the health of epithelial tissue will enhance the ability of bacteria to penetrate and cause infection. Zinc is also required for keratin production. Keratin is a wax-like substance secreted into the teat-end opening. The keratin lining of the teat canal helps in entrapping bacteria and prevents their upward movement into the mammary gland through its bactericidal properties. Approximately 40% of keratin in the teat canal is regenerated after each milking; therefore, the ability of the cow's mammary system to efficiently reproduce keratin is important in the defence against mastitis.

Zn is also involved in facilitating hormonal secretion and function particularly somatomedin-c, osteocalcin, testosterone, thyroid hormones, insulin and growth hormone. In India, commonly fed feeds (roughages as well as concentrates) contain Zn content below the critical level of 40 ppm as recommended by NRC (2001) for dairy cattle. Hence, there is a need to supplement Zn to dairy animals in order to improve their immune status.

7.2.3 Copper (Cu)

Cu is a component of a range of physiologically important metalloenzymes and takes part in (1) antioxidant defence as an integral part of SOD, (2) cellular respiration, (3) cardiac function, (4) bone formation, (5) carbohydrate and lipid metabolism, (6) immune function, (7) connective tissue development, (8) tissue keratinisation and (9) myelination of the spinal cord. The main Cu-containing enzymes have been enlisted in Table 2. Inorganic Cu has a strong pro-oxidant effect and likely to stimulate lipid peroxidation in feed or the intestinal tract (Surai et al. 2003). Dietary Cu may also affect cytokine production in cattle. Mononuclear cells of lactating dairy cows receiving a marginal level of Cu (6–7 mg/kg diet) were observed to produce less IFN-γ

Table 2 Some minerals containing enzymes

Mineral	Enzyme	Function
Cu	Cytochrome oxidase	Transport of electron during aerobic respiration
Cu	Lysyl oxidase	Formation of desmosine cross links in collagen and elastin
Cu	Ceruloplasmin	Iron absorption and transport for haemoglobin synthesis
Cu	Tyrosinase	Melanin production
Cu	Superoxide dismutase	Antioxidant in cells, play role in phagocytic cell Function
Zn	Lactic dehydrogenase	Plays role in glycolytic metabolism
Zn	Carbonic anhydrase	Assists rapid interconversion of carbon dioxide and water into carbonic acid, protons and bicarbonate ions
Zn	Copper zinc superoxide dismutase	Antioxidant in cells
Se	glutathione peroxidase	Antioxidant in cells
	Heme oxidase	Function as antioxidant in protection of biological membranes
Mo	Xanthine oxidase	Generates reactive oxygen species and plays an important role in the catabolism of purines

Source: Modified from National Research Council (2001)

when stimulated with Con A than cells of cows fed on adequate levels of Cu (Torre et al. 1995).

Copper similar to zinc plays a vital role in immune functions of all mammals. The Cu is an integral part of enzymatic activity that affects efficiency complex. It is also involved in proper functioning of immune cells (macrophages and neutrophils). A deficiency of copper has been associated with a decreased ability of these cells to multiply. Failure to multiply quickly may result in a competitive advantage by mastitis-causing bacteria. Diets with 20 ppm supplemental copper have been shown to reduce the severity of mastitis following an *E. coli* challenge compared to diets with 7 ppm (Scaletti et al. 2001). Heifers that received no supplemental copper after weaning and then were fed a diet with no supplemental copper from 84 days prepartum to 108 days postpartum had more infected quarters during lactation than did animals fed 20 ppm supplemental copper from 84 d pre- to 107 d postpartum (Harmon and Torre 1994).

7.2.4 Iron (Fe)

Iron in ferrous form (Fe^{++}) performs a vital role in many biochemical reactions, including (1) antioxidant defence as an essential component of catalase, (2) energy and protein metabolism, (3) as a haem respiratory carrier, (4) oxidation–reduction reactions and (5) in the electron transport system. Reduced Fe is also a catalyst for

lipid peroxidation and radical formation, thus having a strong pro-oxidant effect (Halliwell 1987). The Fe plays role in the immune response, and an insufficient supply causes anaemia in deficient animals due to failure to produce haemoglobin. Fe deficiency is not common in adult cattle as their requirement is low and Fe is ubiquitous in the environment, but it is more frequent in calves as milk Fe content is low (Underwood 1999). In dairy cattle, plasma Fe concentration is decreased during the acute phase response to immunological challenges as are Zn concentrations (Kushner 1982), whereas plasma Cu concentration may increase (Andrieu 2008). These ion changes reflect changes in cation binding of plasma proteins and, more importantly, alterations in cellular uptake mechanisms. During mastitis, increased secretion of binding proteins such as lactoferrin in milk decreases the amount of available Fe and thus reduces the availability of the divalent Fe for growth of Gram-negative bacteria (Todhunter et al. 1990).

7.2.5 Manganese (Mn)

Mn plays an important role in body metabolism as an essential part of a range of enzymes that are involved in (1) antioxidant protection as an integral part of enzyme SOD, (2) bone growth and formation of eggshell, (3) carbohydrate and lipid metabolism, (4) immune and nervous function and (5) reproduction. Like zinc and copper, manganese is involved in detoxifying superoxide radicals (free oxygen radicals) produced by immune cells in response to kill bacteria. Superoxide radicals disrupt cellular membranes and cause cellular damage leaving the mammary gland more susceptible to infection, scarring and loss in milk production. Manganese also enhances macrophage (white blood cell) killing ability. Increasing copper and manganese status of the lactating cow can help reduce SCC as copper and manganese play roles in immune response.

7.2.6 Chromium

Chromium has both humoral and cellular immunomodulatory effects, but the fundamental mechanism of intercellular and intracellular action are still to be fully elucidated. Cr requirements in human and farm animals increase during stress, the heat stress and early lactation could excrete Cr irreversibly through urine (Borel et al. 1984). Heat stress decreases feed intake and adding Cr to the diet help relieve this effect (AL-Saiady et al. 2004; Hayirli et al. 2001). Cows fed the diet with Cr had the highest peak milk production over the control. The increased milk yield could also occur due to higher dry matter intake and reduced rate of mobilisation of fatty acids from adipose tissue (McNamara and Valdez 2005). Cr deficiency can cause insulin resistance, and the ratio of glucose to insulin can act as a crude index of tissue sensitivity to insulin (Evock-Clover et al. 1993). Cows receiving Cr had higher molar ratio of glucose to insulin and lower insulin than cows receiving no Cr indicative of increased insulin sensitivity due to Cr supplementation. Increased insulin sensitivity is likely to stimulate lipogenesis and inhibit lipolysis. A reduced mobilisation of fatty acids from adipose tissue may allow a greater increase in feed intake, stabilise hepatic fat metabolism and reduce hepatic ketogenesis, to increase milk production in cows (Kronfeld 1976).

The immune function may be affected by low Cr levels in association with insulin and/or cortisol activity as corticosteroids suppress immune system. The chromium also mediates production and regulation of certain cytokines (Borgs and Mallard 1998). High-producing dairy cows are

under metabolic and physiological stress during early lactation. State of negative energy balance may increase concentration of nonesterified fatty acids (NEFA) and β-hydroxyl butyric acid (BHBA) in blood which result in ketosis and other metabolic disorders. The supplementation of Cr is beneficial during the early lactation and immediately after parturition in cows. During the periparturient period, insulin resistance may be an important factor in the initiation of catabolic activities (Holstenius 1993). Improved glucose tolerance and milk yield and decreased blood cortisol, NEFA and BHBA were observed in primiparous cows supplemented with 0.5 ppm of organic chromium (Subiyatno et al. 1996). Supplementation of chromium has been observed to reduce blood cortisol concentrations and increase measures of immunological activity in transition dairy cows (Burton et al. 1993; Chang et al. 1996).

8   Effect of Free Radical Production on Immunity and Role of Antioxidants

An increase in oxidative reactions within the cell, or at the cell membrane, produces free radicals or activated molecules with the potential to inhibit cellular functions, damage membranes and even to result in the destruction of the cell. Oxidation–reduction reactions do normally occur in the body, but when the reactions become uncontrolled, the end products of oxidation (i.e. free radicals) accumulate, and tissue damage occurs. In an effort to protect itself, the cell prevents the accumulation of free radicals by the action of several antioxidants present in the body (Table 3).

Various factors facilitate the accumulation of free radicals, but heat stress increases both the metabolic rate of cells and free radicals accumulation (Bernabucci et al. 2002; Lohrke et al. 2005; Chandra 2009). If the levels of antioxidants present are low/suboptimal within the cell, the damage due to accumulation of free radicals is likely to occur. Cows that are under stress of calving, challenged by a high load or experiencing the peak demands of lactation, have a higher level of

Table 3 Antioxidants present in body tissues

Enzymatic	Nutrients involved	Nonenzymatic	Nutrients involved
Superoxide dismutase (cytosol)	Copper and zinc	Ascorbic acid	Vitamin C
Superoxide dismutase (mitochondria)	Manganese	Beta carotene (membranes)	Beta carotene
Catalase (cytosol)	Iron	Ceruloplasmin	Copper
Glutathione peroxidase	Selenium	Uric acid	
Glutathione reductase		Bilirubin	
		Melatonin	
		Isoflavone	
		Methionine	
		α-Tocopherol	Vitamin E

Source: Markesbery et al. (2001), Weiss (2009)

Table 4 Suggested feeding levels of vitamin E and some minerals in total diet

Vitamin E	1,000 IU/day for dry cows
	1,500 IU for dry buffaloes
	500 IU/day for lactating cows
Selenium	0.3 ppm
Copper	20 ppm
Zinc	60–80 ppm

Source: Modified from Scaletti et al. (1999)

free radicals (Bernabucci et al. 2002) and therefore require a greater supplementation of antioxidants. Table 4 shows the suggested feeding levels of vitamin E and minerals for cows.

As has been indicated earlier, free radicals are formed as a normal end product of cellular metabolism arising from either the mitochondrial electron transport chain or from stimulation of NADPH (Valko et al. 2007). The presence of free radicals leading to oxidative reactions in the organism is physiological, and oxidative stress occurs when there is increased production of free radicals and reactive oxygen species, and/or a decrease in antioxidant defence system. Oxidative stress results in damage of biological macromolecules and disruption of normal metabolism and physiology (Trevisan et al. 2001). Heat stress generally increases the production of free radicals that lead to oxidative stress. Under normal physiological conditions, antioxidant defence systems within the body can effectively neutralise the ROS that are produced and eliminate them. Supplementation of antioxidants to the cows before heat stress starts and also during the stress

period may correct the infertility due to heat stress through decreased cortisol secretion and oxidative stress, resulting in enhanced pregnancy rates. Moreover, strong positive correlations between several antioxidant enzymes (e.g. glutathione peroxidase) and vascular adhesion molecules suggest a protective response of antioxidants to an enhanced proinflammatory state in transition dairy cows (Aitken et al. 2009). Antioxidants then could contribute to enhance mechanisms against oxidative stress with various immunity, reproduction and health benefits. Heat stress is associated with reduced activity by antioxidants in the blood plasma. Vitamin E (α-tocopherol), a strong reducing agent that can give electrons to lipids undergoing peroxidation, is a major antioxidant present in plasma membranes (Wang and Quinn 2000).

The treatment of cows with antioxidants to improve fertility in summer has given inconsistent results. Effects of antioxidants on reproductive function may be more pronounced during heat stress because of the increased metabolic rates associated with cellular hyperthermia. High temperature increases liver peroxidation (Ando et al. 1997), and activity of enzymes involved in free radical production such as xanthine oxidase is also increased. Exposure of dairy cows to heat stress decreased total antioxidant activity in blood. Like most cells, preimplantation embryos can produce free radicals (Yang et al. 1998). TrxR1 also can facilitate the gene expression of other cytoprotective antioxidant enzyme factors, such as heme oxygenase in bovine endothelial cells (Trigona et al. 2006). The nonenzymatic antioxidants are tocopherols, ascorbic acid, carotenoids, lipoic acid and GSH. Vitamin E (α-tocopherol acetate) is the predominant antioxidant found in biological membranes. The tocopherols disrupt radical chain reactions that lead to auto-oxidation of adjacent membrane-associated fatty acids. For example, vitamin E can act as a scavenger of both lipid radicals and lipid peroxy radical by donating a hydrogen ion with the formation of a tocopheroxyl radical. The tocopheroxyl radical is then regenerated back to its reduced form by vitamin C (Blokhina et al. 2003). Ascorbic acid (vitamin C) is a water-soluble antioxidant that plays a key role in maintaining the redox state of cells. In addition to recycling vitamin E, ascorbic acid can reduce several other oxidised biomolecules and act as a direct scavenger of free radicals (Blokhina et al. 2003). The carotenoids are another important free radical scavenger and are especially effective at quenching singlet oxygen and can prevent the subsequent formation of secondary ROS. Lipoic acid is a component of the pyruvate dehydrogenase complex and plays a central role in energy metabolism. However, lipoic acid also can function as a metal chelator and ROS scavenger. The reduced form of lipoic acid, dihydrolipoic acid, can further prevent accumulation of ROS by recycling vitamins C and E (Blokhina et al. 2003).

9 Conclusions

Many factors environmental or physiological can influence the immune response of an animal to stress. Stress can suppress or enhance or have a balanced effect on the immune functions of an animal. Many of the conflicting findings reported may be partially explained by the types and durations of the stressors, age, genetics and social status. Moreover, the aspect of the immune system being assessed, the starting point of the immune system and the balance between Th1 and Th2 may contribute to these discrepancies between studies. If the immune system is in a predominantly Th2 state, the animal may have enhanced protection against pathogen to which it

has been previously exposed. However, if the immune system is skewed towards Th2, then viral and early pathogen (innate) immunity would be suppressed, and the animals would be more likely to have an allergic or autoimmune disease. In India, animals are mainly reared on straw-based diets which are low in major nutrients, antioxidant vitamins and trace minerals. On these diets, the cows are under stress especially during periparturient period, this may lead to immunosuppression and can be afflicted with various diseases. Considering these facts, supplementation of antioxidants (combination of micronutrients, i.e. copper/zinc and vitamin E) to the high producers, especially during periparturient period may prove to be beneficial. These antioxidants may improve productivity of cows by reducing the stress, thus protecting against the chances of subclinical/clinical mastitis. A better understanding of the complexity of these relationships in farm animals will improve the animal health and well-being.

References

Abdel-Samee AM (1987) The role of cortisol in improving productivity of heat-stressed farm animals with different techniques. PhD thesis, Faculty of Agriculture, Zagazig University, Zagazig

Aggarwal A, Ashutosh, Chandra G, Singh AK (2012) Heat shock protein 70, oxidative stress and antioxidant status in periparturient crossbred cows supplemented with α-tocopherol acetate. Trop Anim Health Prod. doi:10.1007/s11250-012-0196-z

Aitken SL, Karcher EL, Rezemand P (2009) Evaluation of antioxidant and proinflammatory gene expression in bovine mammary tissue during the periparturient period. J Dairy Sci 92:589–598

AL-Saiady MY, AL-Saiskh MA, AL-Mufarrej SI, Al-Showeimi TA, Mogawer HH, Dirrara A (2004) Effect of chelated chromium supplementation on lactation performance and blood parameters of Holstein cows under heat stress. Anim Feed Sci Technol 117:223–233

Amaral BC, Connor EE, Tao S, Hayen J, Bubolz J, Dahl GE (2010) Heat stress abatement during the dry period influences prolactin signaling in lymphocytes. Domest Anim Endocrinol 38:38–45

Ando M, Katagiri K, Yamamoto S, Wakamatsu K, Kawahara I, Asanuma S, Usuda M, Sasaki K (1997) Age-related effects of heat stress on protective enzymes for peroxides and microsomal monooxygenase in rat liver. Environ Heal Perspect 105:726–733

Andrieu S (2008) Is there a role for organic trace element supplements in transition cow health? Vet J 176: 77–83

Baldwin CL, Sathiyaseelan T, Naiman B, White AM, Brown R, Blumerman S, Rogers A, Black SJ (2002) Activation of bovine peripheral blood γδ T cells for cell division and IFN-γ production. Vet Immunol Immunopathol 87:251–259

Bernabucci U, Ronchi B, Lacetera N, Nardone A (2002) Markers of oxidative status in plasma and erythrocytes of transition dairy cows during hot season. J Dairy Sci 85:2173–2179

Berry MJ, Kedffer JD, Harney JW, Larsen PR (1991) Selenocysteine confers the biochemical properties characteristic of the type I iodothyronine deiodinase. J Biol Chem 266:14155–14158

Blecha F, Baker PE (1986) Effect of cortisol in vitro and in vivo on production of bovine interleukin 2. Am J Vet Res 47:841–845

Blokhina O, Virolainen E, Fagerstedt KV (2003) Antioxidants, oxidative damage and oxygen deprivation stress: a review. Ann Bot (Lond) 91:179–194

Blotta MH, DeKruyff RH, Umestsu DT (1997) Corticosteroids inhibit IL-12 production in human monocytes and enhance their capacity to induce IL-4 synthesis in CD4+ lymphocytes. J Immunol 158:5589–5595

Bluestone JA, Khattri R, Sciammas R, Sperling AI (1995) TCRγδ cells: A specialized T-cell subset in the immune system. Annu Rev Cell Dev Biol 11:307–353

Bonnette ED, Kornegay ET, Lindemann MD, Hammerberg C (1990) Humoral and cell-mediated immune response and performance of weaned pigs fed four supplemental vitamin E levels and housed at two nursery temperatures. J Anim Sci 68:1337–1345

Borel JS, Majerus TC, Polansky MM (1984) Chromium intake and urinary chromium excretion of trauma patients. Biol Trace Elem Res 6:317–326

Borgs P, Mallard BA (1998) Immune-endocrine interactions in agricultural species: chromium and its effect on health and performance. Domest Anim Endocrinol 15:431–438

Bot A, Smith KA, Von Herrath M (2004) Molecular and cellular control of T1/T2 immunity at the interface between antimicrobial defense and immune pathology. DNA Cell Biol 23:341–350

Burdick NC, Carroll JA, Hulbert LE, Dailey JW, Ballou MA, Randel RD, Willard ST, Rhonda CV, Welsh TH Jr (2011) Temperament influences endotoxin-induced changes in rectal temperature, sickness behavior, and plasma epinephrine concentrations in bulls. Innate Immun 4:355–364

Burton JL, Mallard BA, Mowat DN (1993) Effects of supplemental chromium on immune responses of periparturient and early lactation dairy cows. J Anim Sci 71:1532–1539

Calder PC, Kew S (2002) The immune system: a target for functional foods? Br J Nutr 88:165–176

Carr AC, Frei B (1999) Does vitamin C act as pro-oxidant under physiological conditions? FASEB J 13: 1007–1024

Carroll JA, Forsberg NE (2007) Influence of stress and nutrition on cattle immunity. Vet Clin Food Anim 23:105–149

Chandra G (2009) Antioxidative status of high body condition periparturient crossbred cows with and without supplementation of α-Tocopherol acetate during summer and winter seasons. M.V.Sc. thesis, NDRI (deemed University), Karnal

Chandra G, Aggarwal A (2009) Effect of DL-α-tocopherol acetate on calving induced oxidative stress in periparturient crossbred cows during summer and winter seasons. Indian J Anim Nutr 26(3):204–210

Chandra G, Aggarwal A (2010) Effect of DL- α-tocopherol acetate on calving induced oxidative stress in periparturient crossbred cows during summer and winter seasons. Indian J Anim Nutr 26:204–210

Chandra S, Chandra RK (1986) Nutrition, immune response and outcome. Prog Food Nutr Sci 10:1–65

Chang X, Mallard BA, Mowat DN (1996) Effects of chromium on health status, blood neutrophil phagocytosis and in vitro lymphocyte blastogenesis of dairy cows. Vet Immunol Immunopathol 52:37–52

Chaudiere J, Tappel AL (1984) Interaction of gold (I) with the active site of selenium-glutathione peroxidase. J Inorg Biochem 20:313–325

Chawla R, Kaur H (2005) Effect of supplemental vitamin E and β-carotene on cell-mediated immunity and mastitis in crossbred cows. Anim Nutr Feed Technol 5:73–84

Chew BP, Park JS (2004) Carotenoid action on the immune response. J Nutr 134:257–261

Coffman RL (2006) Origins of the Th1-Th2 model: a person perspective. Nat Immunol 7:539–541

Constant SL, Bottomly K (1997) Induction of Th1 and Th2 CD4+ T cell response: the alternative approach. Annu Rev Immunol 15:297–322

Cook NB, Bennett TB, Emery KM, Nordlund KV (2002) Monitoring nonlactating cow intramammary infection dynamics using DHI somatic cell count data. J Dairy Sci 85:1119–1126

Cusack PMV, McMeniman NP, Lean IJ (2005) The physiological and production effects of increased dietary intake of vitamins E and C in feedlot cattle challenged with bovine herpesvirus 1. J Anim Sci 83:2423–2433

Dahlquist SP, Chew BP (1985) Effects of vitamin A and B-carotene on mastitis in dairy cows during the early dry period. J Dairy Sci 68:191

Dalpke K, Heeg H, Bartz Baetz A (2008) Regulation of innate immunity by suppressors of cytokine signaling (SOCS) proteins. Immunobiology 213:225–235

de Jong EC, Vierira PL, Kalinski P, Kapsenberg ML (1999) Corticosteroids inhibit the production of inflammatory mediators in immature monocyte-derived DC and induce development of tolerogenic DC3. J Leukoc Biol 66:201–204

DeKruyff RH, Yang Y, Umetsu DT (1998) Corticosteroids enhance the capacity of macrophages to induce TH2 cytokine synthesis in CD4+ lymphocytes by inhibiting IL-12 production. J Immunol 160:2231–2237

Devaraj C, Upadhyay RC (2007) Effect of catecholamines and thermal exposure on lymphocyte proliferation, IL-1α & β in buffaloes. Ital J Anim Sci 6(suppl 2): 1336–1339

Dhabhar FS, McEwen BS (2006) Bidirectional effects of stress on immune function: possible explanations for salubrious as well as harmful effects. In: Ader R, editor. Psychoneuroimmunology. 4th ed. San Diego (CA): Elsevier

Eicher SD, Morrill JL, Blecha F, Chitko-McKown CG, Anderson NV, Higgins JJ (1994) Leukocyte functions of young dairy calves fed milk replacers supplemented with vitamins A and E. J Dairy Sci 77:1399–1407

Eitam H, Brosh A, Orlov A, Izhaki I, Shabtay A (2009) Caloric stress alters fat characteristics and Hsp70 expression in somatic cells of lactating beef cows. Cell Stress Chaperones 14:173–182

Elenkov IJ (2002) Systemic stress-induced Th2 shift and its clinical implications. Int Rev Neurobiol 52:163–186

Elenkov IJ, Chrousos GP (2002) Stress hormones, pro-inflammatory, and anti-inflammatory cytokines and autoimmunity. Ann N Y Acad Sci 96:290–303

Elenkov IJ, Papanicolaou DA, Wilder RL, Chrousos GP (1996) Modulatory effects of glucocorticoids and catecholamines on human interleukin-12 and interleukin-10 production: clinical implications. Proc Assoc Am Physicians 108:374–381

Elenkov IJ, Chrousos GP, Wilder RL (2000) Neuroendocrine regulation of IL-12 and TNF-α/IL-10 balance. Ann N Y Acad Sci 917:94–105

Elvinger F, Hansen PJ, Natzke RP (1991) Modulation of function of bovine polimorphonuclear leukocytes and lymphocytes by high temperature in vitro and in vivo. Am J Vet Res 52:1692–1698

Enjalbert F, Lebreton P, Salat O (2006) Effects of copper, zinc and selenium status on performance and health in commercial dairy and beef herds, retrospective study. J Anim Physiol Anim Nutr 90:459–466

Erickson KL, Medina EA, Hubbard NE (2000) Micronutrients and innate immunity. J Infect Dis 182 (Suppl 1):5–10

Evock-Clover CM, Polansky MM, Anderson RA (1993) Dietary chromium supplementation with or without somatotropin treatment alters serum hormones and metabolites in growing pigs without affecting growth performance. J Nutr 123:1504–1512

Frankenburg S, Wang X, Milner Y (1998) Vitamin A inhibits cytokines produced by type 1 lymphocytes in vitro. Cell Immunol 185:75–81

Fulford AJ, Harbuz MS (2005) An introduction to the HPA. In: Steckler T, Kalin NH, Reul JMHM (eds) Handbook of stress and the brain. Elsevier, Amsterdam, pp 43–66

Grignard E, Morin J, Vernet P, Drevet JR (2005) GPX5 orthologs of the mouse epididymis-restricted and sperm-bound selenium-independent glutathione peroxidase are not expressed with the same quantitative and spatial characteristics in large domestic animals. Theriogenology 64:1016–1033

Grimble RF (2001) Symposium on evidence-based nutrition. Nutritional modulation of immune function. Proc Nutr Soc 60:389–397. doi:10.1079/PNS2001102

Habeeb AAM (1987) The role of insulin in improving productivity of heat stressed farm animals with different techniques. Ph.D. Thesis, Faculty of Agriculture, Zagazig University, Zagazig, Egypt

Hahn GL, Mader TL, Harrington JA, Nienaber JA, Frank KL (2002) Living with climatic variability and potential global change: climatological analyses of impacts on livestock performance. In: Proceedings of the 16th international congress on biometeorology, Kansas City, pp 45–49

Halliwell B (1987) Oxidants and human disease: some new concepts. Fed Am Soc Biol J 1:398

Hara S (2001) Effects of selenium deficiency on expression of selenoproteins in bovine arterial endothelial cells. Biol Pharm Bull 24:754–759

Harmon RJ, Torre PM (1994) Copper and zinc: do they influence mastitis? In: Proceedings of National Mastitis Council, Orlando, pp 54–56

Hasselmann L, Munchow H, Menzke V, Schneewei TR, Ahrens F (2000) Influence of β-carotene supplementation on cell count of raw milk as well as on reproductive performance of dairy cows under practical conditions. Nutr Abstr Rev (B) 70:708

Hayirli A, Bremmer DR, Bertics SJ, Socha MT, Grummer RR (2001) Effect of chromium supplementation on production and metabolic parameters in periparturient dairy cows. J Dairy Sci 84:1218–1230

Hein WR, Mackay CR (1991) Prominence of gamma delta T cells in ruminant immune system. Immunol Today 12:30–34

Hicks TA, McGlone JJ, Whisnant CS, Kattesh HG, Norman RL (1998) Behavioral, endocrine, immune and performance measures for pigs exposed to acute stress. J Anim Sci 76:474–483

Hogan JS, Smith KL, Weiss WP, Todhunter DA, Schockey WL (1990) Relationship among vitamin E, selenium and bovine blood neutrophils. J Dairy Sci 73:2372–2378

Holstenius P (1993) Hormonal regulation related to the development of fatty liver and ketosis. Acta Vet Scand 89:55–60

Hornef MW, Frisan T, Vandewalle A, Normark S, Richter-Dahlfors A (2002) Toll-like receptor 4 resides in the Golgi apparatus and colocalizes with internalized lipopolysaccharide in intestinal epithelial cells. J Exp Med 195:559–570

Horst RD (1986) Regulation of calcium and phosphorous homeostasis in dairy cows. J Dairy Sci 69:604–616

Irwin M, Vale W, Rivier C (1990) Central corticotropin-releasing factor mediates the suppressive effect of stress on natural killer cell cytotoxicity. Endocrinology 126:2837–2844

Ishizaki H, Kariya Y (1999) Effects of peripheral blood polymorphonuclear leukocyte function and blood components in Japanese Black steers administered ACTH in a cold environment. J Vet Med Sci 61:487–492

Ishizaki H, Hanafusa Y, Kariya Y (2005) Influence of truck-transportation on the function of bronchoalveolar lavage fluid cells in cattle. Vet Immunol Immunopathol 105:67–74

Ismaili J, Olislagers V, Poupot R, Fournié JJ, Goldman M (2002) Human γδ T cells induce dendritic cell maturation. Clin Immunol 103:296–302

Janeway CA, Medzhitov R (2002) Innate immune recognition. Annu Rev Immunol 20:197–216

Janeway CA, Travers P, Walport M, Shlomchik M (2001) Basic concepts in immunology. In: Janeway CA, Travers P, Walport M, Shlomchik M (eds) Immunobiology: the immune system in health and disease. Garland Publishing, New York, pp 1–34

Johnson RW, Von Borell EH, Anderson LL, Kojic LD, Cunnick JE (1994) Intracerebroventricular injection of corticotrophin-releasing hormone in the pig: acute effects on behavior, adrenocorticotropin secretion and immune suppression. Endocrinology 135:642–648

Kamwanja LA, Chase CC, Gutierrez JA, Guerriero V, Olson TA, Hammond AC, Hansen PJ (1994) Responses of bovine lymphocytes to heat shock as modified by breed and antioxidant status. J Anim Sci 72:438–444

Kehri ME Jr, Burton JL, Nonnecke BJ, Lee EK (1999) Effects of stress on leukocytes trafficking and immune responses: implications for vaccination. Adv Vet Med 41:61–81

Kelley KW (1982) Immunobiology of domestic animals as affected by hot and cold weather. In: Proceedings of the second international livestock environment symposium. ASAE Publication No. 3–82, St. Joseph, Michigan, pp 470–483

Kelley KW, Osborne CA, Evermann JF, Parish SM, Gaskins CT (1982a) Effects of chronic heat and cold stressors on plasma immunoglobulin and mitogen-induced blastogenesis in calves. J Dairy Sci 65:1514–1528

Kelley KW, Greenfield RE, Evermann JF, Parish SM, Perryman LE (1982b) Delayed-type hypersensitivity, contact sensitivity, and phytohemagglutinin skin-test responses of heat- and cold-stressed calves. Am J Vet Res 43:775–779

Kidd P (2003) Th1/Th2 balance: the hypothesis, its limitations, and implications for health and disease. Altern Med Rev 8:223–246

Kinjyo I, Hanada T, Inagaki-Ohara K, Mori H, Aki D, Ohishi M, Yoshida H, Kubo M, Yoshimura A (2002) SOCS1/JAB is a negative regulator of LPS-induced macrophage activation. Immunity 17:583–591

Kronfeld DS (1976) The potential importance of the proportions of glucogenic, lipogenic and aminogenic nutrients in regard to the health and productivity of dairy cows. Adv Anim Physiol Anim Nutr 7:5–10

Kruse V (1970) Yield of colostrum and immunoglobulin in cattle at the first milking after parturition. Anim Prod 12:619–626

Kruse-Jarres JD (1989) The significance of zinc for humoral and cellular immunity. J Trace Elem Electrol Health Dis 3:1–8

Kushner I (1982) The phenomenon of the acute phase response. Ann N Y Acad Sci 389:39–48

Lacetera N, Bernabucci U, Ronchi B, Scalia D, Nardone A (2002) Moderate summer heat stress does not modify immunological parameters of Holstein dairy cows. Int J Biometeorol 46:33–37

Lacetera N, Bernabucci U, Scalia D, Ronchi B, Kuzminsky G, Nardone A (2005) Lymphocyte functions in dairy cows in hot environment. Int J Biometeorol 50:105–110

Levine S (2005) Stress: an historical perspective. In: Steckler T, Kalin NH, Reul JMHM (eds) Handbook of stress and the brain. Elsevier, Amsterdam, pp 3–23

Lohrke B, Viergutz T, Kanitz W, Losand B, Weiss DG, Simko M (2005) Hydroperoxides in circulating lipids from dairy cows, implications for bioactivity of endogenous-oxidized lipids. J Dairy Sci 88:1708–1710

Maddox JF, Aherne KM, Reddy CC, Sordillo LM (1999) Increased neutrophil adherence and adhesion molecule mRNA expression in endothelial cells during selenium deficiency. J Leukoc Biol 65:658

Mallard BA, Dekkers JC, Ireland MJ, Leslie KE, Sharif S, Lacey Vankampen C, Wagter L, Wilkie BN (1998) Alteration in immune responsiveness during the periparturient period and its ramification on dairy cows and calf health. J Dairy Sci 81:585–595

Marai IFM, Habeeb AAM, Daader AH, Yousef HM (1995) Effect of Egyptian subtropical conditions and the heat stress alleviation techniques of water spray and diaphoretics on the growth and physiological functions of Friesian calves. J Arid Environ 30:219–225

Marai IFM, Daader AM, Abdel-Samee AM, Ibrahim H (1997a) Winter and summer effects and their amelioration on lactating Friesian and Holstein cows maintained under Egyptian conditions. In: Proceedings of the international conference on animal, poultry, rabbits and fish production and health, Cairo

Marai IFM, Daader AM, Abdel-Samee AM, Ibrahim H (1997b) Lactating Friesian and Holstein cows as affected by heat stress and combination of amelioration techniques under Egyptian conditions. In: Proceedings of the international conference on animal, poultry, rabbits and fish production and health. Cairo

Markesbery WR, Montine TJ, Lovell M (2001) Oxidative alterations in neurodegenerative diseases. In: Mattson MP (ed) Pathogenesis disorders. Humana Press, Totowa

Maurya P (2011) Leptin level in relation to immunity, energy metabolites and cellular adaptations during dry period and early lactation in crossbred cows. M.V.Sc. thesis, NDRI (deemed University), Karnal

McDowell LR (2003) Minerals in animal and human nutrition. Academic, San Diego

McGlone JJ, Lumpkin EA, Norman RL (1991) Adrenocorticotropin stimulates natural killer cell activity. Endocrinology 129:1653–1658

McGlone JJ, Salak JL, Lumpkin EA, Nicholson RI, Gibson M, Norman RL (1993) Shipping stress and social status effects on pig performance, plasma cortisol, natural killer cell activity, and leukocyte numbers. J Anim Sci 71:888–896

McNamara JP, Valdez F (2005) Adipose tissue metabolism and production responses to calcium propionate and chromium propionate. J Dairy Sci 88:2498–2507

Medzhitov R, Preston-Hulburt P, Janeway CA Jr (1997) A human homologue of the drosophila toll protein signals activation of adaptive immunity. Nature 388: 394–397

Michal JJ, Chew B, Wong TS, Heirman LR, Standaert FE (1994) Modulatory effects of dietary β-carotene on blood and mammary leukocyte function in peripartum dairy cows. J Dairy Sci 77:1408–1422

Minton JE (1994) Function of the hypothalamic-pituitary axis and the sympathetic nervous system in models of acute stress in domestic farm animals. J Anim Sci 72:1891–1898

Mishra OK, Hooda Singh G, Meur SK (2011) Influence of induced heat stress on HSP70 in buffalo lymphocytes. J Anim Physiol Anim Nutr 95:540–544

Morrow-Tesch JL, McGlone JJ, Salak-Johnson JL (1994) Heat and social stress effects on pig immune measures. J Anim Sci 72:2599–2609

Mosmann TR, Coffman RL (1989) Th1 and Th2 cells: different patterns of lymphokine secretion lead to different functional properties. Annu Rev Immunol 7:145–173

Mosmann TR, Sad S (1996) The expanding universe of T-cell subsets: Th1, Th2 and more. Immunol Today 17:138–146

Nardone A, Lacetera N, Bernabucci U, Ronchi B (1997) Com- position of colostrum from dairy heifers exposed to high air temperatures during late pregnancy and the early postpartum period. J Dairy Sci 80:838–844

National Research Council (NRC) (2001) Nutrient requirements of dairy cattle. Natl. Acad. Press, Washington, DC

Norrman J, David CW, Sauter SN, Hammon HM, Blum JW (2003) Effects of dexamethasone on lymphoid tissue in the gut and thymus of neonatal calves fed with colostrum or milk replacer. J Anim Sci 81:2322–2332

Panda N, Kaur H (2007) Influence of vitamin E supplementation on spontaneous oxidized flavour of milk in dairy buffaloes. Int J Dairy Technol 60:198–204

Paterson JE, MacPherson A (1990) The influence of low cobalt intake on the neutrophil function and severity of Ostertagia infection in cattle. Br Vet J 146:519–530

Politis I, Hidiroglou N, Batra TR, Gilmore JA, Gorewit RC, Scherf H (1995) Effects of vitamin E on immune function of dairy cows. Am J Vet Res 56:179–184

Pollock JM, Welsh MD (2002) The WC1+ γδ T-cell population in cattle: a possible role in resistance to intracellular infection. Vet Immunol Immunopathol 89:105–114

Reddy PG, Frey RA (1990) Nutritional modulation of immunity in domestic food animals. Adv Vet Sci Comp Med 35:255–281

Reffett JK, Spears JW, Brown TT (1988) Effect of dietary selenium on the primary and secondary immune response in calves challenged with infectious bovine rhinotracheitis virus. J Nutr 118:229–235

Richards DF, Fernandez M, Caulfield J, Hawrylowicz CM (2001) Glucocorticoids drive human CD8(+) T cell differentiation towards a phenotype with high IL-10 and reduced IL-4, IL-5 and IL-13 production. Eur J Immunol 30:2344–2354

Salak JL, McGlone JJ, Lyte M (1993) Effects of in vitro adrenocorticotrophic hormone, cortisol and human recombinant interleukin-2 on porcine neutrophils migration and luminol-dependent chemiluminescence. Vet Immunol Immunopathol 39:327–337

Salak-Johnson JL, McGlone JJ (2007) Making sense of apparently conflicting data: stress and immunity in swine and cattle. J Anim Sci 85:81–88

Salak-Johnson JL, McGlone JJ, Norman RL (1996) In vivo glucocorticoid effects on porcine natural killer cell activity and circulating leukocytes. J Anim Sci 74:584–592

Salak-Johnson JL, McGlone JJ, Whisnant CS, Norman RL, Kraeling RR (1997) Intracerebroventricular porcine corticotrophin-releasing hormone and cortisol effects on pig immune measures and behavior. Physiol Behav 61:15–23

Salem A (1980) Seasonal variations in some body reactions and blood constituents in lactating buffaloes and Friesian cows. J Egypt Vet Med Assoc 40:63

Sapolsky RM, Romero LM, Munck AU (2001) How do glucocorticoids influence stress responses? Integrating permissive, suppressive, stimulatory, and preparative actions. Endocr Rev 21:55–89

Sasaki A, Yasukawa H, Suzuki A, Kamizono S, Syoda T, Kinjyo I, Sasaki M, Johnston JA, Yoshimura A (1999) Cytokine-inducible SH2 protein-3 (CIS3/SOCS3) inhibits Janus tyrosine kinase by binding through the N-terminal kinase inhibitory region as well as SH2 domain. Genes Cells 4:339–351

Scaletti RW, Amaral-Phillips DM, Harmon RJ (1999) Using nutrition to improve immunity against disease in dairy cattle: Copper, Zinc, Selenium, and Vitamin E http://www.ca.uky.edu. Issued 4

Scaletti RW, Trammell DS, Smith BA, Harmon RJ (2001) Role of dietary copper in altering response to intramammary E. coli challenge. Pages 29–33 in Proc. Int. Mastitis Symp, Natl Mastitis Counc/Am Assoc Bovine Practitioners, Vancouver, BC, Canada

Semba RD (1994) Vitamin A, immunity and infection. Clin Infect Dis 19:489–499

Shafferi I, Roussel JD, Koonce KX (1981) Effects of age, temperature, season and breed on blood characteristics of dairy cattle. J Dairy Sci 64:63–68

Shearer J, Mohammed HO, Brenneman JS, Tran TQ (1992) Factors associated with concentrations of immunoglobulins in colostrums at the first milking post calving. Prev Vet Med 14:143–154

Shebaita MK, Kamal TH (1973) In vivo body composition in ruminants. I. Blood volume in Friesian and water buffaloes. Alex J Agric Res 1:329–350

Skjolaas KA, Minton JE (2002) Does cortisol bias cytokine production in cultured splenocytes to a Th2 phenotype? Vet Immunol Immunopathol 87:451–458

Skjolaas KA, Grieger DM, Hill CM, Minton JE (2002) Glucocorticoid regulation of type 1 and type 2 cytokines in cultured porcine splenocytes. Vet Immunol Immunopathol 87:79–87

Smith KL, Todhunter DA, Schoenberger PS (1985) Environmental mastitis: cause, prevalence. Prev J Dairy Sci 68:1531

Soper F, Muscoplat CC, Johnson DW (1978) In vitro stimulation of bovine peripheral blood lymphocytes: analysis of variation of lymphocyte blastogenic response in normal dairy cattle. Am J Vet Res 39:1039–1042

Sordillo LM, Streicher KL (2002) Mammary gland immunity and mastitis susceptibility. J Mammary Gland Biol Neoplasia 7:135–146

Sordillo LM, Shafer-Weaver K, DeRosa D (1997) Immunobiology of the mammary gland. J Dairy Sci 80:1851–1865

Spears JW (2000) Micronutrients and immune function in cattle. Proc Nutr Soc 59:587–594

Starr R, Willson TA, Viney EM, Murray LJL, Rayner JR, Jenkin BJ, Gonda TJ, Alexander WS, Metcalf D, Nicola NA, Hilton DJ (1997) A family of cytokine-inducible inhibitors of signaling. Nature 387:917–921

Stott G (1980) Immunoglobulin absorption in calf neonates with special considerations of stress. J Dairy Sci 63:681–688

Subiyatno A, Mowat DN, Yang WZ (1996) Metabolite and hormonal responses to glucose or propionate infusions in periparturient dairy cows supplemented with chromium. J Dairy Sci 79:1436–1445

Surai PF, Dvorska JE (2002) Strategies to enhance antioxidant protection and implications for the well-being of companion animals. In: Proceedings of the 18th annual symposium on nutritional biotechnology in the feed and food industry, Lexington, pp 521–534

Surai KP, Surai PF, Speake BK, Sparks NHC (2003) Antioxidant–prooxidant balance in the intestine: food for thought. 1. Prooxidants. Nutr Genomics Funct Foods 1:51–70

Swanson LW, Sawchenko PE (1980) Paraventricular nucleus: a site for the integration of neuroendocrine and autonomic mechanisms. Neuroendocrinology 31:410–417

Swanson LW, Sawchenko PE, Wiegand SJ, Price JL (1980) Separate neurons in the paraventricular nucleus project to the median eminence and to the medulla or spinal cord. Brain Res 198:190–195

Tizard IR (2000) Veterinary immunology: an introduction, 6th edn. W.B. Saunders Company, London, p 482

Todhunter D, Smith KL, Hogan JS (1990) Growth of gram-negative bacteria in dry cow secretion. J Dairy Sci 73:363–372

Tomlinson DJ, Socha MT, Rapp CJ, Johnson AB (2002) Summary of twelve trials evaluating the effect of feeding complexed zinc methionine on lactation performance of dairy cattle. J Dairy Sci 85:106 (abstr.)

Torre PM, Harmon RJ, Sordillo LM, Boissonneault GA, Hemken RW, Trammell DS, Clark TW (1995) Modulation of bovine mononuclear cell proliferation and cytokine production by dietary copper insufficiency. J Nutr Immunol 3:3–20

Traber MG, Arai H (1999) Molecular mechanisms of vitamin E transport. Annu Rev Nutr 19:343–355. doi:10.1146/annurev.nutr.19.1.343 (Volume publication date July 1999)

Trevisan M, Browne R, Ram M, Muti P, Freudenheim JA, Carosella M, Armstrong D (2001) Correlates of markers of oxidative status in the general population. Am J Epidemiol 154:348–356

Trigona WL, Mullarky IK, Cao Y, Sordillo LM (2006) Thioredoxin reductase regulates the induction of haem oxygenase-1 expression in aortic endothelial cells. Biochem J 394:207–216

Trout JM, Mashaly MM (1994) The effects of adrenocorticotropic hormone and heat stress on the distribution of lymphocyte populations in immature male chickens. Poult Sci 73:1694–1698

Underwood EJ (1999) In: Underwood EJ, Suttle N (eds) The mineral nutrition of livestock, 3rd edn. CABI Publishing, Wallingford

Upadhyay RC (2011) Impact of climate change on livestock production and health. In: Proceedings international conference on agriculture. ICICCA, Sri Lanka, pp 19–39

Valentino RJ, Van Bockstaele EJ (2005) Functional interactions between stress neuromediators and the locus coeruleusnorepinephrine system. In: Steckler T, Kalin NH, Reul JMHM (eds) Handbook of stress and the brain. Elsevier, Amsterdam, pp 465–468

Valko M, Leibfritz D, Moncol J, Cronin MT, Mazur M, Telser J (2007) Free radicals and antioxidants in normal physiological functions and human disease. Int J Biochem Cell Biol 39:44–84

Visser J, Boxel-Dezaire A, Methorst D, Brunt T, De Kloet ER, Nagelkerken L (1998) Differential regulation of interleukin-10 (IL-10) and IL-12 by glucocorticoids in vitro. Blood 91:4255–4264

Waage S, Sviland S, Odegaard SA (1998) Identification of risk factors for clinical mastitis in dairy heifers. J Dairy Sci 81:1275–1284

Wang X, Quinn PJ (2000) The location and function of vitamin E in membranes (review). Mol Membr Biol 17:143–156

Wall EH, Auchtung-Montgomery TL, Dahl GE, Mc- Fadden TB (2005) Short communication: Short day photoperiod during the dry period decreases expression of suppressors of cytokine signaling in the mammary gland of dairy cows. J Dairy Sci 88:3145–3148

Weber PSD, Madsen SA, Smith GW, Ireland JJ, Burton JL (2001) Pre-translational regulation of neutrophil L-selectin in glucocorticoid-challenged cattle. Vet Immunol Immunopathol 83:213–240

Webster AJF (1983) Environmental stress and the physiology, performance and health of ruminants. J Anim Sci 57:1584–1593

Weiss WP (2009) Antioxidant nutrients and milk quality. http://www.extension.org/pages/antioxidant nutrients and milk quality

Whitaker DA, Eayres HF, Aitchison K, Kelly JM (1997) No effect of a dietary zinc proteinate on clinical mastitis, infection rate, recovery rate, and somatic cell count in dairy cows. Vet J 153:197–204

Wiegers GJ, Reul JMHM (1998) Induction of cytokine receptors by glucocorticoids: functional and pathological significance. Trends Pharmacol Sci 19:317–321

Wiegers GJ, Stec IEM, Sterzer P, Ruel JMHM (2005) Glucocorticoids and the immune response. In: Steckler T, Kalin NH, Reul JMHM (eds) Handbook of stress and the brain. Elsevier, Amsterdam, pp 175–191

Williams RS, Thomas JA, Fina M, German Z, Benjamin IJ (1993) Human heat shock protein 70 (hsp70) protects murine cells from injury during metabolic stress. J Clin Invest 92:503–508

Wrona D, Trojniar W, Borman A, Ciepielewski Z, Tokarski J (2001) Stress-induced changes in peripheral natural killer cell cytotoxicity in pigs may not depend on plasma cortisol. Brain Behav Immun 15:54–64

Wu CY, Fargeas C, Nakajima T, Delespesse G (1991) Glucocorticoids suppress the production of interleukin 4 by human lymphocytes. Eur J Immunol 21:2645–2647

Wu CY, Wang K, McDyer JF, Sedev RA (1998) Prostaglandin E2 and dexamethasone inhibit IL-12 receptor expression and IL-12 responsiveness. J Immunol 161:2723–2730

Yang HW, Hwang KJ, Kwon HC, Kim HS, Choi KW, Oh KS (1998) Detection of reactive oxygen species (ROS) and apoptosis in human fragmented embryos. Hum Reprod 13:998–1002

Yeruham I, Elad D, Friedman S, Perl S (2003) Corynebacterium pseudotuberculosis infection in Israeli dairy cattle. Epidemiol Infect 131:947–955

Yoshimura A, Naka T, Kubo M (2007) SOCS proteins, cytokine signaling and immune regulation. Nat Rev Immunol 7:454–465

Yousef HM (1990) Studies on adaptation of Friesian cattle in Egypt. PhD thesis, Faculty of Agriculture, Zagazig University, Zagazig

Biological Rhythms

Contents

Abstract

A biological rhythm is a cyclical change in the biological or chemical function of body. The biological rhythms are endogenously controlled by self-contained circadian clocks. The daily alternation of light and dark is the main regulatory factor of the pineal hormone melatonin. Alterations in long-term lighting conditions during the year result in metabolic and behavioural changes in most living beings. The suprachiasmatic nuclei (SCN) in the hypothalamus are regarded as the anatomical loci of the circadian pacemaker. The most important synchronising trigger of circadian rhythm is environmental light/dark (LD) cycle. The circadian pacemaker in the mammalian SCN consists of a double complex of circadian genes (Per1/Cry1 and Per2/Cry2), which is able to maintain the endogenous rhythm. Melatonin has the ability to entrain biological rhythms and has important effects on biological function like reproduction of many mammals and livestock. The daily rhythmicity of melatonin is considered to be a very reliable phase marker of the endogenous timing system. The results on the patterns of cortisol levels in

A. Aggarwal and R. Upadhyay, *Heat Stress and Animal Productivity*,
DOI 10.1007/978-81-322-0879-2_6, © Springer India 2013

livestock particularly in ruminants are inconsistent; the levels have been observed to fluctuate episodically, or peaks and troughs have been found at varying times of the day depending on the physiological status and conditions. Increased glucocorticoid secretion at the circadian peak depends on increased hypothalamic–pituitary activity (HPA). Leptin is also a major regulator of neuroendocrine function and has an overall inhibitory effect on HPA activity and suppress the appetite-stimulating effects of glucocorticoid. Leptin secretion is pulsatile but leptin pulses are irregular in cattle. Chronobiologically, to maximise nutrient efficiency and optimise health, nutrient supply to reticulorumen, splanchnic and peripheral tissues needs to be synchronised with endogenous rhythms in hormone production and nutrient metabolism. The circadian system or oscillator coordinates the metabolic and hormonal changes needed to initiate and sustain milk synthesis or lactation. The animal or cow's capacity to produce milk and cope with metabolic stresses in early lactation is related to animal's ability to set circadian rhythms in order particularly during the transition period or early lactation. Circadian variations are also observed in many other biological functions like reactive oxygen species (ROS), defence systems, thermoregulation, the cardiovascular system and other functions in humans and domestic animals.

1   Introduction

A biological rhythm is a cyclic change in the biological or a chemical function. Biological rhythms affect a variety of activities, such as the sleep–wake cycle, migration behaviour in birds and seasonal fattening, hibernation and reproductive cycles in wild animals (Piccione and Caola 2002). Biological rhythms are endogenously controlled by self-contained circadian clocks or oscillators. Circadian rhythm implies that under constant external conditions (without time cues), the rhythms follow a pattern with an endogenous period similar to, but not identical, 24 h. The length of cycle is under the control of a circadian oscillator (Ikonomov et al. 1998). Most living organisms organise their activities coordinating with 24-h light and dark cycle associated with sunrise and sunset. Circadian rhythms like sleeping and waking in animals, brain wave activity, production of hormones and flower closing and opening in angiosperms, tissue growth and differentiation in fungi and other biological activities in plants and animals are related to the daily sunrise and sunset cycle. The circadian rhythms are generated by an internal clock(s) that is synchronised to light–dark cycles and other cues in an organism's environment. This internal clock accounts for waking up at the same time every day spontaneously even without an alarm. The clock also causes nocturnal animals to function at night when diurnal creatures are resting. Circadian rhythms to an extent can be disrupted by changes in daily schedule. In mammals, the circadian clock or oscillator is located within the brain's hypothalamus, and pineal gland helps coordination through release of melatonin in response to the information it receives from retinal photoreceptors in the eye. In the retinal photoreceptors, the light energy perceived is transformed into nerve impulses and transmitted to the hypothalamic nuclei of the central nervous system for further processing. The suprachiasmatic nuclei (SCN) in the hypothalamus are the anatomical loci of the circadian pacemaker (Moore 1997; Fig. 1) and are the generators of circadian rhythms (Klein et al. 1991). The daily alternation in light and dark observed is the regulatory factor of the pineal hormone melatonin. The typical daily pattern of melatonin secretion also depends on the function of the suprachiasmatic circadian clock (Moore and Klein 1974). The alterations or changes in long-term lighting conditions that occur during the year result in metabolic and behavioural changes of mammalians. For instance, the reproductive cycles with concurrent hormonal changes are entrained in many mammalian species and livestock like sheep by annual lighting conditions (Karsch et al. 1984). The SCN–pineal complex is believed to be a probable component in the mechanism

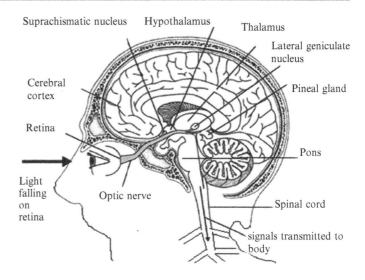

Fig. 1 The biological clock located within the suprachiasmatic nucleus in the brain (Moore 1997; BSCS 2003, Copyright © 2003 BSCS. All rights reserved. Used with permission)

entraining annual rhythms (Zucker et al. 1991; Scott et al. 1995). Pineal gland and its secretion facilitates regulation of seasonal breeding in mammalians by synchronising the hypothalamo–hypophyseal–gonadal axis reproductive functions. The melatonin serves the reproductive system as a messenger of night duration or length (Karsch et al. 1991; Bartness et al. 1993). In addition to the melatonin rhythm and pattern, the daily variation in glucocorticoid levels in blood is also a classical example of circadian rhythms in humans (Bliss et al. 1953; Orth and Island 1969), rats (*Rattus norvegicus*, Guillemin et al. 1959; Moore and Eichler 1972) rhesus monkey (*Macaca mulatta*, Perlow et al. 1981), Syrian hamsters (*Mesocricetus auratus*, de Souza and Meier 1987) and red-backed voles (*Clethrionomys gapperi*, Kramer and Sothern 2001). Though the phenomenon of circadian rhythm of glucocorticoids is classical in many mammalian species, but in ruminants, like cattle (MacAdam and Eberhart 1972; Wagner and Oxenreider 1972; Abilay and Johnson 1973; Lefcourt et al. 1993) and sheep (Simonetta et al. 1991), they do not exhibit any particular pattern. The inconsistent, absence of rhythm or discrepant results in ruminants could be attributed to differences in metabolism of nutrients and to the role of glucocorticoids in the regulation of metabolism in ruminants. The special digestive system of ruminants stores and processes a large amount of undigested, partially digested

Table 1 Periods of biological rhythms

Type of rhythm	Period (t)
Ultradian	$t < 20$ h
Circadian	20 h $\leq t \leq 28$ h
Infradian	$t > 28$ h
Circaseptan	$t = 7 \pm 3$ days
Circannual	$t = 1$ year ± 3 months

Source: Piccione and Caola (2002)

and digested food; nutrients continue to absorb into the blood evenly during the 24-h period. Therefore, variations in metabolites and metabolic hormones are little in ruminants, and characteristic patterns are either absent or inconsistent in ruminants. The lighting conditions affect regulation of lipid metabolism in mammals (Clarke 2001; Bartness et al. 2002; Morgan et al. 2003), and the melatonin is a principal mediator and regulator of photoperiodic information (Lincoln et al. 2003). The variation in the periods of biological rhythm has been presented in Table 1.

Biological rhythms in mammals are of two general types:

- *Exogenous rhythms*: The exogenous rhythms as the name indicates are directly produced by an external influence such as sunlight, food, noise or social interaction and are not generated internally by the organism itself, for example, environmental cue. Since these rhythms are not generated internally, in the absence of environmental cues, the exogenous rhythm is likely to cease or discontinue. The stimuli or cues are called zeitgebers – a German word meaning 'time givers'. Zeitgebers are likely to help to reset the biological clock to a 24-h day.

- *Endogenous rhythms*: The endogenous rhythms are controlled by the internal and self-sustaining biological rhythms, and these biorhythms sustain even in the absence of environmental cue, for example, oscillations in core body temperature and melatonin.

2   Circadian Timekeeping Mechanisms

Several approaches to elucidate the nature of circadian oscillators have emerged over the years (Hastings and Schweiger 1975), aimed at locating the anatomical loci responsible for generating these periodicities and tracing the entrainment pathway for light signals from the photoreceptor(s) to the clock itself. Animals appear to have central clocks that reside in discrete 'pacemaker tissues' in the central nervous system, whose signals direct circadian output responses in peripheral tissue areas (Takahashi 1995). The circadian clock in most mammals and frequently observed in humans has a natural day length of about 24 h. The clock requires resetting to match the day length of the environmental photoperiod, that is, the light/dark, or day/night, cycle. The environmental cue that synchronises the internal biological clock is duration of light or light period. Photoreceptors in the retina perceive and transmit light-dependent signals to the SCN. Receptors in the retina, the rods and cones, are apparently not required for this photoreception (Freedman et al. 1999). Special types of retinal ganglion cells that are photoreceptive and project directly to the SCN appear to provide signals for synchronising the biological clock (Berson et al. 2002).

The first circadian clock gene was identified and isolated from Drosophila, and subsequent analysis of its expression led to the first molecular model of the circadian oscillator, an autoregulatory feedback loop in gene expression. Further discovery of additional clock genes in Drosophila supports the feedback loop model

and add to its mechanistic detail and complexity (Hardin 2009). In mammals and livestock species, circadian oscillators are believed to operate similarly in a variety of tissues, including the brain and numerous other internal organs. The circadian oscillators are primarily photoreceptive, and directly intensity and duration of light modulate these oscillators in peripheral tissues. Therefore, local oscillators and systemic cues control most biorhythms in mammals and livestock.

2.1   Biological Clock

A biological clock acts as an alarm to wake up or to initiate a physiological process at an appropriate phase of the daily activity. The clock also helps to organise an activity and prepare in anticipation of actual need in an organism. The clock also acts as a time management device or an instrument to assess the day and/or night length or duration to regulate them appropriately (Dunlap et al. 2004).

The suprachiasmatic nuclei (SCN) in the hypothalamus are the anatomical loci of the circadian pacemaker (Moore 1997). The SCN is a very small structure consisting of a pair of pinhead-size regions, each containing only about 10,000 neurons out of the estimated 100 billion neurons in the brain. This circadian clock or oscillator, entrained by the light–day cycle via the retinohypothalamic tract, is believed to impose circadian patterns on a wide array of physiological functions and behavioural processes like locomotor activity, body temperature, heart rate, blood pressure, hormone secretion and urinary excretion (Dunlap et al. 2004; Refinetti 2005; Cassone and Stephan 2002). The knowledge on molecular biology and genetics helped in cloning of many mammalian 'clock' genes and to the discovery of new, extracerebral sites which contain circadian oscillators (Yamazaki et al. 2000). Information on circadian clock has revealed that genes and circadian oscillators are expressed in many organs, such as lung, liver, skeletal muscles and kidney (Yoo et al. 2004).

Circadian rhythmicity of body temperature has been considered both in laboratory and farm animals due to the relatively easy monitoring of body temperature and the robustness of its rhythm (Piccione et al. 2003). These studies have advanced knowledge on temperature rhythm, homeothermy and the characteristics of the temperature clock.

3   Daily Rhythms

As has been indicated earlier, the physiological processes are regulated by a circadian oscillator and were first described in the plant leaves movement by the French scientist Jean-Jacques d'Ortous de Mairan (Meijer and Rietveld 1989). The circadian clock is regulated by the light and wavelength, intensity, timing and duration of the light stimulus impacts (Cardinali et al. 1972; Brainard et al. 1983, 1986; Takahashi et al. 1984). The endogenous biological rhythms help to assess and anticipate periodic changes in the environment and are thus important for adaptive behaviour. The circadian rhythm is governed and related to the light/dark cycle, yet the rhythm period can be reset by exposure to a light or dark cycle in individuals or mammalian species. The change in pulse of light or change in the lighting conditions can help in adjustment of the animal gradually to the new pattern provided it does not deviate too much from the species' fixed circadian norm.

Animals confined in total darkness for a long period have been observed to display a 'free-running' rhythm (Redman et al. 1983; Thomas and Armstrong 1988). The sleep cycle of diurnal animals' moves forwards approximately by 1 h a day; their free-running rhythms are about 25 h. In contrast, in nocturnal animals, the free-running rhythm is shorter by about an hour, that is, about 23 h. The biological clock even in total darkness is influenced by events occurring regularly on a daily basis. Continuous light exposure suppresses circadian rhythm of locomotor activity in rats (Homna and Hiroshige 1978; Chesworth et al. 1987). Several other circadian rhythms in rats (e.g., behavioural, temperature and some humoral rhythms) may persist for several weeks depending on the intensity of light (Homna and

Hiroshige 1978; Depres- Brummer et al. 1995). The environmental cues that entrain the circadian rhythm are called zeitgebers or circadian synchronisers as has been indicated earlier. Several environmental and behavioural stimuli act as circadian synchronisers. These include water and food intake, motor activity, sleep–wake rhythm, corticosterone release, activity of pineal N-acetyltransferase enzyme and body temperature (Rusak and Zucker 1979).

The most important synchronising trigger of circadian rhythm is environmental light/dark (LD) cycle associated with sunrise and sunset. In the absence of external cues, the rhythm may become out of phase with, for instance, the ultradian rhythm of digestion. Biological functions, such as hormone production, cell regeneration and brain activation as measured by an electroencephalogram (EEG) and overall behavioural patterns (sleeping, eating) are linked to the circadian cycle. Destruction of the SCN completely abolishes the normal sleep/wake rhythm. Information on day length travels from the SCN to the pineal and in response the pineal gland secretes the melatonin. The melatonin secretion reaches its peak at night and nadir during the day (Zucker et al. 1983). SCN can also transmit its message directly to peripheral organs and tissues through the autonomic nervous system (Bartness et al. 2001; la Fleur 2003; Buijs et al. 2006) to evoke responses.

4   Annual Rhythms

The gradual change in day and night or annual lighting condition has an impact on the behaviour and physiology of most mammalian species (Hastings et al. 1985). The timing of annual reproductive cycle in many mammals is associated with hormonal changes leading to the successful production of offspring (Karsch et al. 1984). In addition to light and dark period, other seasonal alterations also affect the animals. These include variation in temperature, availability of food and metabolism. Similar to circadian rhythm in mammals, the SCN and the pineal gland are the main structures regulating annual rhythms

(Hastings 2001; Schwartz et al. 2001; Zucker 2001). The assessment of time or timing system in the brain regulates annual rhythm such seasonal cycles like sexual behaviour, energy metabolism, food intake and hibernation. The photoperiod is the strongest synchroniser of seasonal functions in most species. For example, in Djungarian hamsters (*Phodopus sungorus*), short days induce reproductive inhibition, inactivity and weight increase, contrary to animals kept in long days that do not display these changes. In sheep, reversal of the annual photoperiodic cycle causes the breeding season to phase shift by 6 months; reduction of its period to 6 months triggers two periods of reproductive activity every year (Malpaux et al. 1993, 2001).

Neurons in the SCN respond to light during subjective night with an expression of the immediate early gene c-*fos*. FOS reactivity in the SCN following a light stimulus depends on the photoperiod history or exposure experience (Sumova et al. 1995; Vuillez et al. 1996). Clock gene expression in the SCN displays photoperiodic variations (Messager et al. 1999, 2000, 2001; Nuesslein-Hildesheim et al. 2000), and the daily profile of arginine vasopressin (VP) messenger ribonucleic acid (mRNA) differs in short and long photoperiods (Jac et al. 2000). There is also an evidence that the thalamic intergeniculate leaflet (IGL), a relay between the retina and SCN, is involved in photoperiod integration (Menet et al. 2001). Some photoperiodic species, for example, ground squirrels (*Spermophilus parryii*), exhibit endogenous circannual rhythms under captivity and seasonally constant conditions (photoperiod and temperature) for long periods (Lee and Zucker 1991; Gorman et al. 2001; Zucker 2001). A morning oscillator (M) adjusted by dawn and an evening oscillator (E) adjusted by dusk in the mammalian circadian system have been suggested (Pittendrigh and Daan 1976). The phase relationship between M and E reflects the day length to which the subject has been exposed. The oscillators control the pineal gland, therefore being able to define not only the time of day but also the time of year (Schwartz et al. 2001). The hypothesis of Daan et al. (2001) suggests that the circadian

pacemaker in the mammalian SCN consists of a double complex of circadian genes (Per1/Cry1 and Per2/Cry2), which is able to maintain the endogenous rhythm. These two types of oscillators are speculated to have slightly different temporal dynamics and light responses. The Per1/Cry1 (or M) oscillator is apparently accelerated by light and decelerated by dark, and the Per2/Cry2 (or E) oscillator is decelerated by light and accelerated by dark. Therefore, changes in the activity of M and E oscillators are likely to have an influence on the adaptation of the endogenous behavioural programme associated with day length.

5   Functions of Sleep

Sleep is essential to all mammals studied and varies widely in species, age, pregnancy, health status, etc. Sleep affects the metabolism, endocrine and immune functions. Lack of sleep increases energy requirements and impairs the immune defence. Studies in cattle on duration of sleep during different stages of lactation in dairy cows show that total sleep time is about 4 h per day and is significantly lower in cows 2 weeks after parturition compared to that during the dry period and peak lactation (Ternman 2011). Studies with rats have demonstrated that while these animals normally live for 2–3 years, rats deprived of REM sleep survive an average of only 5 months. Rats deprived of all sleep survive only about 3 weeks (Rechtschaffen 1998). In humans, extreme sleep deprivation can cause an apparent state of paranoia and hallucinations in otherwise healthy individuals.

Many hypotheses have been advanced to explain the role of this necessary and natural behaviour (Rechtschaffen 1998). The following examples highlight several of these theories:

5.1   Evolution of Sleep

Sleep is ubiquitous among mammals, birds and reptiles, although sleep patterns, habits, postures and places of sleep vary greatly in different subjects.

Table 2 Total sleep time for various species

Species average	Total sleep time (hours/day)
Human infant	16.0
Tiger	15.8
Squirrel	14.9
Golden hamster	14.3
Rat	12.6
Cat	12.1
Mouse	12.1
Rabbit	11.4
Duck	10.8
Dog	10.6
Baboon	10.3
Chimpanzee	9.7
Guinea pig	9.4
Human adolescent	9.0
Human adult	8.0
Goat	5.3
Cow	3.9
Sheep	3.8

Source: Aserinsky (1999), Campbell and Tobler (1984), Kryger et al. (1989), and Tobler (1989)

- *Sleep patterns*: The patterns of sleep are different in all subjects and vary a lot in rhythm and duration. Mammals generally alternate between NREM and (rapid eye movement) REM sleep states in a cyclic fashion, although the length of the sleep cycle and the percentage of time spent in NREM and REM states vary with the animal species. Birds also exhibit NREM/REM cycles, and each phase is very short in duration (NREM sleep is about two and one-half minutes; REM sleep is about 9 s). Birds while asleep do not lose muscle tone during REM sleep similar to mammals.
- *Sleep habits*: In general, mammalian species sleep during night or day. Humans and livestock species sleep primarily at night, while other animals like rats sleep primarily during the day. Small-sized mammals tend to sleep more than large ones (Table 2). In some cases, animals have developed ways to sleep and concurrently satisfy critical life functions. These animals engage in unihemispheric sleep, in which one side of the brain sleeps, while the other side is awake. This phenomenon is observed most notably in birds (like those that make long, transoceanic flights) and aquatic mammals (like dolphins and porpoises). Unihemispheric sleep allows aquatic mammals to sleep and continue to swim and surface to breathe. Also, this type of sleep allows them to keep track of other group members and watch for predators. The mallard ducks increase their use of unihemispheric sleep with the increase in risk of predation (Rattenborg et al. 1999).
- *Sleep postures*: A wide variety of postures are observed in mammals ranging from laying postures to standing. The postural position may be curled up in dogs, cats and many other animals; standing in horses and birds; swimming in aquatic mammals and ducks; hanging upside down in bats; straddling a tree branch in leopard; and lying down (back or abdomen) in humans. Livestock species also exhibit different postures during sleep. During the day, buffalo and cattle usually rest in sternal recumbency. Some of the animals may also be observed in rest and ruminating. The duration of recumbency in animals varies and less than 1 h is likely to be spent in lateral recumbency in one stretch; however, episodes of rest in this position are generally brief and may be several times during a day. The recumbency in large animals may be associated with periods of sleep. The forelimbs are curled under the body, and one hind leg is tucked forwards underneath the body, taking the bulk of weight on an area enclosed by the pelvis above and the stifle and hock joints below. The other hind limb is stretched out to the side of the body with the stifle and hock joints partially flexed. Some of the animals occasionally lie with one or other foreleg stretched out in full extension for a short period. Cattle also occasionally lie fully on their sides for very short periods while holding their heads forwards and upwards in order to facilitate regurgitation and expulsion of gases from the rumen. Adult cattle may also sleep in position similar to calves with their heads extended inwards to their flanks. This posture is considered a normal resting and sleeping position in cattle and buffaloes, but it is also a posture

typically observed in cows suffering from milk fever.

• *Sleep places*: The place of sleep is likely to vary in different mammalian and non-mammal species and depends on habitat of a mammalian species. For example, burrows (rabbits), open spaces (lions), underwater (hippopotami), nests (gorillas) and the bed for comfort (humans).

Sleep may also occur among lower life forms, such as fish and invertebrates, but it is difficult to know because EEG patterns are not comparable to those of vertebrates. Consequently, investigating sleep in species other than mammals and birds depends on the identification of specific behavioural characteristics of sleep: a quiet state, a typical species-specific sleep posture, an elevated arousal threshold (or reduced responsiveness to external stimuli), rapid waking due to moderately intense stimulation (i.e. sleep is rapidly reversible) and a regulated response to sleep deprivation. Fruit fly (*Drosophila melanogaster)* responds similar to mammals when exposed to chemical agents that alter sleep patterns (Hendricks et al. 2000; Shaw et al. 2000). Comparative studies have explored the evolution of sleep. Although REM sleep is thought to have evolved from NREM sleep, studies suggest that NREM and REM sleep may have diverged from a common precursor sate of sleep (Kryger et al. 1989).

6   Pineal Gland and Melatonin

The pineal gland (synonyms – glandula pinealis, epiphysis cerebri) is a small structure located in most mammals between the habenular and posterior commissures. Identification of the pineal gland as a distinct cerebral organ can be traced back to the third and fourth centuries BC (Kappers 1960; Hoffman and Reiter 1965). At the end of the nineteenth century, Ahlborn and Rabl-Ruckhardt, then Graaf, Korschelt and Spencer, described the anatomy, histology, innervation and embryology of the mammalian pineal gland and noted its resemblance to the epiphysis organ of lower vertebrates (Simonneaux and Ribelayga 2003).

Phylogenetically the pineal gland is derived from a photoreceptor organ, but its function remained unknown for long (Simonneaux and Ribelayga 2003). Bioassay techniques enabled the discovery of an active pineal extract capable of lightening the colour of frog skin (McCord and Allen 1917) followed by the isolation of the pineal hormone melatonin in 1958 (Lerner et al. 1958, 1959). Fluorescent techniques allowed the measurement of melatonin and serotonin concentrations, which led to the discovery of circadian variations in their levels (Quay 1963, 1964). The mammalian pineal gland is innervated by peripheral sympathetic and parasympathetic fibres and those originating from the central nervous system. The pineal gland receives relatively scarce afferent innervation from the brain. The most important afferents are postganglionic sympathetic fibres which originate from the superior cervical ganglia (SCG) and form the bilateral nervi conarii, which enter the pineal gland posteriorly (Kappers 1979). The neurons of this noradrenergic pathway receive regulatory input from the suprachiasmatic nucleus of the hypothalamus, which receives direct input from retinal ganglion cells (Kappers 1960, 1979) via the monosynaptic retinohypothalamic tract. The transmission in this tract is predominantly modulated both pre- and postsynaptically by neuropeptide Y (NPY) in many mammals (Simonneaux et al. 1994; Mikkelsen et al. 2000). Retinohypothalamic fibres synapse in the suprachiasmatic nuclei (SCN), and connections from the SCNs to the intermediolateral grey column in the spinal cord have been observed. Preganglionic neurons pass from the spinal cord to the superior cervical ganglion (SCG), and the postganglionic neurons of this ganglion project to the pineal gland (Andersson 1978; Ganong 1997).

The pineal gland receives afferent fibres anteriorly that travel through the commissural peduncles, possibly originating from the hypothalamus (Vollrath 1984). A third pathway, the ventrolateral pineal tract, has also been observed (Sparks 1998). The myelinated fibres of this tract originate from the pretectal region, posterior and lateral to the posterior commissure. Central nerve fibres originating from the hypothalamic, limbic forebrain and visual cortex also innervate the pineal gland in non-human mammals (Kappers 1960). The parasympathetic innervation of the pineal gland with fibres containing the primary neurotransmitter of parasympathetic neurons, acetylcholine (Ach), has also been observed in some mammalian species, including the cow and rat (Phansuwan-Pujito et al. 1991; Korf et al. 1996; Weihe et al. 1996). There is also an evidence of parasympathetic fibres innervating pineal arising from the mammalian pterygopalatine ganglia and containing vasoactive intestinal peptide (VIP) and other neuropeptides (Moller 1992). Besides NPY, which is located in the sympathetic fibres, and VIP, many other peptides have been found in nerve fibres terminating in perivascular and intraparenchymal areas in mammals. These include substance P, vasopressin, oxytocin and luteinising hormone-releasing hormone (Barry 1979; Ronnekleiv 1988).

6.1   Synthesis and Metabolism of Melatonin

The pineal gland synthesises and secretes melatonin, a structurally simple hormone that communicates information about photoperiodic information or environmental lighting to various parts of the body. Melatonin has the ability to entrain biological rhythms and has important effects on reproductive rhythm of many animals. The light-transducing ability of the pineal gland and the role pineal plays, it is believed as the 'third eye'.

Melatonin (5-methoxy-N-acetyltryptamine) is a small-sized (molecular weight 232.3) indoleamine secreted rhythmically, and its synthesis increases at night. Tryptophan is a precursor molecule for the biosynthesis of melatonin. Pinealocytes hydroxylate and decarboxylate tryptophan to serotonin, which is then acetylated to N-acetyl-serotonin by the rate-limiting enzyme N-acetyltransferase (NAT). This is methylated by hydroxyindole-O-methyltransferase (HIOMT) to melatonin (Reiter 1991). After its biosynthesis, the highly lipophilic melatonin is released into capillaries, where most of it binds to albumin

(Cardinali et al. 1972). Melatonin is metabolised by hydroxylation and conjugation with sulphate or glucuronic acid, mainly in the liver and also in the kidney. The degraded product of melatonin is excreted into urine as 6-sulphatoxymelatonin. Functional disorders of pineal have been shown to affect the elimination rate (Lane and Moss 1985; Viljoen et al. 1992; Kunz et al. 1999). The half-life of melatonin in blood after intravenous administration is about 30 min (Mallo et al. 1990).

The biosynthesis of melatonin is initiated by the uptake of the essential amino acid tryptophan into pineal parenchymal cells. Tryptophan is converted to 5-hydroxytryptophan, through the action of the enzyme tryptophan hydroxylase and then to 5-hydroxytryptamine (serotonin) by the enzyme aromatic amino acid decarboxylase. Serotonin concentrations are higher in the pineal than in any other organ or in any brain region and exhibit a striking diurnal rhythm remaining at a maximum level during the daylight hours and falling by more than 80% soon after the onset of darkness as the serotonin is converted to melatonin, 5-hydroxytryptophol and other methoxyindoles. Serotonin's conversion to melatonin involves the following two enzymes that are characteristic of the pineal:

1. SNAT (serotonin-N-acetyltransferase) which converts the serotonin to N-acetylserotonin
2. HIOMT (hydroxyindole-O-methyltransferase) which transfers a methyl group from S-adenosylmethionine to the 5-hydroxyl of the N-acetylserotonin (Lerner et al. 1959)

The activities of both enzymes rise soon after the onset of darkness because of the increased release of norepinephrine from sympathetic neurons terminating on the pineal parenchymal cells. Another portion of the serotonin liberated from pineal cells after the onset of darkness is deaminated by the enzyme monoamine oxidase (MAO) and then either oxidised to form 5-hydroxyindole acetic acid or reduced to form 5-hydroxytryptophol. Both of these compounds are also substrates for HIOMT and can thus be converted in the pineal to 5-methoxyindole acetic acid 5-methoxytryptophol. The level of 5-methoxytryptophol, like that of melatonin, rises in the pineal with the onset of darkness. Since 5-methoxytryptophol synthesis does not require the acetylation of serotonin, the nocturnal increase in pineal SNAT activity may not be related to the rise in pineal methoxyindole levels. The intraparenchymal release of stored pineal serotonin which then becomes accessible to both SNAT and MAO ultimately controls the rates at which these three major pineal methoxyindoles are synthesised and generates the nocturnal increases in pineal melatonin and 5-methoxytryptophol. The rates of methylation of all three 5-hydroxyindoles formed from pineal serotonin depend on HIOMT activity in pineal gland. The proportion of available serotonin acetylated at any particular time of day or night depends on the relative activities of pineal SNAT and MAO at that specific time.

Melatonin has been found in blood, urine and saliva, also in the cerebrospinal fluid (CSF), at a concentration much higher than in blood, and in the anterior chamber of the eye (Martin et al. 1992). Melatonin is also found in semen, amniotic fluid, urine and breast milk (Cagnacci 1996). Melatonin in plasma, CSF, saliva and urine is eliminated by pinealectomy, indicating that it is mainly synthesised in the pineal gland (Nelson and Drazen 1999). There is, however, evidence that melatonin is also synthesised at other sites which include, in humans, the retina, gut and bone marrow (Cagnacci 1996; Conti et al. 2000), indicating a localised action of melatonin besides a central regulatory function (Fjaerli et al. 1999).

6.2   Melatonin Receptors

In humans, there are two types of melatonin receptors (Mel1a and Mel1b) with different binding affinity and chromosomal localisation (Reppert et al. 1995). Melatonin receptors have been found in the SCN of the hypothalamus, which controls the rhythmic production of melatonin by the pineal gland (Reppert et al. 1988; Weaver and Reppert 1996). In addition, it is also located in the cerebellum (Al-Ghoul et al. 1998), retinal rods, horizontal amacrine and ganglion cells (Reppert et al. 1995; Scher et al. 2002). Besides the CNS, human melatonin receptors have been found in lymphocytes (Lopez-Gonzalez

et al. 1992), prostate epithelial cells (Zisapel et al. 1998), granulose cells of preovulatory follicles (Yie et al. 1995), spermatozoa (van Vuuren et al. 1992), the mucosa layer of the colon (Poon et al. 1996) and blood platelets (Vacas et al. 1992). In the absence of receptors, melatonin molecules exert systemic effects also at the basic cellular levels (Benitez- King 1993; Fjaerli et al. 1999).

6.3   Daily Rhythm of Melatonin

The daily alternation of light and dark is the most important regulatory element in the synthesis of pineal hormone melatonin. In all mammals studied, whether they exhibit nocturnal or diurnal activity, melatonin levels are higher at night than during the day. The melatonin level starts to rise during the evening and is at its highest in the middle of the night and starts to decrease reaching low levels in the morning. The daily rhythm of melatonin is considered to be a very reliable phase marker used by the endogenous timing system. The study on zebu-cross heifers reported that plasma melatonin level was observed to be the lowest at 12.30 h (23.33 ± 5.78 pg/ml) during day, and thereafter, an increase in plasma melatonin level was observed (after 20.30 h) with increase in darkness and the mean peak level (124.33 ± 16.16 pg/ml) occurred at 00.30 h during night (Aggarwal et al. 2005; Fig. 2). The level declined and was low during daytime. There was a significant ($P < 0.05$) variation in plasma melatonin levels during different times of the day.

In the absence of the light–dark cycle, melatonin rhythms do not exhibit characteristic pattern and observed to free run with a period slightly different from 24 h (Aschoff 1965). In rats, Syrian hamster and Siberian hamster (*P. sungorus*), pharmacological doses of exogenous melatonin are capable of synchronising the circadian rhythms of locomotor activity and melatonin synthesis in free-running circadian rhythms (Redman et al. 1983; Schuhler et al. 2002). Lesioning of the SCN abolishes pineal melatonin rhythm (Klein and Moore 1979). Therefore, input from the main endogenous circadian pacemaker, located in the SCN, is essential for the circadian rhythm of

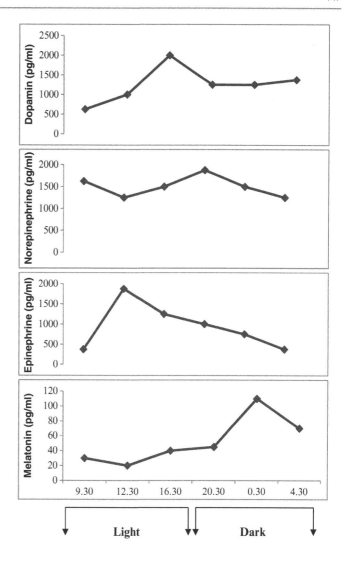

Fig. 2 Circadian levels of dopamine, norepinephrine, epinephrine and melatonin in crossbred cattle during summer (Aggarwal et al. 2005)

melatonin and its synchronisation with the external light–dark cycle. Light not only entrains the circadian rhythm but also directly suppresses nocturnal melatonin synthesis and levels in plasma. Suppression in melatonin synthesis is dependent on the intensity and dose of light. In humans, a decrease in nocturnal melatonin levels has been observed at relatively low light intensities, similar to normal indoor lighting (Lewy et al. 1980; Brainard et al. 2001). Light is also able to phase shift the nocturnal melatonin rhythm. In endogenous circadian mammalian rhythms, melatonin acts as a synchroniser (Armstrong 1989). The synchronising effect is likely to occur at a particular circadian time and may vary in different species. Application of

exogenous melatonin directly into the SCN has been found to advance the endogenous melatonin peak but also increases its peak (Bothorel et al. 2002). Some in vitro studies have also demonstrated a local effect of melatonin on SCN metabolism, electrical activity and circadian rhythm (Cassone et al. 1988; McArthur et al. 1991). Small doses of exogenous melatonin induce a phase-shifting effect in rats (Warren et al. 1993) and humans (Lewy et al. 2005). Melatonin may exert its synchronising properties indirectly on circadian clock inputs and outputs or directly on the clock via melatonin MEL-R receptors (Song et al. 1999) or other binding sites (Pévet et al. 2002). This property of melatonin helps, along with several circadian signals, in setting circadian rhythm of the fetus (Reppert et al. 1979; Reppert and Weaver 2001).

In both sheep breeding during short days and hamsters breeding during long days, short-duration melatonin infusions result in long-day responses typical of the species, whereas long-duration infusions elicited responses associated with short days. This indicated that the signal about day length is encoded in the duration of nocturnal melatonin secretion, or in other words, the melatonin pattern serves as a humoral signal to convey day length information (Schwartz et al. 2001).

In humans, this 'chronobiotic' function of melatonin helps to resynchronise the rhythms of individuals with disrupted circadian rhythms. A disrupted rhythm can, for example, be due to 'delayed sleep phase' syndrome, jet lag, night-shift work or blindness (Arendt et al. 1984, 1988, 1997; Takahashi et al. 2000).

6.4 Annual Rhythm of Melatonin

Accumulating evidence indicates that in mammals, the SCN and the pineal gland are the main neural structures involved in the regulation of annual cycles (Goldman 2001; Zucker 2001). The pineal gland is a major structure in the endocrine system allowing mammals to respond to annual changes in the photoperiod by adaptive alterations of their physiological state. The cyclicity of reproductive behaviour of most mammals is an example of adaptive behaviour that is entrained by alterations of day length during the year. The pineal gland and its melatonin rhythm are essential triggers of this cyclicity of reproduction; numerous studies have demonstrated that the pineal gland is a neuroendocrine transducer responding to photoperiodic information from the retina and transmitting this to the reproductive system via a particular dynamic pattern of melatonin secretion (Bittman 1984; Goldman 2001). The periodic synthesis and secretion of melatonin by the pineal gland is governed by oscillations in the SCN connected via a complex multisynaptic pathway to the pineal gland (Teclemariam-Mesbah et al. 1999; Kalsbeek and Buijs 2002). The disruption of any portion of this pathway abolishes melatonin rhythm.

The duration of melatonin secretion has been observed to be inversely related to day length, and therefore, the melatonin signal encodes information about gradual changes in day length during the year. Several hypotheses have been proposed concerning which parameters of the melatonin secretion pattern (duration, amplitude, phase or total quantity) convey the photoperiodic message to target structures. The hypotheses have been based on the analyses of endogenous melatonin patterns in different experimental conditions and on experiments investigating the effects of acute injections or chronic infusions of exogenous melatonin (Pitrosky et al. 1991; Bartness et al. 1993). Studies of pinealectomised Siberian and Syrian hamsters and sheep administered daily infusions of melatonin indicate that photoperiodic information is indeed encoded in the melatonin signal (Bartness et al. 1993). In both sheep breeding during short days and hamsters breeding during long days, short-duration melatonin infusions resulted in long-day responses typical of their pattern, whereas long-duration infusions elicited responses associated with short days. These results indicate that the signal of the day length is encoded in the duration of nocturnal melatonin secretion or the melatonin pattern serves as a humoral signal conveying day length information (Schwartz et al. 2001). Observations of the melatonin secretion pattern in various species which were kept under different photoperiodic conditions have shown that the duration of the nocturnal melatonin peak is positively related to length of the night in sheep (Rollag and Niswender 1976) in Siberian hamster (Ribelayga et al. 2000) and in Syrian hamster (Skene et al. 1987).

7 Cortisol

Cortisol is a glucocorticoid hormone that is involved in the response to stress; it increases blood pressure and blood sugar levels and suppresses the immune system. Synthetic cortisol, also known as hydrocortisone, is used as a drug mainly to suppress allergic reactions and inflammatory responses (Greco 2002). The synthesis of cortisol in the adrenal cortex is stimulated by adrenocorticotrophic hormone (ACTH) from the anterior lobe of the pituitary gland which is in turn stimulated by corticotrophin-releasing hormone (CRH) released from the hypothalamus. Cortisol inhibits the secretion of CRH resulting in feedback inhibition of ACTH secretion. Chronic stress is likely to break down this normal feedback system (Greco 2002).

7.1 Daily Rhythm of Cortisol

The diurnal variation of glucocorticoid levels in blood occurs and may be regarded as a classical circadian rhythm similar to melatonin as the concentrations are low during rest, and there is a rapid rise just before the active period in many mammalian species, performing activities during the day or night. In the rats, the secretion of glucocorticoids has been shown to exhibit rhythmicity and is observed to be under the control of the main body clock, that is, the SCN (Moore and Eichler 1972). The rhythm is maintained by other mechanisms and is not exclusively under the control of SCN at least in some mammals as is evidenced in suprachiasmatic region-lesioned primates that the daily cortisol rhythm does not disappear (Reppert et al. 1981). The morning rise of cortisol in animals active during daytime is endogenous. In humans, the rise in cortisol may be enhanced by bright light after awakening (Scheer and Buijs 1999) and also by sleep deprivation (Leproult et al. 2001). In pigs, exposure to supplementary artificial light in the morning after sunrise increased the cortisol level (Andersson et al. 2000), and in male Creole goats (*Capra hircus*), an abrupt exposure to sunlight in the middle of the day resulted in enhanced cortisol concentrations (Sergent et al. 1985). There was no effect of intensity of light on blood cortisol levels in pigs (Griffith and Minton 1992) or in bulls (Leining et al. 1980). The pattern of cortisol level in blood of ruminants is inconsistent; the levels fluctuate episodically or peaks and troughs do not have any specific pattern indicative of rhythmicity and have been observed at different times of the

day depending on the physiological status and conditions. The absence of circadian variation in cortisol levels was found in sheep (Kennaway et al. 1981; Lincoln et al. 1982; Simonetta et al. 1991), Eld's deer (Ingram et al. 1999) and cattle (Hudson et al. 1975; Fulkerson et al. 1980; Lefcourt et al. 1993).

Serum cortisol levels have been reported to increase either during the night or early morning in goats (Kokkonen et al. 2001), sheep (Kennaway et al. 1981) and cattle (Fulkerson et al. 1980; Thun et al. 1981; Lefcourt et al. 1993; Lyimo et al. 2000). Mean plasma level of cortisol in cattle and buffalo has been observed to be higher early in the morning and lower during evening hours. However, in general the increasing trend in cortisol levels from evening to morning was observed with a peak in early hours (Table 3). Lefcourt et al. (1993) found weak circadian rhythms of cortisol in lactating cows acclimated to a rigidly controlled environment and reported that these could easily be obscured under less rigidly controlled conditions.

The daily rhythm of adrenocorticoid activity has been shown to be as a result of a diurnal rhythm of corticotrophin-releasing factor secretion from the hypothalamus (Takebe et al. 1972). The timing of this cycle is dependent on the sleep–wake activity of the subject rather than on daily environmental variation. Lack of diurnal rhythm or weak circadian rhythm in cows and buffaloes may be because of the dependence of cortisol on the sleep–wake schedule. Ruminants do not enter the deep sleep similar to that observed in man and other animals (Balch 1955). Sleep periods in the ruminant have been related to the digestive needs of the animals since rumination requires both time and consciousness. This was supported by Morag (1967) who found that sheep fed only with finely chopped grass entered states of deep sleep but that when their diet was changed to unlimited amounts of hay, they did not sleep at all.

Alterations of cortisol levels have been observed to be associated with the feeding times in goats (Eriksson and Teravainen 1989), and a circadian rhythm of plasma cortisol concentrations has been observed in pregnant ewes fed once a day, but not in ewes fed throughout the

Table 3 Plasma cortisol concentration (ng/ml) in cattle and buffalo during 24 h

| Time (h) | Cattle | | | | | | Buffalo | | | | | |
| | Winter | | Summer | | | | Winter | | Summer | | | |
	Female	Male	Female	Male			Female	Male	Female	Male		
8:30 AM	1.80±0.29	1.78±0.39	1.39±0.33	1.43±0.55			2.51±0.10	2.32±0.62	2.17±0.26	1.42±0.08		
12:30 PM	1.34±0.23	1.15±0.11	1.33±0.32	1.13±0.32			1.71±0.28	1.69±0.64	1.57±0.17	1.08±0.05		
4:30 PM	1.90±0.12	1.21±0.08	0.99±0.14	1.20±0.35			1.45±0.35	1.53±0.24	1.52±0.29	1.24±0.17		
8:30 PM	1.17±0.07	1.07±0.19	1.18±0.09	0.96±0.18			1.52±0.24	1.61±0.21	1.55±0.25	1.36±0.26		
12:30 AM	1.21±0.10	1.13±0.12	1.32±0.12	1.36±0.20			2.67±0.18	2.03±0.47	2.02±0.24	1.47±0.22		
4:30 AM	1.86±1.17	1.79±0.32	1.44±0.32	1.55±0.40			2.75±0.27	2.62±0.42	2.23±0.37	1.70±0.11		

Source: Aggarwal et al. (2001)

day (Simonetta et al. 1991). This indicates that ruminants exhibit only a weak and low-amplitude intrinsic daily cortisol rhythm that is likely to be masked by other external environmental factors.

Increase in glucocorticoid secretion at the circadian peak depends on increased hypothalamic–pituitary activity (HPA) and on increased sensitivity of the adrenal cortex to ACTH (Jacobson 2005). The stress response is a classical example of rhythmic response. HPA feedback loops occur at different time domains, referred as slow that occurs in response to chronic exposure to glucocorticoids; intermediate and fast feedback are observed both in response to stress and to circadian events (Keller-Wood and Dallman 1984). In response to the fast and intermediate feedbacks, cortisol levels have been observed to follow a circadian variability, in which maximum levels are observed at 2–4 h around the time of awakening and subsequently decrease to a minimum or a nadir at 2–4 h around the time of sleeping. This activity is most likely driven by the hypothalamic SCN (light) and by ventromedial hypothalamus (food) which regulate the expression of the CRH gene. In a cycle of 24 h, there are about 15–18 pulses of corticotrophin of various amplitudes representing quiescent period and an acrophase (Buckley and Schatzberg 2005; Veldhuis et al. 1989).

The role of the cortisol in awakening response is characterised by a sharp increase of cortisol release by about 38–75% of awakening levels, reaching a maximum approximately 30 min after awakening (Pruessner et al. 1999). The cortisol awakening response (CAR) is a distinct phenomenon superimposing the circadian rhythm of cortisol and adding a significant incremental effect to the linear trend of increasing cortisol concentrations in the early morning hours (Wilhelm et al. 2007). Stress is also one of the important factors that influence CAR. Other factors likely to have effect are sex, health status and health behaviour (Fries et al. 2008). The CAR is associated with the stress-related physical and mental symptoms, particularly in anticipation of upcoming demands (Fries et al. 2008).

7.2 Annual Rhythm of Cortisol

Long-term lighting conditions play an important role in the seasonality displayed in behaviour and physiology of many mammalian species (Hastings et al. 1985). The cyclic hormonal changes required for the reproduction and successful production of offspring (Karsch et al. 1984) and seasonal alterations in metabolism result due to variation in environmental conditions (temperature, humidity) and the availability of feed resources and nutrients.

Seasonal changes in the activity and responsiveness of the hypothalamic–pituitary–adrenal (HPA) axis have been reported in most mammalian species particularly in rodents (Boswell et al. 1994), primates (Schiml et al. 1996) and humans (Walker et al. 1997). These seasonal changes in physiology and behaviour are associated with the breeding season, for example, changes in the secretion of hormones involved in the reproduction (Suttie et al. 1992) and growth and metabolic axes. Increased aggressive behaviour and reduced voluntary food intake result in weight loss and an acute rise in cortisol and testosterone levels in male goats (Howland et al. 1985). HPA axis activity can be modulated by changes in reproductive function (Verkerk and Macmillan 1997), metabolic and growth demands and social factors.

Annual changes or photoperiodic modulations of glucocorticoid levels have been observed in ruminants including sheep (Brinklow and Forbes 1984) and bulls (Leining et al. 1980). The variability of patterns depends on the age and sex of animals or on the social structure of the herd (Feher et al. 1994).

7.3 Altered Circadian HPA Activity and Rhythmicity

In order to avoid the genesis of stress-associated glucocorticoid excess, it is important that the circadian rhythm not be disrupted. Physiologically, the increase in glucocorticoid after stressful stimuli is beneficial to improve chances of survival. However, prolonged glucocorticoid excess leads to myriad prejudicial effects. Changes in the glucocorticoid nadir that occurs close to the onset of sleep alter the regulation of HPA function and other glucocorticoid-sensitive endpoints, leading to inadequate glucocorticoid secretion. To avoid glucocorticoid excess, the nadir must be maintained for 4–6 h (Jacobson 2005). However, both the levels of glucocorticoid and the patterns of pulsatility play an important role, as studies have indicated that different pulse sizes and frequencies determine different effects on mineralocorticoid (MR) and glucocorticoid (GR) binding and also probably on MR and GR homodimer and heterodimer formation (Lightman 2008; Reul and de Kloet 1985).

Metabolic syndrome is associated with the disruption of the glucocorticoid nadir, and the elevated 24-h glucocorticoid levels or peak is not the main factor (Dallman et al. 2000). The effects of elevated glucocorticoid secretion at the circadian peak are, however, less well defined. In humans, increased cortisol responses to stress also have been correlated with diminished amplitude of the circadian rhythm (Rosmond et al. 1998).

Stress, either physical and/or psychological, is capable of disrupting the HPA axis homeostasis. These stimuli can originate from extra-hypothalamic sites, such as the catecholaminergic cell groups throughout brainstem, the spinohypothalamic–spinothalamic–spinoreticulothalamic pain pathways, proinflammatory cytokines related or originated from the immune system and psychogenic inputs from the medial prefrontal cortex and from the hippocampus. The intrahypothalamic sites that may regulate HPA axis are the arcuate complex/melanocortin/NPY systems and the periventricular hypothalamic network. These may function as a hypothalamic visceromotor pattern generator constituting of nodes of five preoptic nuclei and the dorsomedial nucleus in the hypothalamus. This intrahypothalamic pattern generator plays an important role in coordinating neuroendocrine, autonomic and behavioural outflows to circadian, immune and psychogenic stimuli (Pecoraro et al. 2006).

Animals under chronic stress have a markedly altered regulation of the circadian and ultradian rhythms. The circadian rhythm is either flattened or lost, and the frequency of corticosterone pulses becomes almost double (Windle et al. 2001).

Leptin, an adipocyte hormone, is a major regulator of neuroendocrine function. Leptin has an overall inhibitory effect on HPA activity (Pralong et al. 1998) and suppressing the appetite-stimulating effects of glucocorticoid. Leptin's effects are explained by its ability to inhibit CRH release, during normal or stressful situations (Heiman et al. 1997). The decrease in leptin levels caused by starvation, a stressful condition, leads to increases in HPA activity (Schwartz and Seeley 1997). In humans with deficiency of leptin, cortisol dynamics is characterised by a higher number of smaller peaks, with smaller morning rise, increased relative variability and increased irregularity in pattern has been observed. Replacement treatment with recombinant human methionyl leptin (r-metHuLeptin) has been observed to organise the dynamics of human HPA (Licinio et al. 2004). In patients with anorexia nervosa, or in highly trained athletes, elevated cortisol levels are probably a consequence of a relative, functional leptin deficiency induced by the paucity of adipose tissue in conjunction with food restriction and intensive energy expenditure.

Stress-induced changes in HPA axis are important to facilitate survival against external or internal stressors. The changes vary in individuals due to myriad factors, such as genetic make-up, neonatal and early-life environment and previous experiences. In addition, many factors regarding the type of stress also play a major role on eliciting the HPA axis response.

The stress-induced responses of the HPA axis overcome the negative feedback provided by endogenous cortisol levels. It is not only the glucocorticoid levels but also their pulsatility and rhythmicity that influence the outcomes determined by hypercortisolism. Stress disrupts the HPA axis rhythmicity by altering the hormonal pulses' amplitude and frequency as well as its circadian cycles.

8   Leptin

Leptin (from the Greek word leptos, meaning thin) is a 16-kDa protein hormone (Zhang et al. 1994), and it is mainly produced by adipose tissues. A small amount of leptin has been shown to be produced by other cells and organs such as bone marrow and placenta (Hoggard et al. 1997; Laharrague et al. 1998; Margetic et al. 2002; Zhao et al. 2004). Leptin is involved in the regulation of body weight, food intake, energy balance and reproduction (Ahima et al. 1997). It is one of the hormones that are closely associated with lipid metabolism (Unger et al. 1999). The rates of leptin secretion and leptin plasma concentration are correlated with total fat mass (Klein et al. 1996). The leptin receptor is expressed in various tissues such as the cerebral cortex, cerebellum, choroid plexus, lung, kidney, skeletal muscles, liver, pancreas, adipose tissue, adrenal medulla and especially in the ventromedial nucleus of the hypothalamus (VHM), also known as the 'satiety centre' (Reidy and Weber 2000). Circulating leptin levels communicate with the brain about the energy supply to regulate appetite and metabolism. It is postulated that leptin's primary role is to provide information to the hypothalamus to regulate energy level based on the amount of energy stored in the adipose tissue. Daily administration of recombinant leptin has been observed to decrease food intake, increase energy expenditures and promote weight loss in mice (Halaas et al. 1997). Leptin acts by inhibiting the activity of neurons containing neuropeptide Y (NPY) and agouti-related peptide (AgRP) and by increasing the activity of neurons expressing α-melanocyte-stimulating hormone (α-MSH). AgRP functions as an endogenous antagonist of the anorectic effect of α-MSH at melanocortin receptors (Arora and Anubhuti 2006). Short-term food deprivation or fasting in cattle reduces leptin mRNA expression in adipose tissue and circulating leptin concentrations (Amstalden et al. 2000). Therefore, leptin plays an important role in regulating energy balance in cattle similar to humans and rodents.

8.1 Daily Rhythm of Leptin

Leptin levels fluctuate in serum of rats (Kalsbeek et al. 2001), mice (Ahima et al. 1998) and humans (Sinha et al. 1996a; Langendonk et al. 1998), with a nocturnal increase and decrease at the end of the dark period and during light. A daily rhythm in plasma leptin levels has been observed in fed Cosmina ewes (Bertolucci et al. 2005) with a minimum level during the light phase and a peak during the dark phase. However, in Syrian and Siberian hamsters, the results have been inconsistent (Gunduz 2002), and no circadian rhythm was observed in sheep and Blackface ewes (Blache et al. 2000). Ruminants have a distinct multistomach digestive tract and functions. The sleeping time is short and transient in cattle (Arave and Albright 1981). Therefore, the circadian or pulsatile pattern of plasma leptin levels in ruminants may be different from that in monogastric animals. Plasma leptin levels show a pulsatile secretion pattern, and about 15 pulses during the 24 h are observed in cattle (Kawakita et al. 2001). Plasma leptin levels reach to a maximum during the daytime and decline to a minimum around the midnight. Leptin secretion showed a pulsatile pattern, but leptin pulses showed irregular periodicity in cattle.

8.2 Annual Rhythm of Leptin

In rodents, leptin has been reported to be a powerful annual regulator of food intake and energy expenditure (Friedman and Halaas 1998) and also affects their reproduction (Ahima et al. 1996). Investigations of the photoperiodic regulation of leptin levels in ruminants are inconclusive and contradictory. Plasma leptin levels and leptin gene expression in perirenal adipose tissue were observed to decrease in ovariectomised ewes exposed to short days, independently of the feeding regime (Bocquier et al. 1998). It was concluded that leptin is modulated by day length independently of food intake, fatness and gonadal activity. Similarly, in ovariectomised cows, serum leptin levels were observed to be lower in winter than in summer, without changes in body weight (Garcia et al. 2002). In the Soay ram, the leptin levels were also found to be lower under short days than long days, indicating that the difference was due to the photoperiod-induced changes in food intake and adiposity rather than on the direct effects of lighting on leptin secretion (Marie et al. 2001). In another study in Soay rams, no difference was found in serum leptin levels between long- and short-day exposed animals, although the mean body weight was somewhat higher under long days (Clarke et al. 2003). In a study, the serum leptin level decreased over 16 weeks under short days and increased under long days, roughly paralleling the changes in body weight of the rams (Lincoln et al. 2001). In seasonal breeding mammals, there are fluctuations in leptin gene expression. In the Djungarian hamster, adipose tissue leptin gene expression was greatly reduced during winter or during exposure to a short photoperiod (Klingenspor et al. 1996). The decrease observed in leptin expression with decreasing photoperiod is attributed to an adaptive behaviour to decrease energy expenditure.

9   Lipid Metabolism

Fatty acids are an important source of energy in body of many organisms. Triglycerides (triacylglycerols or triacylglycerides) are glycerides of the glycerol esterified with three fatty acids. The breakdown of fat stored in fat cells is referred as lipolysis that occurs under the influence of lipases. Triglycerides are broken down into glycerol and free or nonesterified fatty acids (FFA or NEFA) by lipases with the help of bile salts. The hormones adrenaline, noradrenaline, glucagon and adrenocorticotrophin help in inducing lipolysis (Voet and Voet 2002). Triglycerides yield more than twice as much energy for the same mass as carbohydrates or proteins. All cell membranes are made of phospholipids, each of which contains two fatty acids, proteins and cholesterol. Fatty acids are also commonly used for protein modification, and all steroid hormones are derived from fatty acids. The metabolism of fatty acids, therefore, consists of catabolic processes which generate energy and primary metabolites from fatty acids (Heidemann 2002).

9.1   Lipid Metabolism in Ruminants

Cattle and buffaloes differ from monogastric animals in that dietary carbohydrates are degraded in rumen to hexoses and pentoses,

which are then fermented by microorganisms to produce short-chain volatile fatty acids, namely, acetate, propionate and butyrate (Hocquette and Bauchart 1999). These fatty acids get absorbed in the rumen by simple diffusion. Acetate, the major constituent of VFAs, is not greatly metabolised by the liver and is distributed to other tissues to be used as an energy source and substrate for lipogenesis in the adipocytes. However, 80–90% of propionate and butyrate are removed by liver for energy requirements. A rise in the level of the volatile fatty acids in peripheral plasma occurs within 2–4 h of feeding concentrate to the animal (Blum et al. 2000). Triacylglycerols are also hydrolysed by rumen lipases, but most of the absorption of medium- and long-chain fatty acids occurs in the jejunum after interaction with bile and pancreatic lipase. Plasma concentration of FFA in ruminants increases before the morning feeding and decreases rapidly after the feeding (Marie et al. 2001). Rumination occurs mainly during times of rest, and the time spent in ruminating depends on the type of diet and range from almost negligible for diets high in grain to about 10 h per day for diets high in forage.

9.2   Daily Rhythms of Free Fatty Acids (FFA) and Glycerol

Regular daily fluctuations of plasma concentrations of FFA have been observed in rats (Dallman et al. 1999), sheep (Marie et al. 2001) and cattle (Blum et al. 2000). No significant daily rhythm of FFA levels has been observed in dairy cows with free access to feed (Bitman et al. 1990). The daily changes in FFA concentrations occurring during voluntary feeding behaviour have been attributed to the time intervals elapsed between eating due to lipolysis during fasting increases the plasma FFA levels. Rapid changes in the lipolysis/lipogenesis balance have been observed in rats (Escobar et al. 1998; Dallman et al. 1999), sheep (Marie et al. 2001) and cattle (Blum et al. 2000) with restricted feeding. Other factors that are likely to contribute to maintenance of the daily rhythm of lipid metabolism are time and frequency of feeding. Extended fasting in rats abolished the 24-h rhythmicity of FFA concentrations, but after 48 h of fasting, the rhythm reappeared (Escobar et al. 1998). In ruminants fed twice daily, the rise and fall of plasma FFA levels were observed to be associated with the morning feed only, or the peak was more pronounced in the morning than in the afternoon (Blum et al. 2000; Marie et al. 2001). The FFA response to adrenaline administration in cows was observed to be more pronounced in the morning than in the evening (Frohli and Blum 1988). Therefore, the rhythm in FFAs in ruminants is likely to occur only during early part of the day and is associated with feeds and feeding.

9.3   Seasonal Variation of Lipid Metabolism

In mammals, there are profound seasonal fluctuations in metabolism in addition to daily variations (Bartness et al. 2002). These fluctuations are mainly triggered by the photoperiod, but the exact mechanisms underlying this phenomenon are not known, and considerable species differences exist in the regulatory systems. In ruminants in seminatural conditions, plasma concentrations of FFA and glycerol of the reindeer (Rangifer tarandus tarandus) were found to be relatively low in summer, suggesting high lipogenic activity (Larsen et al. 1985), whereas in the Alaskan reindeer, FFA concentrations fluctuated throughout the year without any clear rhythm or trend (Bubenik et al. 1998). In artificial lighting conditions, sheep with freely available food gained about 10 kg during 16 weeks in a long-day condition after an equal period under short days, and their plasma FFA levels were significantly lower at the end of long-day exposure than short-day exposure (Marie et al. 2001). FFA concentrations were found to be significantly lower in long than in short-day exposed underfed ewes, indicating enhanced fat mobilisation in underfed ewes during short days (Bocquier et al. 1998). Variations in enzyme activity of adipose and muscle tissues of sheep exposed to short or long photoperiods demonstrated that sheep have the ability to anticipate future seasonal changes associated with food

resources scarcity even when the food intake is kept constant (Faulconnier et al. 2001).

10   Feeding-Related Circadian Rhythms

Feeding timing has been observed to alter ruminant chronobiology. The time of feed delivery and the time of most intense eating activity affect postprandial rhythms and daily amount of intake (Nikkhah et al. 2011). In order to maximise nutrient efficiency and improve health, nutrient supply to reticulorumen, splanchnic and peripheral tissues should be synchronised with endogenous rhythms in ruminants (Piccione and Caola 2002). Greatest portion of the cyclic pattern of rumination activity in cattle and sheep occurs at night (Woodford and Murphy 1988) and rest. Cattle and buffaloes also display a distinct period of rumination during the day, and several episodes of ruminations may be observed. The circadian basis of rumination may be partially derived from a biological clock influenced by environmental factors such as feeding, milking schedules and patterns of lighting each day (Metz 1975).

A feeding-entrained circadian rhythm, independent of the light–dark period has been described in animals (Stephan 2002). Temporal restrictions of feeding (RF) can phase shift behavioural and physiological circadian rhythms in mammals. These changes in biological rhythms associated with feeding have been postulated to be caused by a food-entrainable oscillator (FEO) not under the control of the SCN (Mieda et al. 2006). Restriction of food or food availability to a single-period schedule at a fixed time of the day in mice induces adaptive response within a few days, and feeding during the period of food availability occurs (Lowrey and Takahashi 2004). Phase advances of circadian rhythms associated with food-anticipatory activity have also been observed in gene expression. The liver, kidney, heart, pancreas and some brain structures participate, independently from the control of the SCN, whose entrainment to light remains intact (Mendoza 2006). Feeding–fasting signals may be involved in the entrainment of the peripheral circadian oscillators (Damiola et al. 2000; Stokkan et al. 2001). The dorsomedial hypothalamic nucleus (DMH) is believed to be a key structure for FEO expression (Gooley et al. 2006; Mieda et al. 2006). The existence of the FEO is not unequivocal. A study on rats with electrolytic DMH lesions does not support (Laundry et al. 2006), and the circadian mechanisms of FEO at the molecular levels are not yet clear (Mendoza 2006). The evidence supporting the existence of FEO feeding-entrained circadian system has been obtained only during restriction of feeding and yet to be fully understood.

The greater appetite for evening versus morning feeding observed in some mammals could partly be due to photoperiod effects. Experiments related to feeding time and lights turned on at 0345 h and turned off at 2245 h or at 1:45 h after evening feeding revealed that the possible anticipation of the light-off by evening-fed cows might contribute to their rapid eating upon feeding (Nikkhah et al. 2011). Since melatonin is known to regulate glucose metabolism in humans and rats (Picinato et al. 2002), the melatonin may contribute to feed intake regulation in ruminants. Exogenous melatonin consumed via drinking water has been found to boost postprandial insulin response (La Fleur et al. 2001). Also, melatonin secretion is known to be induced by dark exposure. In 2100 h fed cows, feeding was observed shortly before lights were put off (2245 h) when melatonin secretion was expected to be high (Nikkhah et al. 2011). In humans, reduced glucose tolerance is linked to elevated melatonin secretion and due to reduced insulin responsiveness and peripheral glucose uptake, which reflects a reduction in glucose and insulin demands (Nikkhah et al. 2006). With such simultaneously reduced glucose tolerance and elevated melatonin secretion in cows, evening eating is likely to increase blood insulin and glucose. Therefore, a decline in peripheral glucose uptake and insulin turnover is expected to occur (Picinato et al. 2002).

Peripheral metabolites such as glucose and VFA are likely to depress feed intake through cell entry and not only by staying in the blood (Forbes 1995). Thus, factors reducing peripheral metabolite uptake can consequently attenuate the metabolite-driven satiety. As such, the expected rise in blood melatonin in the evening may affect and reduce peripheral metabolites partitioning in favour of milk synthesis. Therefore, a higher melatonin level may affect or weaken the feed-driven satiety in evening-fed cows. This may have led evening-fed cows to consume more feed shortly post-feeding compared to morning-fed cows (Nikkhah et al. 2011). An altered regulation of feed intake may cause night-time glucose intolerance in lactating cows (Furedi et al. 2006). Milk energy output was observed to be higher in 2100 h fed cows than in 0900 h fed cows, indicating that at times of high feed intake and hepatic metabolite output, the nutrients were partitioned in favour of milk synthesis by the mammary gland. The greater rumen VFA concentration shortly post-feeding by evening versus morning feeding supports the increased milk output.

11 Circadian Clocks as Mediators of the Homeorhetic Response

The transition from pregnancy to lactation in cows is the most stressful period as during this period homeorhetic adaptations are coordinated across almost every organ of the body and are marked by tremendous changes in hormones and metabolism to cope the increased energetic demands of milk production or lactation. Changes in circadian clocks have been observed to occur in multiple tissues during the transition period in rats indicating that the circadian system regulates the activities in transition phase and similar changes also occur in the cow's physiology to support lactation in transition period. Circadian rhythms coordinate the timing of physiological processes and synchronise these activities in tune with the animal's environment. The effect of the circadian clock on lactation may be through the photoperiod effect, which is accompanied by coordinated changes in the cow's endocrine system and metabolism in response to changes in light and dark period. The exposure to long days (LD) of 16–18 h of light and 6–8 h of darkness increase milk yield by 2–3 kg/day, regardless of the stage of lactation. This LD effect is attributed to increased circulating insulin-like growth factor-I that is independent of any effect on growth hormone concentrations (Dahl et al. 2011). Cows housed under SD during the dry period have higher mammary growth and produce 3–4 kg/day more milk in the subsequent lactation compared with cows on LD when dry. While on SD, circulating prolactin (PRL) diminishes, but expression of PRL receptor increases in mammary, liver and immune cells. The PRL-signalling pathways within these tissues are affected by photoperiod. The replacement of PRL to cows on SD is able to partially reverse the effects of SD on production in the next lactation. Therefore, effects on cows are likely mediated through a PRL-dependent pathway. In prepubertal stage, LD improves mammary parenchymal accumulation and lean body growth which lead to higher yields and production in the first lactation (Dahl et al. 2011). Therefore, photoperiod manipulation can be one of the important tools to improve the efficiency of production across the life cycle of the dairy cow particularly under temperate climate.

Intracellular circadian rhythm generation occurs through an auto-regulatory transcription–translation feedback loop (Reppert and Wever 2001). The positive loop consists of ARNTL (aka BMAL1) and CLOCK gene products (and NPAS2 outside the SCN), and the negative loop consists of the PER and CRY gene products (Gekakis 1998). ARNTL expression is also regulated by Rev-erbα (NR1D1) and Rorα (RORA) that respectively repress or activate ARNTL transcription (Preitner et al. 2002; Guillaumond et al. 2005). The genes that RORA and NR1D1 regulate are often coordinately regulated by these two molecules, and crosstalk between RORA and NR1D1 likely acts to fine-tune their target physiologic networks, such as circadian rhythms, metabolic homeostasis and inflammation (Forman et al. 1994).

Bovine mammary epithelial cells have been postulated to possess a functional clock that can be synchronized by external stimuli and the expression of ARNTL (ARNTL and CLOCK gene products make up core clock elements to generate circadian rhythms) (Casey et al. 2009). A positive limb of the core clock is responsive to prolactin in bovine mammary explants. Others showed that 7% of genes expressed in breasts of lactating women had circadian patterns of expression, and the diurnal variation of composition of cow's milk is associated with changes in expression of mammary core clock genes. These studies indicate that the circadian system coordinates the metabolic and hormonal changes required for lactation initiation and sustaining it. Thus, cow's capacity to produce milk and cope with metabolic stresses in early lactation depends on its ability to set circadian rhythms during the transition period (Casey et al. 2009).

12 Oxidative Stress

Living organisms and their systems attempt to maintain homeostasis by adjusting to the continuous state of flux in the external environment (Crawford and Davies 1994). Circadian synchronisation of physiological events and cellular processes is essential for survival and welfare of the organism. The synchronisation of gene expression to external stimulus can help minimise the damage caused by external stressors (Langmesser and Albrecht 2006) and adaptive process. The redox balance, the balance between oxidants and antioxidants, is one example of such a system that helps maintaining reactive oxygen species (ROS) homeostasis (Krishnan et al. 2008). The rhythmic expression of genes involved in various metabolic pathways and stress resistance has been reported in flies (Ceriani et al. 2002) and in mammals (Miller et al. 2007). Daily rhythms are likely to occur in the expression of some antioxidants (catalase, SOD or GST) and may be involved in protecting the organism from excessive levels of ROS that may damage biological macromolecules (Hardeland et al. 2003; Kondratov 2007). ROS are generated by normal physiological process (Sugino 2006), but when produced in excessive quantities or at rates faster than they are removed by antioxidant mechanisms, ROS may cause oxidative stress (Nordberg and Arner 2001) that, in turn, can result in udder oedema, milk fever, retained placenta and mastitis in cows (Miller and Brezezinska-Slebodzinska 1993). Therefore, the circadian organisation is important for avoiding excessive oxidative stress (Hardeland et al. 2000), the periodical formation of free radicals and other potentially harmful oxidants. Thus, rhythmicities in free radical formation is related to the oxygen consumption in mammals active during the day and with the circadian rhythms of locomotor activity (Hardeland et al. 2000). The stress response varies daily and is governed by rhythmic gene expression under the control of the circadian clock (Simonetta et al. 2008). The circadian clock gene period is essential for maintaining a robust antioxidative defence. Both the ROS production and ROS defence processes are controlled (partly or fully) by the circadian clock. Since ROS production occurs in a circadian manner, circadian variations have been found in many ROS defence systems (Langmesser and Albrecht 2006).

13   Circadian Rhythm of Cardiovascular Parameters

Rhythmic patterns in the activities of cardiovascular system have been well documented in humans (Ruesga and Estrada 1999) and in domestic animals (Piccione et al. 2005). Rhythmic variations in arterial blood pressure and electrocardiographic patterns in the athletic horse indicate that exercise acts as an exogenous synchroniser to modulate the periodicity of the cardiovascular system (Piccione et al. 2001, 2002).

The diurnal variability in cardiovascular functions occurs to match the requirements of different level of activities at different times of the day. Heart rate, stroke volume and blood pressure vary with changes in activity, posture and other external stimuli to adjust oxygen or blood perfusion of specific tissue or organ performing activity. Cardiovascular parameters, such as heart rate and blood pressure, are higher during the day and lower at night (Veerman et al. 1995), parallel

with activation of the adrenergic system and adrenal cortex (Cornelissen and Fagard 2005). These diurnal changes result from both external stimuli and endogenous homeostatic control mechanisms.

14   Conclusions

The melatonin is synthesised in a rhythmic manner with high levels during the dark phase of a natural or any imposed light–dark cycle. The rhythmic levels of the circulating melatonin hormone provide information concerning the temporal position and duration of darkness (i.e. the clock or the oscillator information). Thus, melatonin acts as a coordinator of internal physiological rhythm under the SCN. The widely accepted role of melatonin as one of the most important endogenous mediators of photoperiodic messages governs central or peripheral circadian clock. There is an increasing interest in beneficial effects of melatonin in the therapy of circadian rhythm disorders. An increasing body of evidence indicates that treatment with melatonin is an effective and safe way of manipulating breeding of some photoperiod-sensitive animals such as sheep, goat and deer. The role of melatonin also exists in buffalo, domestic pig, wild pig and horse and may be exploited to increase production of milk, meat and their productivity. The synchronisation of feeding and eating behaviours in ruminants with nutrient utilisation efficiency will help the optimisation of production. Such chronobiological insights into modern ruminant production will offer prospects to improve animal production and health. Ruminant feed intake is an evolutionary bioscience interfacing viable ruminant longevity, adequately safe and secure food supply and quality environment. Knowledge on feeding and eating rhythm and times affect diurnal and postprandial intake patterns will enable predicting diurnal patterns in rumen, post-rumen and peripheral nutrient assimilation. These will suggest optimal, suboptimal and unfavourable times of nutrient supply to mammary cells and milk synthesis.

References

Abilay TA, Johnson HD (1973) Plasma steroids during the ovarian cycle at 18.2 °C temperature. J Anim Sci 37:298–299

Aggarwal A, Kumar P, Upadhyay RC, Singh SV (2001) Circadian variation in plasma levels of cortisol in cattle and buffalo during different seasons. Int J Anim Sci 16:95–98

Aggarwal A, Upadhyay RC, Singh SV, Kumar P (2005) Adrenal-thyroid pineal interaction and effect of exogenous melatonin during summer in crossbred cattle. Indian J Anim Sci 75:915–921

Ahima RS, Prabakaran D, Mantzoros C, Qu D, Lowell B, Maratos-Flier E, Flier JS (1996) Role of leptin in the neuroendocrine response to fasting. Nature 382, 250–252

Ahima RS, Dushay J, Flier SN, Prabakaran D, Flier JS (1997) Leptin accelerates the onset of puberty in normal female mice. J Clin Invest 99:391–395

Ahima RS, Prabakaran D, Flier JS (1998) Postnatal leptin surge and regulation of circadian rhythm of leptin by feeding. Implications for energy homeostasis and neuroendocrine function. J Clin Invest 101:1020–1027

Al-Ghoul WM, Herman MD, Dubocovich ML (1998) Melatonin receptor subtype expression in human cerebellum. Neuroreport 9:4063–4068

Amstalden M, Garcia MR, Williams SW, Stanko RL, Nizielski SE, Morrison CD, Keisler DH, Williams GL (2000) Leptin gene expression, circulating leptin, and luteinizing hormone pulsatility are acutely responsive to short-term fasting in prepubertal heifers: relationships to circulating insulin and insulin-like growth factor I. Biol Reprod 63:127–133

Andersson B (1978) Regulation of water intake. Physiol Rev 58:598

Andersson H, Lillpers K, Rydhmer L, Forsberg M (2000) Influence of light environment and photoperiod on plasma melatonin and cortisol profiles in young domestic boars, comparing two commercial melatonin assays. Domest Anim Endocrinol 19:261–274

Arave CW, Albright JL (1981) Cattle behavior. J Dairy Sci 64:1318–1329

Arendt J, Borbely AA, Franey C, Wright J (1984) The effects of chronic small doses of melatonin given in the late afternoon on fatigue in man: a preliminary study. Neurosci Lett 45:317–321

Arendt J, Aldhous M, Wright J (1988) Synchronisation of a disturbed sleep-wake cycle in a blind man by melatonin treatment. Lancet 1:772–773

Arendt J, Skene DJ, Middelton B, Lockley SW, Deacon S (1997) Efficacy of melatonin treatment in jet lag, shift work and blindness. J Biol Rhythms 12:604–617

Armstrong SM (1989) Melatonin: the internal zeitgeber of mammals. J Pineal Res 7:157–202

Arora S, Anubhuti S (2006) Role of neuropeptides in appetite regulation and obesity – a review. Neuropeptides 40:375–401

Aschoff J (1965) Circadian rhythms in man. Science 148:1427–1432

Aserinsky E (1999) Eyelid condition at birth: relationship to adult mammalian sleep-waking patterns. In: Mallick BN, Inoue S (eds) Rapid eye movement sleep. Naroca Publishing, New Delhi

Balch CC (1955) Sleep in ruminants. Nature 175:940–942

Barry J (1979) Immunofluorescence study of the preoptico-terminal LHRH tract in the female squirrel monkey during the estrous cycle. Cell Tissue Res 198:1–13

Bartness TJ, Powers JB, Hastings MH, Bittman EL, Goldman BD (1993) The timed infusion paradigm for melatonin delivery: what has it taught us about the melatonin signal, its reception, and the photoperiodic control of seasonal responses. J Pineal Res 15: 161–190

Bartness TJ, Song CK, Demas GE (2001) SCN efferents to peripheral tissues: implications for biological rhythms. J Biol Rhythms 16:196–204

Bartness TJ, Demas GE, Song CK (2002) Seasonal changes in adiposity: the roles of the photoperiod, melatonin and other hormones, and sympathetic nervous system. Exp Biol Med 227:363–376

Benitez- King G (1993) Calmoduling mediates melatonin cytoskeletal effects. Experentia 49:635–641

Berson DM, Dunn FA, Takao M (2002) Phototransduction by retinal ganglion cells that set the circadian clock. Science 295:1070–1073

Bertolucci C, Caola G, Foa A, Piccione G (2005) Daily rhythms of serum leptin in ewes. Effect of feeding, pregnancy and lactation. Chronobiol Int 22:817–827

Bitman J, Wood DL, Lefcourt AM (1990) Rhythms in cholesterol, cholesteryl esters, free fatty acids, and triglycerides in blood of lactating dairy cows. J Dairy Sci 73:948–955

Bittman EL (1984) Melatonin and photoperiodic time measurement: evidence from rodents and ruminants. In: Reiter RJ (ed) The pineal gland. Raven, New York, pp 155–191

Blache D, Tellam RL, Chagas LM, Blackberry MA, Vercoe PE, Martin GB (2000) Level of nutrition affects leptin concentrations in plasma and cerebrospinal fluid in sheep. J Endocrinol 65:625–637

Bliss EL, Sandberg AA, Nelson DH, Eik-Nes K (1953) The normal levels of 17-hydroxycorticosteroids in the peripheral blood in man. J Clin Invest 32:818–823

Blum JW, Bruckmaier RM, Vacher PY, Munger A, Jans F (2000) Twenty-four-hour patterns of hormones and metabolites in week 9 and 19 of lactation in high-yielding dairy cows fed triglycerides and free fatty acids. J Vet Med 47:43–60

Bocquier F, Bonnet M, Faulconnier Y, Guerre-Millo M, Martin P, Chilliard Y (1998) Effects of photoperiod and feeding level on perirenal adipose tissue metabolic activity and leptin synthesis in the ovariectomized ewe. Reprod Nutr Dev 38:489–498

Boswell T, Woods SC, Kenagy GJ (1994) Seasonal changes in body mass, insulin, and glucocorticoids of free-living golden-mantled ground squirrels. Gen Comp Endocrinol 96:339–346

Bothorel B, Barassin S, Saboureau M, Perreau S, Vivien-Roels B, Malan A, Pévet P (2002) In the rat, exogenous melatonin increases the amplitude of pineal melatonin secretion by a direct action on the circadian clock. Eur J Neurosci 16:1090–1098

Brainard GC, Richardson BA, King TS, Matthews SA, Reiter RJ (1983) The suppression of pineal melatonin content and N-acetyltransferase activity by different light irradiances in the Syrian hamster: a dose–response relationship. Endocrinology 113:293–296

Brainard GC, Podolin PL, Leivy SW, Rollag MD, Cole C, Barker FM (1986) Near-ultraviolet radiation suppresses pineal melatonin content. Endocrinology 119:2201–2205

Brainard GC, Hanifin JP, Greeson JM, Byrne B, Glickman G, Gerner E, Rollag MD (2001) Action spectrum for melatonin regulation in humans: evidence for a novel circadian photoreceptor. J Neurosci 21:6405–6412

Brinklow BR, Forbes JM (1984) Effect of pinealectomy on the plasma concentrations of prolactin, cortisol and testosterone in sheep in short and skeleton long photoperiods. J Endocrinol 100:287–294

BSCS (2003) Sleep, sleep disorders, and biological rhythms, NIH publication no. 04-4989. National Institutes of Health, Bethesda

Bubenik GA, Schams D, White RG, Rowell J, Blake J, Bartos L (1998) Seasonal levels of metabolic hormones and substrates in male and female reindeer (Rangifer tarandus). Comp Biochem Physiol 120:307–315

Buckley TM, Schatzberg AF (2005) On the interactions of the hypothalamic-pituitary-adrenal (HPA) axis and sleep: normal HPA axis activity and circadian rhythm, exemplary sleep disorders. J Clin Endocrinol Metab 90:3106–3114

Buijs RM, Scheer FA, Kreier F, Yi C, Bos N, Goncharuk VD, Kalsbeek A (2006) Organization of circadian functions: interaction with body. Prog Brain Res 153:341–360

Cagnacci A (1996) Melatonin in relation to physiology in adult humans. J Pineal Res 21:200–213

Campbell SS, Tobler I (1984) Animal sleep: a review of sleep duration across phylogeny. Neurosci Biobehav Rev 8:269–300

Cardinali DP, Larin F, Wurtman RJ (1972) Control of the rat pineal gland by light spectra. Proc Natl Acad Sci USA 69:2003–2005

Casey T, Patel O, Dykema K, Dover H, Furge K, Plaut K (2009) Molecular signatures reveal circadian clocks May orchestrate the homeorhetic response to lactation. PLoS One 4(10):e7395. doi:10.1371/journal.pone.0007395

Cassone VM, Stephan FK (2002) Central and peripheral regulation of feeding and nutrition by the mammalian circadian clock: implications for nutrition during manned space flight. Nutrition 18:814–819

Cassone VM, Roberts MH, Moore RY (1988) Effects of melatonin on 2-deoxy- [1-14C] glucose uptake within rat suprachiasmatic nucleus. Am J Physiol 255:332–337

Ceriani MF, Hogenesch JB, Yanovsky M, Panda S, Straume M, Kay SA (2002) Genome-wide expression analysis in Drosophila reveals genes controlling circadian behaviour. J Neurosci 22:9305–9319

Chesworth MJ, Cassone VM, Armstrong SM (1987) Effects of daily melatonin injections on activity rhythms of rats in constant light. Am J Physiol 253:101–107

Clarke IJ (2001) Sex and season are major determinants of voluntary food intake in sheep. Reprod Fertil Dev 13:577–582

Clarke IJ, Rao A, Chilliard Y, Delavaud C, Lincoln GA (2003) Photoperiod effects on gene expression for hypothalamic appetite-regulating peptides and food intake in the ram. Am J Physiol 284:101–115

Conti A, Conconi S, Hertens E, Skwarlo- Sonta K, Markowska M, Maestroni GJM (2000) Evidence for melatonin synthesis in mouse and human bone marrow cells. J Pineal Res 28:193–202

Cornelissen VA, Fagard RH (2005) Effects of endurance training on blood pressure, blood pressure-regulating mechanisms, and cardiovascular risk factors. Hypertension 46(4):667–675

Crawford DR, Davies KJ (1994) Adaptive response and oxidative stress. Environ Health Perspect 102:25

Daan S, Albrecht U, Van der Horst T, Illnerova H, Roenneberg T, Wehr TA, Schwartz WJ (2001) Assembling a clock for all seasons: are there M and E oscillators in the genes? J Biol Rhythms 16:105–116

Dahl GE, Tao S, Thompson IM (2011) Effects of photoperiod on mammary gland development and lactation. doi: 10.2527/jas.2011-4630

Damiola F, Le Minh N, Preitner N, Kornmann B, Fleury- Olela F, Schibler U (2000) Restricted feeding uncouples circadian oscillators in peripheral tissues from the central pacemaker in the suprachiasmatic nucleus. Genes Dev 14:2950–2961

Dallman MF, Akana SF, Bhatnagar S, Bell ME, Choi SJ, Chu A, Horsley C, Levin N, Meijer O, Soriano LR, Strack AM, Viau V (1999) Starvation: early signals, sensors, and sequelae. Endocrinology 140:4015–4023

Dallman MF, Akana SF, Bhatnagar S, Bell ME, Strack AM (2000) Bottomed out: metabolic significance of the circadian trough in glucocorticoid concentrations. Int J Obes Relat Metab Disord 24(Suppl 2):40–46

de Souza CJ, Meier AH (1987) Circadian and seasonal variations of plasma insulin and cortisol concentrations in the Syrian hamster, Mesocricetus auratus. Chronobiol Int 4:141–151

Depres- Brummer P, Levi F, Metzger G, Touitou Y (1995) Light-induced suppression of the rat circadian system. Am J Physiol 37:1111–1116

Dunlap JC, Loros JJ, DeCoursey PJ (2004) Fundamental properties of circadian rhythms. In: Chronobiology–biological timekeeping. Sinauer Associates Inc., Sunderland

Eriksson L, Teravainen TL (1989) Circadian rhythm of plasma cortisol and blood glucose in goats. Asian-Australas J Anim Sci 2:202–203

Escobar C, Diaz-Munoz M, Encinas F, Aguilar-Roblero R (1998) Persistence of metabolic rhythmicity during fasting and its entrainment by restricted feeding schedules in rats. Am J Physiol 274:1309–1316

Faulconnier Y, Bonnet M, Bocquier F, Leroux C, Chilliard Y (2001) Effects of photoperiod and feeding level on adipose tissue and muscle lipoprotein lipase activity and mRNA level in dry non-pregnant sheep. Br J Nutr 85:299–306

Feher T, Zomborszky Z, Sandor E (1994) Dehydroepiandrosterone, dehydroepiandrosterone sulphate, and their relation to cortisol in red deer (Cervus elaphus). Comp Biochem Physiol 109:247–252

Fjaerli O, Lund T, Osterud B (1999) The effect of melatonin on cellular activation processes in human blood. J Pineal Res 26:50–55

Forbes JM (1995) Voluntary food intake and diet selection in farm animals. CABI Int, Wallingford

Forman BM, Chen J, Blumberg B, Kliewer SA, Henshaw R, Ong ES, Evans RM (1994) Cross-talk among ROR alpha 1 and the Rev-erb family of orphan nuclear receptors. Mol Endocrinol 8:1253–1261

Freedman MS, Lucas RJ, Soni B, von Schantz M, Munoz M, David-Gray Z, Foster R (1999) Regulation of mammalian circadian behavior by non-rod, non-cone, ocular photoreceptors. Science 284:502–504

Friedman J, Halaas JL (1998) Leptin and the regulation of body weight in mammals. Nature 395:763–770

Fries E, Dettenborn L, Kirschbaum C (2008) The cortisol awakening response (CAR): facts and future directions. Int J Psychophysiol 72:67–73

Frohli DM, Blum JW (1988) Nonesterified fatty acids and glucose in lactating dairy cows: diurnal variations and changes in responsiveness during fasting to epinephrine and effects of beta-adrenergic blockade. J Dairy Sci 71:1170–1177

Fulkerson WJ, Sawyer GJ, Gow CB (1980) Investigations of ultradian and circadian rhythms in the concentration of cortisol and prolactin in the plasma of dairy cattle. Aust J Biol Sci 33:557–561

Furedi C, Kennedy AD, Nikkhah A et al (2006) Glucose tolerance and diurnal variation of circulating insulin in evening and morning fed lactating cows. Adv Dairy Technol 18:356

Ganong WF (1997) Review of medical physiology. A lange medical book, 18th edn. Appleton and Lange, Stamford, p 433

Garcia MR, Amstalden M, Williams SW, Stanko RL, Morrison CD, Keisler DH, Nizielski SE, Williams GL (2002) Serum leptin and its adipose gene expression during pubertal development, the estrous cycle, and different seasons in cattle. J Anim Sci 80:2158–2167

Gekakis N (1998) Role of the CLOCK protein in the mammalian circadian mechanism. Science 280:1564–1569

Goldman BD (2001) Mammalian photoperiodic system: formal properties and neuroendocrine mechanisms of photoperiodic time measurement. J Biol Rhythms 16:283–301

Gooley JJ, Schomer A, Saper CB (2006) The dorsomedial hypothalamic nucleus is critical for the expression of food-entrainable circadian rhythms. Nat Neurosci 9:398–407

Gorman MR, Goldman BD, Zucker I (2001) Mammalian photoperiodism. In: Handbook of behavioral neurobiology, vol 12. Plenum/Kluwer, New York, pp 481–508

Greco D (2002) Endocrine glands and their function. In: Cunningham textbook of veterinary physiology, 3rd edn. W.B. Saunders Company, Philadelphia, pp 341–372

Griffith MK, Minton JE (1992) Effect of light intensity on circadian profiles of melatonin, prolactin, ACTH, and cortisol in pigs. J Anim Sci 70:492–498

Guillaumond F, Dardente H, Giguère V, Cermakian N (2005) Differential control of Bmal1 circadian transcription by REV-ERB and ROR nuclear receptors. J Biol Rhythms 20:391–403

Guillemin R, Dear WE, Liebelt RA (1959) Nychthemeral variations in plasma free corticosteroid levels of the rat. Proc Soc Exp Biol Med 101:394–395

Gunduz B (2002) Daily rhythm in serum melatonin and leptin levels in the Syrian hamster (Mesocricetus auratus). Comp Biochem Physiol 132:393–401

Halaas JL, Boozer C, Blair-West J, Fidahusein N, Denton DA, Friedman JM (1997) Physiological response to long-term peripheral and central leptin infusion in lean and obese mice. Proc Natl Acad Sci 94:8878–8883

Hardeland R, Coto-Montes A, Burkhardt S, Zsizsik BK (2000) Circadian rhythms and oxidative stress in nonvertebrate organisms. In: Vanden Driessche T (ed) The redox state and circadian rhythms. Kluwer Academic Publishers, Dordrecht, pp 121–126

Hardeland R, Coto-Montes A, Poeggeler B (2003) Circadian rhythms, oxidative stress, and antioxidative defense mechanisms. Chronobiol Int 20:921–962

Hardin P (2009) Molecular mechanisms of circadian timekeeping in Drosophila. Sleep Biol Rhythms 7:235–242

Hasting MH (2001) Adaptation to seasonal change: photoperiodism and its mechanism. J Biol Rhythms 16:283–430

Hastings JW, Schweiger HG (1975) The molecular basis of circadian rhythms. Dahlem Workshop Report

Hastings MH, Herbert J, Martensz ND, Roberts AC (1985) Annual reproductive rhythms in mammals: mechanisms of light synchronization. Ann N Y Acad Sci 453:182–204

Heidemann SR (2002) The molecular and cellular bases of physiologic regulation. In: Cunningham textbook of veterinary physiology, 3rd edn. W.B. Saunders Company, Philadelphia, pp 2–29

Heiman ML, Ahima RS, Craft LS, Schoner B, Stephens TW, Flier JS (1997) Leptin inhibition of the hypothalamic-pituitary-adrenal axis in response to stress. Endocrinology 138:3859–3863

Hendricks JC, Finn SM, Panckeri KA, Chavkin J, Williams JA, Sehgal A, Pack AI (2000) Rest in Drosophila is a sleep-like state. Neuron 25:129–138

Hocquette JF, Bauchart D (1999) Intestinal absorption, blood transport and hepatic and muscle metabolism of fatty acids in preruminant and ruminant animals. Reprod Nutr Dev 39:27–48

Hoffman R, Reiter RJ (1965) Pineal gland: influence on gonads of male hamsters. Science 148:1609–1611

Hoggard N, Hunter L, Duncan JS, Williams LM, Trayhurn P, Mercer JG (1997) Leptin and leptin receptor mRNA and protein expression in the murine fetus and placenta. Proc Natl Acad Sci USA 94:11073–11078

Homna K, Hiroshige T (1978) Endogenous ultradian rhythms in rats exposed to prolonged continuous light. Am J Physiol 235:250–256

Howland BE, Sanford LM, Palmer WM (1985) Changes in serum levels of LH, FSH, prolactin, testosterone, and cortisol associated with season and mating in male pygmy goats. J Androl 6:89–96

Hudson S, Mullord M, Whittlestone WG, Payne E (1975) Diurnal variations in blood cortisol in the dairy cow. J Dairy Sci 58:30–33

Ikonomov OG, Stoynev AG, Shisheva AC (1998) Integrative coordination of circadian mammalian diversity: neuronal networks and peripheral clocks. Prog Neurobiol 54:87–97

Ingram JR, Crockford JN, Matthews LR (1999) Ultradian, circadian and seasonal rhythms in cortisol secretion and adrenal responsiveness to ACTH and yarding in unrestrained red deer (Cervus elaphus) stags. J Endocrinol 162:289–300

Jac M, Kiss A, Sumova A, Illnerova H, Jezova D (2000) Daily profiles of arginine vasopressin mRNA in the suprachiasmatic, supraoptic and paraventricular nuclei of the rat hypothalamus under various photoperiods. Brain Res 887:472–476

Jacobson L (2005) Hypothalamic-pituitary-adrenocortical axis regulation. Endocrinol Metab Clin North Am 34:271–292

Kalsbeek A, Buijs RM (2002) Output pathways of the mammalian suprachiasmatic nucleus: coding circadian time by transmitter selection and specific targeting. Cell Tissue Res 309:109–118

Kalsbeek A, Fliers E, Romijn JA, La Fleur SE, Wortel J, Bakker O, Endert E, Buijs RM (2001) The suprachiasmatic nucleus generates the diurnal changes in plasma leptin levels. Endocrinology 142:2677–2685

Kappers JA (1960) The development, topographical relations and innervation of the epiphysis cerebri in the albino rat. Z Zellforsch 52:163–215

Kappers JA (1979) Short history of pineal discovery and research. In: Kappers JE, Pévet P (eds) The pineal gland of vertebrates including man, Progress in brain research 52. Elsevier, Amsterdam, pp 3–22

Karsch FJ, Bittman EL, Foster DL, Goodman RL, Legan SJ, Robinson JE (1984) Neuroendocrine basis of seasonal reproduction. Rec Prog Horm Res 40:185–232

Karsch FJ, Woodfill CJI, Malpaux B, Robinson JE, Wayne NL (1991) Melatonin and mammalian photoperiodism: synchronization of annual reproductive cycles. In: Klein DC, Moore RY, Reppert SM (eds) Suprachiasmatic nucleus: the mind's clock. Oxford University Press, New York, pp 217–232

Kawakita Y, Abe H, Hodate K (2001) Twenty four hour variation of plasma leptin concentration and pulsatile leptin secretion in cattle. Asian Aust J Anim Sci 14:1209–1215

Keller-Wood ME, Dallman MF (1984) Corticosteroid inhibition of ACTH secretion. Endocr Rev 5:1–24

Kennaway DJ, Obst JM, Dunstan EA, Friesen HG (1981) Ultradian and seasonal rhythms in plasma gonadotropins, prolactin, cortisol, and testosterone in pinealectomized rams. Endocrinology 108:639–646

Klein KC, Moore RY (1979) Pineal N-acetyltransferase and hydroxyindole-O methyltransferase: control by the retinohypothalamic tract and the suprachiasmatic nucleus. Brain Res 174:245–262

Klein DC, Moore RY, Reppert SM (1991) Suprachiasmatic nucleus: the mind's clock. Oxford University Press, New York

Klein S, Coppack SW, Mohamed-Ali V, Landt M (1996) Adipose tissue leptin production and plasma leptin kinetics in humans. Diabetes 45:984–987

Klingenspor M, Dickopp A, Heldmaier G, Klaus S (1996) Short photoperiodic reduces leptin gene expression in white and brown adipose tissue of Djungarian hamsters. FEBS Lett 16:290–294

Kokkonen UM, Riskila P, Roihankorpi MT, Soveri T (2001) Circadian variation of plasma atrial natriuretic peptide, cortisol and fluid balance in the goat. Acta Physiol Scand 171:1–8

Kondratov RV (2007) A role of the circadian system and circadian proteins in aging. Ageing Res Rev 6:12–27

Korf HW, Schomerus C, Maronde E, Stehle JH (1996) Signal transduction molecules in the rat pineal organ: Ca2+, pCREB, and ICER. Naturwissenschaften 83:535–543

Kramer MK, Sothern RB (2001) Circadian characteristics of corticosterone secretion in red-backed voles (Clethrionomys gapperi). Chronobiol Int 18:933–945

Krishnan N, Davis AJ, Giebultowicz JM (2008) Circadian regulation of response to oxidative stress in *Drosophila melanogaster*. Biochem Biophys Res Commun 374:299–303

Kryger MH, Roth Y, Dement WC (1989) Principles and practice of sleep medicine. W.B. Saunders, Philadelphia

Kunz D, Schmitz S, Mahlberg R, Mohr A, Stoter C, Wolf KJ, Hermann WM (1999) A new concept for melatonin deficit: on pineal calcification and melatonin excretion. Neuropsychopharmacology 21:765–772

la Fleur SE (2003) Daily rhythms in glucose metabolism: suprachiasmatic nucleus output to peripheral tissue. J Neuroendocrinol 15:315–322

La Fleur SE, Kalsbeek A, Wortel J et al (2001) Role for the pineal and melatonin in glucose homeostasis: pinealectomy increases night-time glucose concentrations. J Neuroendocr 13:1025–1032

Laharrague P, Larrouy D, Fontanilles AM, Truel N, Campfield A, Tenenbaum R, Galitzky J, Corberand JX, Pénicau L, Casteilla L (1998) High expression of leptin in human bone marrow adipocytes in primary culture. FASEB J 12:747–752

Lane EA, Moss HB (1985) Pharmacokinetics of melatonin in man: first pass hepatic metabolism. J Clin Endorinol Metab 61:1214–1216

Langendonk JG, Pijl H, Toornvliet AC, Burggraaf J, Frolich M, Schoemaker RC, Doornbos J, Cohen AF, Meinders AE (1998) Circadian rhythm of plasma leptin levels in upper and lower body obese women: influence of body fat distribution and weight loss. J Clin Endocrinol Metab 83:1706–1712

Langmesser S, Albrecht U (2006) Life time-circadian clock, mitochondria and metabolism. Chronobiol Int 23:151–157

Larsen TS, Lagercrantz H, Riemersma RA, Blix AS (1985) Seasonal changes in blood lipids, adrenaline, noradrenaline, glucose and insulin in Norwegian reindeer. Acta Physiol Scand 124:53–59

Laundry GJ, Simon MM, Webb IC, Mistlberger RE (2006) Persistence of a behavioral food-anticipatory circadian rhythm following dorsomedial hypothalamic ablation in rats. Am J Physiol 290:R1527–R1534

Lee T, Zucker I (1991) Suprachiasmatic nucleus and photic entrainment of circannual rhythms in ground squirrels. J Biol Rhythms 6:315–330

Lefcourt AM, Bitman J, Kahl S, Wood DL (1993) Circadian and ultradian rhythms of peripheral cortisol concentrations in lactating dairy cows. J Dairy Sci 76:2607–2612

Leining KB, Tucker HA, Kesner JS (1980) Growth hormone, glucocorticoid and thyroxine response to duration, intensity and wavelength of light in prepubertal bulls. J Anim Sci 51:932–942

Leproult R, Colecchia EF, L'Hermite-Baleriaux M, Van Cauter E (2001) Transition from dim to bright light in the morning induces an immediate elevation of cortisol levels. J Clin Endocrinol Metab 86:151–157

Lerner AB, Case JD, Takahashi Y, Lee TH, Mori N (1958) Isolation of melatonin, the pineal gland factor that lightens melanocytes. J Am Chem Soc 80:2587

Lerner AB, Case JD, Heinzelmann RV (1959) Structure of melatonin. J Am Chem Soc 81:6084–6085

Lewy AJ, Wehr TA, Goodwin FK, Newsome DA, Markey SP (1980) Light suppresses melatonin secretion in humans. Science 210:1267–1269

Lewy AJ, Emens JS, Lefler BJ, Yuhas K, Jackman AR (2005) Melatonin entrains free-running blind people according to a physiological dose–response curve. Chronobiol Int 22:1093–1106

Licinio J, Caglayan S, Ozata M, Yildiz BO, de Miranda PB, O'Kirwan F, Whitby R, Liang L, Cohen P, Bhasin S, Krauss RM, Veldhuis JD, Wagner AJ, DePaoli AM, McCann SM, Wong ML (2004) Phenotypic effects of leptin replacement on morbid obesity, diabetes mellitus, hypogonadism, and behavior in leptin-deficient adults. Proc Natl Acad Sci U S A 101:4531–4536

Lightman SL (2008) The neuroendocrinology of stress: a never ending story. J Neuroendocrinol 20:880–884

Lincoln GA, Almeida OFX, Klandorf H, Cunningham RA (1982) Hourly fluctuations in the blood levels of melatonin, prolactin, luteinizing hormone, follicle stimulating hormone, testosterone, tri-iodothyronine, thyroxine and cortisol in rams under artificial photoperiods, and the effects of cranial sympathectomy. J Endocrinol 92:237–250

Lincoln GA, Rhind SM, Pompolo S, Clarke IJ (2001) Hypothalamic control of photoperiod-induced cycles in food intake, body weight, and metabolic hormones in rams. Am J Physiol 281:76–90

Lincoln GA, Anderson H, Loudon A (2003) Clock genes in calendar cells as the basis of annual timekeeping in mammals – a unifying hypothesis. J Endocrinol 179:1–13

Lopez-Gonzalez MA, Calvo JR, Osuma C, Guerrero JM (1992) Interaction of melatonin with human lymphocytes: evidence for binding sites coupled to potentation of cyclic AMP stimulated by vasoactive intestinal peptide and activation of cyclic GMP. J Pineal Res 12:97–104

Lowrey PL, Takahashi JS (2004) Mammalian circadian biology: elucidating genome-wide levels of temporal organization. Annu Rev Genomics Hum Genet 5:407–441

Lyimo ZC, Nielen M, Ouweltjes W, Kruip TAM, van Eerdenburg FJCM (2000) Relationship among estradiol, cortisol and intensity of estrous behavior in dairy cattle. Theriogenology 53:1783–1795

MacAdam WR, Eberhart RJ (1972) Diurnal variation in plasma corticosteroid concentration in dairy cattle. J Dairy Sci 55:1792–1795

Mallo C, Zaidan R, Galy G, Vermeulen E, Brun J, Chazot G, Claustrat B (1990) Pharmacokinetics of melatonin in man after intravenous infusion and bolus injection. Eur J Clin Pharmacol 38:297–301

Malpaux B, Daveau A, Maurice F, Gayrard V, Thierry JC (1993) Short-day effects of melatonin on luteinizing hormone secretion in the ewe: evidence for central sites of action in the mediobasal hypothalamus. Biol Reprod 48:752–760

Malpaux B, Migaud M, Tricoire H, Chemineau P (2001) Biology of mammalian photoperiodism and the critical role of the pineal gland and melatonin. J Biol Rhythms 16:336–347

Margetic S, Gazzola C, Pegg GG, Hill RA (2002) Leptin: a review of its peripheral actions and interactions. Int J Obes 26:1407–1433

Marie M, Findlay PA, Thomas L, Adam CL (2001) Daily patterns of plasma leptin in sheep: effects of photoperiod and food intake. J Endocrinol 170:277–286

Martin XD, Malina HZ, Brennan MC, Hendrikson PH, Lichter PR (1992) The ciliary body-the third organ found to synthesize idoleamines in humans. Eur J Ophthalmol 2:67–72

McArthur AJ, Gillette MU, Prosser RA (1991) Melatonin directly resets the rat suprachiasmatic circadian clock in vitro. Brain Res 565:158–161

McCord CP, Allen FP (1917) Evidence associating pineal gland function with alterations in pigmentation. J Exp Zool 23:207–243

Meijer JH, Rietveld WJ (1989) Neurophysiology of the suprachiasmatic circadian pacemaker in rodents. Physiol Rev 69:671–707

Mendoza J (2006) Circadian clocks: setting time by food. J Neuroendocrinol 19:127–137

Menet J, Vuillez P, Jacob N, Pevet P (2001) Intergeniculate leaflets lesion delays but does not prevent the integration of the photoperiodic change by the suprachiasmatic nuclei. Brain Res 906:176–179

Messager S, Ross AV, Barret PJ, Morgan PJ (1999) Decoding photoperiodic time through Per1 and ICER gene amplitude. Proc Natl Acad Sci USA 96:9938–9943

Messager S, Hazlerigg DG, Mercer JG, Morgan PJ (2000) Photoperiod differentially regulates the expression of Per1 and ICER in the pars tubelaris and the suprachiasmatic nucleus of the Siberian hamster. Eur J Neurosci 12:2865–2870

Messager S, Garabette ML, Hastings MH, Hazlerigg DG (2001) Tissue-specific abolition of Per1 expression in the pars tubelaris by pinealectomy in the Syrian hamster. Neuroreport 12:579–582

Metz JHM (1975) Time patterns of feeding and rumination in domestic cattle, Landbouwhogeschool Wageningen; Mededelingen. H. Veenman, Wageningen, 75–12 274 pp

Mieda M, Williams SC, Richardson JA, Tanaka K, Yanagisawa M (2006) The dorsomedial hypothalamic nucleus as a putative food-entrainable pacemaker. Proc Natl Acad Sci USA 103:12150–12155

Mikkelsen JD, Hauser F, Olcese J (2000) Neuropeptide Y (NPY) and NPY receptors in the rat pineal gland. In: Olcese J (ed) Melatonin after four decades. Kluwer Academic/Plenum Press, New York, pp 95–107

Miller JK, Brezezinska-Slebodzinska E (1993) Oxidative stress and antioxidants in disease: oxidative animal function. J Dairy Sci 76:502–511

Miller BH, McDearmon EL, Panda S, Hayes KR, Zhang J, Andrews JL, Antoch MP, Walker JR, Esser KA, Hogenesch JB, Takahashi JS (2007) Circadian and CLOCK-controlled regulation of the mouse transcriptome and cell proliferation. Proc Natl Acad Sci USA 104:3342–3347

Moller M (1992) The structure of the pinealopetal innervation of the mammalian pineal gland. Microsc Res Tech 21:188–204

Moore RY (1997) Circadian rhythms: basic neurobiology and clinical applications. Annu Rev Med 48:253–266

Moore RY, Eichler VB (1972) Loss of a circadian adrenal corticosterone rhythm following suprachiasmatic lesions in the rat. Brain Res 13:201–206

Moore RY, Klein DC (1974) Visual pathways and the central neural control of a circadian rhythm in pineal serotonin N-acetyltransferase activity. Brain Res 71:17–33

Morag M (1967) Influence of diet on the behavior pattern of sheep. Nature (London) 213:110–115

Morgan PJ, Ross AW, Mercer JG, Barrett P (2003) Photoperiodic programming of body weight through the neuroendocrine hypothalamus. J Endocrinol 177:27–34

Nelson RJ, Drazen DL (1999) Melatonin mediates seasonal adjustments in immune function. Reprod Nutr Dev 39:383–398

Nikkhah A, Plaizier JC, Furedi CJ et al (2006) Response in diurnal variation of circulating blood metabolites to nocturnal vs diurnal provision of fresh feed in lactating cows. J Anim Sci 84:111

Nikkhah A, Furedi CJ, Kennedy AD et al (2011) Feed delivery at 2100 h vs. 0900 h for lactating dairy cows. Can J Anim Sci 91:113–122

Nordberg J, Arner ESJ (2001) Reactive oxygen species, antioxidants, and the mammalian thioredoxin system. Free Radic Biol Med 31:1287–1312

Nuesslein-Hildesheim B, O'Brien JA, Ebling FJP, Maywood ES, Hastings MH (2000) The circadian cycle of mPER clock gene products in the suprachiasmatic nucleus of the Siberian hamster encodes both daily and seasonal time. Eur J Neurosci 12:2856–2864

Orth DN, Island DP (1969) Light synchronization of the circadian rhythm in plasma cortisol (17-OHCS) concentration in man. J Clin Endocrinol Metabol 29:479–486

Pecoraro N, Dallman MF, Warne JP, Ginsberg AB, Laugero KD, la Fleur SE, Houshyar H, Gomez F, Bhargava A, Akana SF (2006) From Malthus to motive: how the HPA axis engineers the phenotype, yoking needs to wants. Prog Neurobiol 79:247–340

Perlow MJ, Reppert SM, Boyar RM, Klein DC (1981) Daily rhythms in cortisol and melatonin in primate cerebrospinal fluid. Effects of constant light and dark. Neuroendocrinology 32:193–196

Pévet P, Bothorel B, Slotten H, Saboureau M (2002) The chronobiotic properties of melatonin. Cell Tissue Res 309:183–191

Phansuwan-Pujito P, Mikkelsen JD, Govitrapong P, Moller M (1991) A cholinergic innervation of the bovine pineal gland visualized by immunohistochemical detection of choline acetyltransferase-immunoreactive nerve fibers. Brain Res 545:49–58

Piccione G, Caola G (2002) Review: biological rhythms in livestock. J Vet Sci 3:145–157

Piccione G, Assenza A, Attanzio G, Fazio F, Caola G (2001) Chronophysiology of arterial blood pressure and heart rate in athletic horses. Slov Vet Res 38:243–248

Piccione G, Caola G, Refinetti R (2002) Maturation of the daily body temperature rhythm in sheep and horse. J Therm Biol 27:333–336

Piccione G, Caola G, Refinetti R (2003) Daily and estrous rhythmicity of body temperature in domestic cattle. http://www.biomedcentral.com/1472-6793/3/7

Piccione G, Grasso F, Giudice E (2005) Circadian rhythm in the cardiovascular system of domestic animals. Res Vet Sci 79:155–160

Picinato MC, Haber EP, Carpinelli, AR (2002) Daily rhythm of glucose-induced insulin secretion by isolated islets from intact and pinealectomized rat. Journal of Pineal Research 33:172–177

Pitrosky B, Masson-Pevet M, Kirch R, Vivien-Roels B, Canguilhem B, Pevet P (1991) Effects of different doses and duration of melatonin infusions on plasma melatonin concentrations in pinealectomized Syrian hamster: consequences at the level of sexual activity. J Pineal Res 11:149–155

Pittendrigh CS, Daan S (1976) A functional analysis of circadian pacemakers in nocturnal rodents. V. Pacemaker structure: a clock for all seasons. J Comp Physiol A106:333–355

Poon AMS, Mak ASY, Luk HT (1996) Melatonin and iodomelatonin binding sites in the human colon. Endocrinol Res 25:77–94

Pralong FP, Roduit WG, Castillo E, Mosimann F, Thorens B, Gaillard RC (1998) Leptin inhibits directly glucocorticoid secretion by normal human and rat adrenal gland. Endocrinology 139:4264–4268

Preitner N, Damiola F, Luis Lopez M, Zakany J, Duboule D, Albrecht U, Schibler U (2002) The orphan nuclear receptor REV-ERB[alpha] controls circadian transcription within the positive limb of the mammalian circadian oscillator. Cell 110:251–260

Pruessner JC, Hellhammer DH, Kirschbaum C (1999) Burnout, perceived stress, and cortisol responses to awakening. Psychosom Med 61:197–204

Quay WB (1963) Circadian rhythm in rat pineal serotonin and its modification of estrous cycle and photoperiod. Gen Comp Endocrinol 3:1473–1479

Quay WB (1964) Circadian and estrous rhythms in pineal melatonin and 5-hydroindole- 3-indole acetic acids. Proc Soc Exp Biol Med 115:710–714

Rattenborg NC, Lima SL, Amlaner CJ (1999) Half-awake to the risk of predation. Nature 397:397–398

Rechtschaffen A (1998) Current perspectives on the function of sleep. Perspect Biol Med 41:359–390

Redman J, Armstrong S, Ng KT (1983) Free-running activity rhythms in the rat: entrainment by melatonin. Science 219:1089–1091

Refinetti R (2005) Circadian physiology, 2nd edn. CRC Press, Boca Raton

Reidy SP, Weber J (2000) Leptin: an essential regulator of lipid metabolism. Comp Biochem Physiol 125:285–298

Reiter RJ (1991) Melatonin: the chemical expression of darkness. Mol Cell Endocrinol 79:C153–C158

Reppert SM, Weaver DR (2001) Molecular analysis of mammalian circadian rhythms. Annu Rev Physiol 63:647–676

Reppert SM, Perlow MJ, Tamarkin L, Klein DC (1979) A diurnal rhythm in primate cerebrospinal fluid. Endocrinology 104:295–301

Reppert SM, Perlow MJ, Ungerleider LG, Mishkin M, Tamarkin L, Orloff DG, Hoffman HJ, Klein DC (1981) Effects of damage to the suprachiasmatic area of the anterior hypothalamus on the daily melatonin and cortisol rhythms in the rhesus monkey. J Neurosci 1:1414–1425

Reppert SM, Weaver DR, Rivkees SA, Stopa EG (1988) Putative melatonin receptors in a human biological clock. Science 242:78–81

Reppert SM, Godson C, Mahle CD, Weaver DR, Slaugenhaupt S, Gusella JF (1995) Molecular characterization of a second melatonin receptor expressed in human retina and brain: the Mel 1b melatonin receptor. Proc Natl Acad Sci USA 92:8734–8738

Reul JM, de Kloet ER (1985) Two receptor systems for corticosterone in rat brain: microdistribution and differential occupation. Endocrinology 117:2505–2511

Ribelayga C, Pevet P, Simonneaux V (2000) HIOMT drives the photoperiodic changes in the amplitude of the melatonin peak of the Siberian hamster. Am J Physiol 278:1339–1345

Rollag MD, Niswender GD (1976) Radioimmunoassay of serum concentrations of melatonin in sheep exposed to different lighting regimens. Endocrinology 106:231–236

Ronnekleiv OK (1988) Distribution in the macaque pineal of nerve fibers containing immunoreactive substance P, vasopressin, oxytocin, and neurophysins. J Pineal Res 5:259–271

Rosmond R, Dallman MF, Bjorntorp P (1998) Stress-related cortisol secretion in men: relationships with abdominal obesity and endocrine, metabolic and hemodynamic abnormalities. J Clin Endocrinol Metab 83:1853–1859

Ruesga ZE, Estrada GJ (1999) Chronobiology its application in cardiology. Rev Mex de Cardiol Issue 10:143–145

Rusak B, Zucker I (1979) Neural regulation of circadian rhythms. Physiol Rev 59:449–526

Scheer FA, Buijs RM (1999) Light affects morning salivary cortisol in humans. J Clin Endocrinol Metabol 84:3395–3398

Scher J, Wankiewicz E, Brown GM, Fujieda H (2002) Melatonin receptor in the human retina: expression and localization. Investig Ophthalmol Vis Sci 43:889–897

Schiml PA, Mendoza SP, Saltzman W, Lyons DM, Mason WA (1996) Seasonality in squirrel monkeys (Saimiri sciureus): social facilitation by females. Physiol Behav 60:1105–1113

Schuhler S, Pitrosky B, Kirsch R, Pevet P (2002) Entrainment of locomotor activity rhythm in pinealectomized Syrian hamster by daily melatonin infusion under different conditions. Behav Brain Res 133:343–350

Schwartz MW, Seeley RJ (1997) Seminars in medicine of the Beth Israel Deaconess Medical Center. Neuroendocrine responses to starvation and weight loss. N Engl J Med 336:1802–1811

Schwartz WJ, De la Iglesia HO, Zlomanczuk P, Illnerova H (2001) Encoding Le Quattro Stagioni with the mammalian brain: photoperiodic orchestration through the suprachiasmatic nucleus. J Biol Rhythms 16:302–311

Scott CJ, Jansen HT, Kao CC, Kuehl DE, Jackson GL (1995) Disruption of reproductive rhythms and patterns of melatonin and prolactin secretion following bilateral lesions of the suprachiasmatic nuclei in the ewe. J Neuroendocrinol 7:429–443

Sergent D, Berbigier P, Kann G, Fevre J (1985) The effect of sudden solar exposure on thermophysiological parameters and on plasma prolactin and cortisol concentrations in male Creole goats. Reprod Nutr Dev 25:629–640

Shaw PJ, Cirelli C, Greenspan R, Tononi G (2000) Correlates of sleep and waking in *Drosophila melanogaster*. Science 287:1834–1837

Simonetta G, Walker DW, McMillen IC (1991) Effect of feeding on the diurnal rhythm of plasma cortisol and adrenocorticotrophic hormone concentrations in the pregnant ewe and sheep fetus. Exp Physiol 76:219–229

Simonetta SH, Romanowski A, Minniti AN, Inestrosa NC, Golombek DA (2008) Circadian stress tolerance in adult Caenorhabditis elegans. J Comp Physiol Neuroethol Sens Neural Behav Physiol 194:821–828

Simonneaux V, Ribelayga C (2003) Generation of the melatonin endocrine message in mammals: a review of the complex regulation of melatonin synthesis by norepinephrine, peptides, and other pineal transmitters. Pharmacol Rev 55:325–395

Simonneaux V, Quichou A, Craft C, Pevet P (1994) Presynaptic and postsynaptic effect of neuropeptide Y in the rat pineal gland. J Neurochem 62:2464–2471

Sinha MK, Ohannesian JP, Heiman ML, Kriauciunas A, Stephens TW, Magosin S, Marco C, Caro JF (1996a) Nocturnal rise of leptin in lean, obese, and noninsulin-dependent diabetes mellitus subjects. J Clin Invest 97:1344–1347

Skene DJ, Pevet P, Vivien-Roels B, Masson-Pevet M, Arendt J (1987) Effect of different photoperiods on concentrations of 5-methoxytryptophol and melatonin in the pineal gland of Syrian hamster. J Endocrinol 114:301–309

Song CK, Bartness TJ, Petersen SL, Bittman EL (1999) SCN cells expressing mt1 receptor mRNA coexpress AVP mRNA in Syrian and Siberian hamsters. Adv Exp Med Biol 460:229–232

Sparks DL (1998) Anatomy of a new paired tract of the pineal gland in humans. Neurosci Lett 248:179–182

Stephan FK (2002) The "other" circadian system: food as a Zeitgeber. J Biol Rhythms 17:284–292

Stokkan KA, Yamazaki S, Tei H, Sakaki Y, Menaker M (2001) Entrainment of the circadian clock in the liver by feeding. Science 291:490–493

Sugino N (2006) Roles of reactive oxygen species in the corpus luteum. Anim Sci J 77:556–565

Sumova A, Travnickova Z, Illnerova H (1995) Memory on long but not on short days is stored in the rat suprachiasmatic nucleus. Neurosci Lett 200:191–194

Suttie JM, White RG, Littlejohn RP (1992) Pulsatile growth hormone secretion during the breeding season in male reindeer and its association with hypophagia and weight loss. Gen Comp Endocrinol 85:36–42

Takahashi JS (1995) Molecular neurobiology and genetics of circadian rhythms in mammals. Annu Rev Neurosci 18:531–553

Takahashi JS, DeCoursey PJ, Bauman L, Menaker M (1984) Spectral sensitivity of a novel photoreceptive system mediating entrainment of mammalian circadian rhythms. Nature 308(5955):186–188

Takahashi T, Sasaki M, Itoh H, Ozone M, Yamadera W, Hayshida K, Ushijima S, Matsunaga N, Obuchi K, Sano H (2000) Effect of 3 mg melatonin on jet lag syndrome in an 8-h eastward flight. Psychiatry Clin Neurosci 54:377–378

Takebe K, Sakakura M, Mashimo K (1972) Continuance of diurnal rhythmicity of CRF activity in hypophysectomized rats. Endocrinology 91:1515–1519

Teclemariam-Mesbah R, Ter Horst GJ, Postema F, Wortel J, Buijs RM (1999) Anatomical demonstration of the suprachiasmatic nucleus-pineal pathway. J Comp Neurol 406:171–182

Ternman E, Agenäs S, Nielsen PP (2011) Sleep and drowsing in dairy cows; do Swedish cows need more sleep. http://www.slu.se/en/faculties/vh/departments/department-of-animal-nutrition-and-management/research/on-going-projects/sleep-and-drowsing-in-dairy-cows

Thomas EMV, Armstrong SM (1988) Melatonin administration entrains female rat activity rhythms in constant darkness but not in constant light. Am J Physiol 255:237–242

Thun R, Eggenberger E, Zerobin K, Luscher T, Vetter W (1981) Twenty-four-hour secretory pattern of cortisol in the bull: evidence of episodic secretion and circadian rhythm. Endocrinology 109:2208–2212

Tobler I (1989) Napping and polyphasic sleep in mammals. In: Dinges DF, Broughton RJ (eds) Sleep alertness: chronological, behavioral, and medical aspects of napping. Raven, New York, pp 9–31

Unger R, Zhou YT, Orci L (1999) Regulation of fatty acid homeostasis in cells: novel role of leptin. Proc Natl Acad Sci 96:2327–2332

Vacas MI, Del Zar MM, Martinuzzo M, Cardinali DP (1992) Binding sites for [3H]-melatonin in human platelets. J Pineal Res 13:60–65

Van Vuuren RJ, Pitout MJ, Van Aswegen CH, Theron JJ (1992) Putative melatonin receptor in human spermatozoa. Clin Biochem 25:125–127

Veerman DP, Imholz BPM, Wieling W, Wesseling KH, Van Montfrans GA (1995) Circadian profile of systemic hemodynamics. Hypertens 26:55–59

Veldhuis JD, Iranmanesh A, Lizarralde G, Johnson ML (1989) Amplitude modulation of a burstlike mode of cortisol secretion subserves the circadian glucocorticoid rhythm. Am J Physiol 257:E6–E14

Verkerk GA, Macmillan KL (1997) Adrenocortical responses to an adrenocorticotropic hormone in bulls and steers. J Anim Sci 75:2520–2525

Viljoen M, Steyn ME, Van Rensburg BW, Reinach SG (1992) Melatonin in chronic renal failure. Nephron 60:138–143

Voet D, Voet JG (2002) Lipid metabolism. In: Biochemistry. 3rd ed. Wiley, Hoboken, New Jersey, pp 909–984

Vollrath L (1984) Functional anatomy of the human pineal gland. In: Reiter RJ (ed) The pineal gland. Raven, New York, pp 285–322

Vuillez P, Jacob N, Teclemariam-Mesbah R, Pevet P (1996) In Syrian and European hamsters, the duration of sensitive phase to light of the suprachiasmatic nuclei depends on the photoperiod. Neurosci Lett 208:37–40

Wagner WC, Oxenreider SL (1972) Adrenal function in the cow. Diurnal changes and the effects of lactation and neurohypophyseal hormones. J Anim Sci 34:630–635

Walker BR, Best R, Noon JP, Watt GCM, Webb DJ (1997) Seasonal variation in glucocorticoid activity in healthy men. J Clin Endocrinol Metab 82:4015–4019

Warren WS, Hodges DB, Cassone VM (1993) Pinealectomized rats entrain and phase-shift to melatonin injections in a dose-dependent manner. J Biol Rhythms 8:233–245

Weaver DR, Reppert SM (1996) The Mel1a melatonin receptor gene is expressed in human suprachiasmatic nucleus. Neuroreport 8:109–112

Weihe E, Tao-Cheng JH, Schafer MKH, Ericson JD, Eiden LE (1996) Visualization of the vesicular acetylcholine transporter in cholinergic nerve terminals and its targeting to the specific populations of small synaptic vesicles. Proc Natl Acad Sci 93:3547–3552

Wilhelm I, Born J, Kudielka BM, Schlotz W, Wust S (2007) Is the cortisol awakening rise a response to awakening? Psychoneuroendocrinology 32:358–366

Windle RJ, Wood SA, Kershaw YM, Lightman SL, Ingram CD, Harbuz MS (2001) Increased corticosterone pulse frequency during adjuvant-induced arthritis and its relationship to alterations in stress responsiveness. J Neuroendocrinol 13:905–911

Woodford ST, Murphy MR (1988) Effect of forage physical form on chewing activity, dry matter intake, and rumen function of dairy cows in early lactation. J Dairy Sci 71:674

Yamazaki S, Numano R, Abe M, Hida A, Takahashi R, Ueda M, Block GD, Sakaki Y, Menaker M, Tei H (2000) Resetting central and peripheral circadian oscillators in transgenic rats. Science 288:682–685

Yie SM, Niles LP, Youglai EV (1995) Melatonin receptors on human granulosa cell membranes. J Clin Endocrinol Metab 80:1747–1749

Yoo SH, Yamazaki S, Lowrey PL, Ko CH, Buhr ED, Siepka M, Hong H, Oh WN, Yoo OJ, Menaker M, Joseph S (2004) Period2: luciferase real-time reporting of circadian dynamics reveals persistent circadian oscillations in mouse peripheral tissues. Proc Natl Acad Sci U S A 101(15):5339–5346

Zhang Y, Proenca R, Maffei M, Barone M, Leopold L, Friedman JM (1994) Positional cloning of the mouse obese gene and its human homologue. Nature 372:425–432

Zhao J, Townsend KL, Schulz LC, Kunz TH, Li C, Widmaier EP (2004) Leptin receptor expression increases in placenta, but not in hypothalamus, during gestation in Mus musculus and Myotis lucifungus. Placenta 25:712–722

Zisapel N, Matzkin H, Gilad E (1998) Melatonin receptors in human prostate epithelial cells. In: Touitou Y (ed) Biological clocks, mechanisms and application. Elsevier, Amsterdam, pp 295–299

Zucker I (2001) Circannual rhythms: mammals. In: Takahashi JS, Turek FW, Moore RY (eds) Handbook of behavioral neurobiology. Circadian clocks, vol 12. Plenum/Kluwer, New York, pp 509–528

Zucker I, Boshes M, Dark J (1983) Suprachiasmatic nuclei influence circannual and circadian rhythms of ground squirrels. Am J Physiol 244:472–480

Zucker I, Lee TM, Dark J (1991) The suprachiasmatic nucleus and annual rhythms of mammals. In: Klein DC, Moore RY, Reppert SM (eds) Suprachiasmatic nucleus: the mind's clock. Oxford University Press, New York

Shelter Management for Alleviation of Heat Stress in Cows and Buffaloes

Contents

Abstract

Air temperature above the thermoneutral zone of tolerance exchanges heat from the environment to body, and body temperature of livestock species like cattle and buffaloes increases during heat. Solar exposure increases heat load of animals by increasing the core and surface temperature. In addition to increasing heat load, heat exchange at the body surface is reduced. Radiant energy that strikes a surface can be absorbed, reflected or transmitted through the material. If the material is nontransparent, the fraction of transmission will be zero, that is, all the radiant energy is either absorbed or reflected. It is necessary to find effective methods to manage heat stress in order to sustain/increase milk production by reducing heat load on the animal. In hot and humid weather, it is important that the cooling system functions to maintain appropriate microclimatic conditions, that is, temperature, humidity and wind flow. Evaporative cooling is the main means for heat stress relief at higher temperatures which is implemented in a variety of modes that differ conceptually and technically. However, the forced ventilation increases convective heat loss and is mostly effective in the lower range of stressing air temperatures. The ventilation in the animal house should be considered for animal comfort improvement, and heat and moisture need to be removed. High gaseous concentrations in animal houses affect the animals, workers and the life span of the buildings themselves.

A. Aggarwal and R. Upadhyay, *Heat Stress and Animal Productivity*,
DOI 10.1007/978-81-322-0879-2_7, © Springer India 2013

There are several methods available that can alleviate animal's heat stress by lowering air temperature or the enthalpy of the ambient air. A number of animal cooling options exist based on combinations of the principles of convection, conduction, radiation and evaporation. Air movement (fans), wetting the cow, evaporation to cool the air and shade to minimise transfer of solar radiation are used to enhance heat dissipation from high-producing cows. Animals in pond lose heat fast to cool water primarily by conduction and coefficient of heat transfer to water from skin. Passive solar designs make use of the sun's energy for the heating and cooling of the animal buildings.

1　Introduction

Thermal stress is a major limiting factor in livestock production under tropical climate and also during summer season in temperate climates. Heat stress occurs when the ambient temperature lies above thermoneutral zone. Meteorological variables which influence the ambient temperature significantly are dry bulb temperature, wet bulb temperature, wind velocity and intensity of solar radiation and radiations from surrounding structures in the animal byre. Ambient temperature and humidity are used to assess thermal comfort, and derivation based on both is termed the temperature–humidity index (THI). High-producing cows have been shown to be in heat stress if the THI exceeds 72 and in severe heat stress at a THI above 80 (Armstrong 1994). Milk production of dairy cows declines at air temperature above 25°C and with relative humidity above 50% (Chiappini and Christiaens 1992). The severity increases as the temperature rise beyond the thermoneutral zone or above the capacity to handle heat effectively.

Lactating dairy cows exposed to high ambient temperature and high relative humidity or radiant energy (direct sunlight) tends to decrease milk yield and may exhibit other signs of heat stress. Lactating cows and buffalo produce large amounts of heat due to digestion and metabolic processes, and this heat needs to be exchanged with the environment to maintain normal body temperature. Animal can give and/or receive heat energy from the environment. Solar radiation increases heat load by increasing the surface temperature of cattle and buffaloes, and air temperature above normal body temperature also increases the heat load. At high temperature, in addition to increasing heat load, heat exchange at the body surface is reduced. Heat stress increases maintenance cost and decreases dry matter intake (DMI) and milk yield on exposure to hot or hot and humid environments for long with THI above 80 (Huber et al. 1994). High-producing cattle and buffaloes are more vulnerable to heat stress. The problem of heat stress becomes more acute as the production level increases (Armstrong 1994).

Two types of radiation, that is, direct radiation from the sun and diffused radiation from the sky, influence the microclimatic conditions around the animal. The diffused solar radiation is scattered at least once before it reaches the surface of animal and influences mainly on cloudy days. Even on clear cloudless day, a certain diffused radiation appears. Therefore, heat is transferred into the animal house or building from solar radiation on cloudy day. The intensity of the direct solar radiation can reach more than $1,000 \text{ W/m}^2$, while diffused radiation may be about $50–100 \text{ W/m}^2$ (Gustafsson 1988). The total solar radiation on animal also sometimes called global radiation is the sum of the direct radiation and the diffused radiation (Duffie and Beckman 1974). Radiant energy that strikes a surface may be absorbed, reflected or transmitted. The relationship between reflectance and absorption is based on the assumption that a black surface has zero reflectance and the absorption is one. The factor for reflectance and absorption varies for different materials. If the material is non-transparent, the fraction of transmission will be zero, that is, all the radiant energy is either absorbed or reflected. To achieve minimum heat load on the animal building from solar radiation, the reflectance factor should be high, and absorption factor should be low for the exposed areas of the materials, that is, roof top or the upper surface of the roof. In direct radiation when the surface is perpendicular to the sun, the rate of solar radiation received is maximum

(Gustafsson 1988). The rate of heat conduction varies greatly with nature of material. Thermal conductivity of skin (0.8–3.5) influences the microclimatic conditions around the animal (Hafez 1969). Therefore, a quantitative understanding of how environmental variables affect the heat budget of animals in particular environment is likely to help in proper livestock management. The heat budget, information based on heat exchanges, may suggest how the ambient environment needs to be manipulated to protect from wind, sun and precipitation.

It is particularly important to find effective methods to manage heat stress to help sustain or increase milk production by reducing heat load on the animal particularly in tropical climate. In hot and humid weather, it is important that the heat alleviation or animal cooling system functions to maintain appropriate microclimatic conditions in the animal sheds, that is, temperature, humidity and wind flow. Several approaches have been used successfully to improve milk productivity in the tropical climates. Publications emanating on the effects of heat stress in dairy cattle report environmental temperatures lower than those observed in region of tropical climate and may not be truly representing conditions typically observed under tropical conditions. The body water plays a central role in evaporative cooling and the mechanisms used for heat dissipation. Evaporative cooling is the main means for heat stress relief at higher temperatures which is implemented in a variety of modes that differ conceptually and technically. However, the forced ventilation increases convective heat loss and is mostly effective in the lower range of stressing air temperatures. Provision of adequate shade reduces the radiant heat load.

2 Heat Transfer and Concentration and Emission of Harmful Gases in Dairy Buildings

Adequate ventilation is a prerequisite in animal shelters for removing harmful gases and to ensure an acceptable indoor microclimate. Microclimate parameters such as the concentration of greenhouse gases, temperature, velocity, dust and humidity affect the welfare of animals, humans and the buildings themselves. In practice, the ventilation rate need not be continuously measured for a good microclimate but needs to be planned in advance so that it is high enough under all environmental situations or can be regulated when required to ensure adequate indoor air quality (Brockett and Albright 1987).

Solar radiation is composed primarily of visible and near-infrared radiation. Far-infrared radiation has longer wavelengths than near-infrared and is largely outside the solar band (both near- and far-IR are felt as heat). Therefore, to understand how a material behaves when exposed to solar radiation, its emissivity in each of these three bands is more important than its overall value summed over the entire spectrum. To maintain a 'cool roof', there is need to reflect solar radiation, and whatever heat is absorbed needs to emit back to the surroundings. Therefore, high reflectivity in the visible and near-infrared bands is required, but low reflectivity (i.e., high emissivity) in the far-infrared bands. White roofs have high emissivity in the far-infrared band, while unpainted metal roofs have low emissivity in this band.

Heat and energy transfer from and into a building occurs in different ways: transmission, radiation and change of air (Hamrin 1996). For designing of heating or mechanical cooling systems, each of these three ways is important. To calculate temperature and moisture within a closed animal shelter, the simplest way is to use steady-state conditions which means that the building do not have any storage component. This means that the inflow of heat plus the heat that is generated in the building must equal the outflow of heat. In steady state, it is assumed that the temperatures on both sides of the building are at a constant level for a sufficient period of time, so that the heat that is leaving on one side of the building equals the heat entering on the other side. This never happens in real sense, the temperatures changes constantly, so the steady-state concept is a simplification (CHPS 2002).

The steady-state energy balance can be expressed as in following equation:

$$P_s + P_{cow} + M_1 e_1 = Pt + M_2 e_2$$

where

P_s = extra heat load from equipment

P_{cow} = heat from cows

$M_1 e_1$ = mass of incoming air multiplied by its enthalpy per unit of mass

Pt = heat conducted through the exterior surfaces of the building

$M_2 e_2$ = mass of outgoing air multiplied by its enthalpy per unit of mass

(Esmay 1978).

At the time of planning, the animal shelter the orientation of the building must be considered. If the shelter is located in north–south orientation, the building is exposed to greater solar radiation than with an east–west orientation. This is because sunlight can enter north–south-orientated buildings in the mornings and afternoon. In the afternoon, the sun is low in the sky and therefore most detrimental. Also the heat stress on the animals is usually maximum or at its greatest magnitude in the afternoon. The preferred orientation of the buildings is an east–west orientation to avoid solar radiation and to achieve maximum shade (Shearer et al. 1999). The wall height and eave extension of the building should also be considered. With greater sidewall heights, more direct afternoon sun can hit the building (Brouk et al. 2001). The direct irradiation on the sidewalls can be reduced with greater eave extensions. A big eave extension can provide shade to the building (Gralla Architects 2000–2001). The eave extension should be one-third the height of the sidewall (Palmer 2004). The size of the holding area should be designed in relation to the size of the milking parlour, so the holding area can hold one group of animals. When determining the size of the holding pen, it should be sized to a minimum of 1.5 m² per cow. The floor of the holding area should be sloped between 3 and 5%. The slope makes the cows facing the parlour, because cows do not like to stand downhill (Van Lieu 2003). The holding pen can be divided into wash and drip pens. If using holding pen washing,

cow cleaning in parlour can be reduced. In that case, each pen should be sized to hold one group of cows (Smith et al. 1997).

For ventilation of the milk centre, both operator and cow comfort should be considered, and heat and moisture need to be removed. In smaller and large parlours, a combination of natural and mechanical ventilation can be used. Fresh air from openings in the sidewall is forced into the parlour above the operator pit and then forced out of the parlour into the holding area. Also a tunnel ventilation system can be used to ventilate the milk centre. Air is drawn by fans through large inlets at the front of the parlour and downwards the holding area (Van Lieu 2003).

The generation and emission of gases associated with livestock production have been extensively studied. High gaseous concentrations in animal buildings affect the welfare of animals, workers and the life span of the buildings themselves (Auvermann and Rogers 2000; De Belie et al. 2000a, b, c; Radon et al. 2002; Zähner et al. 2004). Carbon dioxide, methane, ammonia, hydrogen sulphide and nitrous oxide are the most prominent gases found in dairy buildings. When gases produced in concentrated dairy production escape from the buildings, they contribute to environmental problems such as global warming, acid rain and upsetting the nutrient balance in the environment (Anderson et al. 2003; Erisman et al. 2003). Global estimates show that animal production facilities emit about 536 Mt NH_3-N (Bouwman et al. 1997) and 689 Mt CH_4 (Moss et al. 2000) annually. The main source of carbon dioxide in dairy buildings is contributed by respiration. Some proportions (6.1%) are produced through the degradation of manure and urea (Kinsman et al. 1995). The average concentration of CO_2 in dairy buildings is 1,900 ppm (Phillips et al. 1998). And the rate of production of CO_2 per cow has been estimated to be 330 g/h (CIGR 1994). Methane is generated through enteric fermentation in ruminants and, to a lower extent, through the anaerobic degradation of manure (CIGR 1994). Dairy cows produce approximately 9 g/h of CH_4 per cow, which is 94.2% of the total amount produced in a dairy building (Jungbluth et al. 2001; Kinsman et al. 1995). The average concentration of CH_4 in dairy buildings has been reported to be 70 ppm (Jungbluth et al. 2001). The emission of CH_4 per dairy production animal is 194 g to 390 g/day (Jungbluth et al. 2001; Kinsman et al. 1995; Sneath et al. 1997). Ammonia causes acid deposition and eutrophication when suspended NH_3 from dairy and other animal production facilities is deposited on land and in bodies of water (Anderson et al. 2003; Erisman et al. 2003). The sources of ammonia in dairy buildings are mainly dairy manure, urine, bedding materials and animal feed. The transformation of organic nitrogen to ammonia in dairy buildings is well documented (Sommer et al. 2006). The average concentration of NH_3 in dairy buildings is 10 ppm (Phillips et al. 1998), and the emission of NH_3 per animal is 6.2–31.7 g/day (Demmers et al. 1998; Zhang et al. 2005).

Hydrogen sulphide is very toxic and has been reported to contribute to the acidification of the soil and water in the environment (Sakamotoa et al. 2006). Hydrogen sulphide is formed when manure remains in the dairy building for a period of over 5 days (CIGR 1994). The average concentration of hydrogen sulphide in dairy buildings is usually around 14 ppb, with a rate of emission per area between 0.016 and 0.084 $g/m^2/$ day (Zhu et al. 2000).

3   Methods for Alleviation of Heat Stress

There are several methods available that can help alleviate animal's heat stress by reducing dry bulb temperature or the enthalpy in the ambient air. Increasing air velocity around the animals may facilitate heat dissipation provided that the ambient temperature is below body temperature. Previous research targeted prevention (Silanikove 2000) of heat stress in lactating cows and the application of cooling systems for lactating cows, which minimises decreases in milk production during summer (Igono et al. 1985; Her et al. 1988; Verbeck et al. 1996; Aggarwal 2004; Aggarwal and Singh 2007). To reduce heat stress in hot climate, handling of the animals should be kept at a minimum. The air movement caused

naturally and mechanically helps in improving heat dissipation by means of convection and evaporation (Bucklin et al. 1988). Mechanical ventilation is very effective in providing air movement, but it is expensive and should be combined with other methods (Bucklin et al. 1991). Different methods for reducing heat stress in dairy animals are as follows:

3.1   Provision of Shade

Lowering or preventing heat stress in lactating animals in summer requires protection from radiant heat and increase of heat loss from the body to the environment. Animal shade may reduce more than 30% of all the heat radiated on animal and is the single most important contribution for lowering heat stress (Bond et al. 1967). Animal shade is one of the more easily implemented and economical methods to minimise heat from solar radiation.

Cows in a shaded versus no shade environment had lower rectal temperatures (38.9 and 39.4°C) and reduced respiratory rate (54 and 82 breaths/min) and yielded 10% more milk when shaded (Roman-Ponce et al. 1977). Cattle with no shade had reduced ruminal contractions, higher rectal temperature and reduced milk yield compared with shaded cows (Collier et al. 1981). Armstrong (1994) reviewed shade and cooling for cows and discussed the benefits and deficiencies of various types of shade. The author suggested differing shade orientations, depending on whether the application was in a dry or wet climate. In the humid Southeast, cows should be allocated 4.2–5.6 m^2 of space beneath the shade and a north–south orientation to allow for penetration of sunlight beneath the shade for drying the ground beneath if earthen floors are used. In hot and humid climate, the cattle and buffalo prefer to use shade when exposed to high temperature and solar radiation. This is more obvious at high temperatures and especially in the middle of the day (Tucker et al. 2008; Kendall et al. 2006). Fisher et al. (2008) reported that cows start using shade when the temperature is above 25°C and the critical temperature when the cows

Fig. 1 Cows under tree during summer

stopped grazing and searched shade was 28°C (Langbein and Nichelmann 1993). However, since cows also use shade at lower temperatures, this demonstrates the importance of shade even more (Schütz et al. 2009; Tucker et al. 2008). Cows also prefer to use shade that offer a greater protection from solar radiation.

Animal shade has also been shown to be a successful way to reduce the heat when compared to other techniques. Comparative study on shade and sprinklers indicated that shaded heifers had a higher feed intake, increased average daily gain and increased feed efficiency than the heifers with sprinklers. Shaded heifers were also observed to have a lower respiration rate than unshaded (Marcillac-Embertson et al. 2008).

Numerous types of shades are available and may vary from natural tree shade (Fig. 1) to metallic and synthetic material shade. There are various types of roofing materials (Buffington et al. 1983), and the most effective is a reflective roof made of aluminium or galvanised with insulation beneath the metal. The floor of shed should be of concrete floor to keep the shaded area as clean as possible.

3.2   Ceiling Fans

If air is flowing over the animal with a velocity between 2 and 3 m/s, it increases convective heat loss during stressful conditions. In holding areas and free stall shelters, the fans should be installed longitudinally, spaced no more than ten times their blade diameter. They should be located vertically and just high enough, so they are out of reach of the cattle and do not interfere with alley scraping or bedding operations (Shearer et al. 1991). A different kind of fan is 'high-volume low-speed fans', HVLS. The 7.3-m diameter operates at a speed of 50 rpm. Air volume moved by a 7.3-m HVLS fan ranges from 0.20 to 0.25×10^6 cubic metre/h or m^3/h (Kammel et al. 2003). Air movement in the velocity range of 200–400 ft per minute across the cows is needed. However, the air movement at the cow level depends on the discharge characteristics of the fan and the distance of the cow from the fan. Another type of fan often used, particularly in holding pens or other areas with higher eave heights, is a 1/2-hp 36″ fan. Such fans will blow about 10,000–11,000 cfm with a 'throw' distance of about 30 ft. A 1-hp 48″ fan provides 21,000 cfm with an effective throw distance of about 40 ft. These fans should be mounted out of reach of the cows and angled downwards slightly. Overhead paddle fans can also be used to provide airflow, but they offer limited air movement unless cows are directly under a fan. The paddle fans do have much higher efficiencies of operation, with ratings of 100 cfm/W of power input, as compared to 15–20 cfm/W for conventional propeller-type fans.

For a given airflow, a larger diameter fan is more energy efficient than several small diameter fans. When two fans have equal diameter and rpm, the fan with lowest motor current rating is usually more efficient. When two fans have equal airflow, the fan with slower speed is usually quieter and more efficient.

Reasonable Fan Goals
36″ fan = 11,000 cfm
48″ fan = 20,000 cfm

3.3   Evaporative Cooling

Animal shades provide shield from solar radiation, but there is no effect on air temperature or relative humidity, and additional cooling may be necessary for lactating cows in a hot and humid climate. A number of cooling options exist for lactating dairy cows based on combinations of the principles of convection, conduction, radiation and evaporation. Air movement (fans), wetting the cow, evaporation to cool the air and shade to minimise transfer of solar radiation are used to enhance heat dissipation. Any cooling system that is to be effective must take into consideration the intense solar radiation, high ambient temperature and the typically high daytime relative humidity, which increases to almost saturation at night. These challenging conditions tax the ability of any cooling system to maintain a normal body temperature of any animal. Evaporative cooling utilises energy from the air to evaporate water, that is, adiabatic cooling. This cooling system lowers the temperature of the air and increases the relative humidity, making evaporative cooling most effective in dry environment (Bucklin et al.

1991). The body surface exposed to moving air is reduced when cows huddle, as well as when cows adopt a recumbent posture during resting or rumination. Mature Holstein cows in stall housing systems spend about 15 h/day lying, and the duration and frequency of lying probably are indicators of cow comfort (Haley et al. 2000). A smaller body surface reduces both the convective and the evaporative components of heat loss by skin. The time spent lying was reduced in cows exposed to heat stress (Frazzi et al. 2000). Ambient temperature has been observed to negatively impact on the percentage of cows lying (Shultz 1984; Overton et al. 2002). Animals standing were observed to have higher respiration rate and body temperatures (Frazzi et al. 2000). Spray cooling of lying cows increased their lying time (Hillman et al. 2005), supporting the contention that a lying cow is more sensitive to heat stress than a standing cow. Respiratory heat loss is recruited when the heat loss by skin is insufficient to maintain thermal stability. A high demand for heat loss by respiration may, by itself, be a stressing factor because it reduces the time spent lying.

3.3.1   Evaporative Cooling Pads and Fan Systems (Desert Cooler)

Evaporative cooling through water evaporation in the incoming airflow is an economical method resulting in a temperature reduction of 8–10°C, but it causes an increased relative humidity up to 90% (Chiappini and Christiaens 1992). This system uses pads and pumps to pour water through the pads (Kelly and Bond 1958). Cooled air must have a short way through the building to avoid high temperature and high humidity. It should be installed in an adequate insulated building (Chiappini and Christiaens 1992).

3.3.2   High-Pressure Foggers

Foggers disperse very fine droplets of water which quickly evaporate and cool the surrounding air, but they also raise the relative humidity. A ring of fog nozzles is attached to the exhaust side of the fan, and then the cooled air is blown down over the animal's body (Jones and Stallings 1999). This system is expensive and requires a lot of maintenance, and the water must be kept very

Fig. 2 Buffaloes under mist cooling system

clean; otherwise, fogging nozzles get clogged or plug (Worely 1999). A fogging system is effective in hot areas with low humidity. In an environment which is saturated with water, the droplets cannot evaporate, and a 'steam bath' effect is likely to be created (Bucklin et al. 1991). Atomisers of different diameter help in conservation of water by regulating the water dispersion.

3.3.3 Misters

A misting system consists of a series of misting nozzles placed in a sequence in tubing and pressurised to provide a fine spray of water droplets between 15 and 50 μm in size. A mist droplet is larger than a fog droplet. The animal is primarily cooled by breathing in the cooler air. This system is not very effective for humid environments, because the mist droplets are too large to evaporate before they reach the floor and the bed or feed becomes wet (Shearer et al. 1999). Another complication with misters is that if the system does not wet the hair coat through the skin, an insulating layer of air can be trapped between the water layer and the skin. Therefore, the natural evaporative

heat loss from the skin is inhibited (Jones and Stallings 1999). Use of mist system for cooling cows and buffaloes was found to be economical also (Aggarwal and Singh 2007). Figure 2 shows a shed with mist cooling facilities.

If the misting system application is in an open area, the cooling capability may be diminished due to a lack of shade and the inability to mount the mist system so that complete evaporation is achieved. In these cases, adding fans to the misting system design will improve its cooling capacity. If there is no structure to mount to, the mist system can be laid on the ground with the misting system nozzles pointing directly upwards. Although this is not the ideal design for a mist system, it can provide relatively good results.

3.3.4 Sprinkler-cum-Fan Cooling

An alternative to mist and fog systems is the sprinkling system. This method does not attempt to cool the air but instead uses a large droplet size to wet the hair coat to the skin of the cow or buffalo, and then water evaporates and cools the hair and skin. Sprinkling is most effective when combined with air movement. Fans should provide an airflow of about 11,000 cfm and should be tilted downwards at 20–30° angles. At least one 36-in. fan is needed for each 40 animals which will move air effectively for about 30 ft. Other fan sizes can be used (a 48-in. fan at 40 ft. intervals). Nozzles (10 psi, 180° spray) are spaced above cows approximately every 8 ft. Sprinklers should be located immediately below the fans so that water is thrown just under the bottom of the fans which run continuously. Cows are sprinkled for 1–2 min at 15-min intervals. Concrete floors must be sloped to handle water run-off rates of 50–100 gal./cow/day. The sprinkling system can be used in holding areas, shade structures and feed barns. Avoid wetting feed and free stalls. In hot–humid conditions, sprinklers alone do not effectively cool animals. The sprinkler system combined with forced air movement help increase the loss of body heat up to three or fourfolds (Shearer et al. 1999).

The evaporative cooling system increases animal feed intake (7–10%) and milk yield (8–13 kg/day) and decreases body temperature (0.2–0.5°C) and breathing rate by 20–25% (Strickland et al. 1989; Igono et al. 1987; Turner et al. 1992; Aggarwal and Singh 2005). In a humid climate, the recommended velocities are 2.9–4.0 m/s depending on airflow direction (Chastain and Turner 1994).

A drawback with the evaporative cooling system is the large quantity of water used (Turner et al. 1992; Aggarwal and Singh 2005). It is important that the water is clean to avoid spreading of diseases. Research trials of fan and sprinkler cooling system have not shown increase in mastitis for cooled cows. If the area of the shed where the water is supplied does not have good drainage, cows may lie in water and could have an increased likelihood of mastitis.

3.3.5 Exit Lane Sprinklers

Animals can be cooled by activated nozzles which deliver water when the animal exits the parlour (Verbeck et al. 1995). Soaking the cows' skin as they leave the milking parlour is a system to prolong the cooling period at milking time

(Bray and Bucklin 1997). When the cows return to the free stall or corral, moisture from the wet hair coat evaporates (Armstrong 1994) and may be effective in hot–dry or arid climatic conditions (Verbeck et al. 1995).

3.3.6 Cooling Ponds

Cooling animal ponds are man-made and may vary in size as per the size of herds (Fig. 3). The size of average pond may be 50×80 and 4–6 ft. deep. Ponds need to be drained regularly and dredged every 1–2 years. Animal cooling ponds should be located close to a water supply as freshwater needs to enter the ponds constantly. There must be a proper way to dispose of run-off water. Some slope from the ponds to the waste management may be necessary. Cooling ponds should in close proximity of the milking parlour to avoid long walks in the sun.

Concrete Ponds: Concrete ponds have been used, but they are expensive to build. However, they are easier to maintain, as the entrances and exits do not have to be rebuilt. The finish of the concrete on exit and entrance slopes should be cross grooved to prevent cows from falling. A 1:8 slope would be a safe slope if concrete ponds are constructed.

Pond Shapes and Depths: Using the sample of $5,000/\text{ft}^2/100$ cows, a pond could be 50′ wide × 100′ long, or a circle 80′ in diameter ($A = (\text{Pi}) \ r^2$; $A = 3.14 \times 40 = 5,024$ ft^2). If a 50′ × 100′ pond is used and the sides are fenced so the cows can only enter or exit at each end of the pond, the amount of dirt falling into the pond is reduced. If a circular pond is made, cow carpet should be used to eliminate the problem of dirt falling in the pond. As cows can swim, pond depth may be kept reasonable. Deeper ponds may allow the settling of organic matter on the bottom of the pond. Preferably depths of 3′ or 4′ is sufficient to cool the cows and let them stand and walk.

The dairy animals are cooled by dipping themselves in the water with head open. Animals in pond lose heat primarily by conduction, but during 5–10 min after exiting the pond, a small amount of heat is lost by evaporative cooling

Fig. 3 Wallowing pond

(Shearer et al. 1999). The man-made pond has a continuous inflow of water with an overflow at the end of the pool. Buffaloes, because of their black skin and less number of sweat glands than cows, especially like to wallow in water for alleviation of heat stress (Chauhan et al. 1998; Aggarwal and Singh 2008).

3.3.7 Tunnel Ventilation
The advantage of tunnel ventilation is that the cows get access to fresh airflow in the whole building, while the disadvantage is the electricity consumption (Stowell et al. 2001; Bray et al. 2003). The system provides a combination of high air exchange rates and high airflow speeds to help the cows to remove body heat by increasing the heat loss by convection. In tunnel-ventilated barns, large exhaust fans are placed at one end of the barn blast the airflow and large openings are placed at the other end. The fans pull outside fresh air through the inlet openings. All windows, doors and other openings remain closed along the sidewalls. To provide a uniform air movement along the barn, it

is important that the tunnel ventilation system is properly designed (Tyson et al. 1998).

3.3.8 Zone Cooling
Inspired air or zone cooling applies a jet of cooled air onto the head and neck of the animal. The air may be cooled by evaporative cooling or by mechanical refrigeration. Because of the installation costs and lack of compatibility with housing systems, this has not become a common cooling system (Bucklin et al. 1991). Air cooled by refrigeration to around 15°C, supplied at a rate of 0.7–0.85 m^3/min, has been observed to increase milk production.

3.3.9 Underground Pipes
If the ground temperature is lower than the atmospheric temperature, air can be passed through underground pipes and the air can be cooled. The pipes are normally placed at a depth of 1.5–2 m under the ground. A temperature reduction of 8–10°C can occur in the peak summer day. The main effect of the system is to reduce extremes of the temperatures (Chiappini and Christiaens

Table 1   Tested roofing material solar reflectances and emissivity

Sample	Solar reflectance	Long-wave emissivity
Black asphalt shingle	0.27	0.90
White asphalt Shingle	0.21	0.91
White tile, concrete	0.75	0.88
Red tile, concrete	0.19	0.91
White metal	0.67	0.83
Limestone paves	0.53	0.89

Source: Modified from Parker and Sherwin (1998)

1992), but it is also a way of reducing the energy consumption for cooling by using the relatively stable soil temperature (Nilsson and Kangro 1998).

3.3.10 Roof Cooling

The radiant heat load on animals can be reduced by cooling surrounding surfaces by evaporation of water. If the roof is sprinkled, the temperature of the roof can be reduced up to 28 °C by application of 1.5-l water per hour and per square metre roof area. If a wall or a roof is wet, energy and therefore heat will be used to evaporate the water. Therefore, radiated sun energy will be reduced (Nevander and Elmarsson 1994).

3.3.11 Mechanical Cooling

To elaborate the cooling design of a building, the total heat load must be calculated. That can be done in a similar way to designing air exchange of the ventilation system for barns in cold weather. The following four aspects needs to be considered (Esmay 1978):

1. Heat transmission through surfaces that are exposed to temperature difference and solar radiation
2. Heat and moisture production by the animals
3. Ventilation air and air that infiltrate the building
4. Supplementary heat from equipment and lights

   The mechanical cooling can change temperature inside to considerable lower temperature compared to outside (Chiappini and Christiaens 1992). An insulated building with this system needs a close attention to air filtration, adequate ventilation and maintenance. Air conditioning of dairy housing depends on the local design condi-

tions and the individual situation of the cow, but approximately the energy needed is 2,500 J/s (W) or more.

3.3.12 Passive Solar Cooling

Heat and light from the sun is freely available and is renewable. Ironically, the majority of modern buildings are not designed to harness solar power. Instead, buildings are often constructed to block sunlight in favour of insulated, opaque partition walls. Passive cooling is based on the interaction of the building and its surroundings. Passive solar design refers to the use of the sun's energy for the heating and cooling of the building. In this approach, the building itself or some element of it takes advantage of natural energy characteristics in materials and air created by exposure to the sun. Passive systems are simple, have few moving parts and require minimal maintenance and require no mechanical systems. In temperate and hot climates, solar heat infiltrating the building has typically been the most costly thing to mitigate (i.e. air conditioning). However, effective passive solar cooling design can eliminate much of this conventional operating cost with proper building design. Passive cooling includes overhangs for south-facing windows, few windows on the west, shade trees, thermal mass and cross ventilation. Table 1 shows solar reflectances and emittances for some roofing material. The coolness of a roof is determined by two radiative properties and their combined effect on temperature:

1. Solar reflectance: The fraction of sunlight that is reflected by a roofing material's surface. Energy that is not reflected by the roof is potentially absorbed by it; this is where thermal emittance comes into play.

2. Thermal emittance: The relative ability of a roofing material with which a surface cools itself by emitting thermal radiation.

Both properties are measured on a scale of 0–1. The higher the value, the cooler the roof. Solar reflectance index allows actual measured solar reflectance and thermal emittance values to be combined into a single value by determining how hot a surface would get relative to standard black and standard white surface. The standard black roofing material has a high emittance value of 90%, and the reflectance value is 5%. The standard black roof has SRI value of 0. The standard white roofing material is 80% reflective, and emittance is 90%. Surface of white roofing material is much cooler, and SRI value is 100.

Preventing excess solar heat from entering the building can help in reducing heat load on the animal. Appropriate window glazing and shading devices may be used to avoid or reduce the need for mechanical cooling. One or more shading strategies may be used including fixed shading devices as part of building design (porches, overhangs, extrusions), trees or other vegetation that provide seasonal shading, awnings that can be extended or removed, operable shades or blinds. In general, east and west glazings to avoid low solar angle exposures may be used. Other cooling strategies that may be used for animal shelters include taking advantage of natural ventilation, radiative cooling in regions that have significant differences in day and night temperatures, ground-coupled cooling and dehumidification in humid climates (Green Building Encyclopedia).

3.3.13 Night Grazing

Although air temperature and the intensity of solar radiation begin to fall after about 2 pm, the temperature of the roof, if metallic, still remains high. As a result, the body temperature and respiration rate of animal are high. Cattle kept in a hot shed have a rapid heartbeat during the day and night also. However, when the cattle are allowed out into a pasture or in open during night, the physiological responses are low due to both of the reduction in radiatio n heat during night during night from the surroundings and the rise in heat loss from the cattle (Shinde et al. 1994).

Considerations in Choosing Cooling Systems

1. Shade the cow from solar radiation. This should always be the first step in any cooling system.
2. Consider average temperature and relative humidity of location during each hour of the day. Determine when during the day evaporative cooling would be effective. Even in humid environments, afternoon humidity may be low enough to benefit from evaporative cooling.
3. If environmental temperature is near or above the normal body temperature of the cow for a significant portion of the summer, some form of evaporative cooling will likely benefit.
4. Do not depend upon evaporative cooling alone, except in very arid environments. In most environments, feed line soaking will provide cooling over and above the evaporative system.
5. Consider all costs associated with evaporative cooling and feed line soaking. While additional benefits are realised by combination systems, additional milk production may not offset expenses.

---

4   Conclusions

The tropical environmental conditions necessitate reduction of heat stress due to solar radiation and heat. The solar radiation on animals may be reduced by shelter; provision of tree shelter and animal shelter may substantially reduce solar radiation and heat load on animals. High-producing animals experience heat stress, and thermal stress should be alleviated. The most common way to alleviate stress of cows and buffalo and their environment is by evaporative cooling. The effectiveness of evaporative cooling depends on the capacity of the air to take up moisture. Consequently, in a tropical climate with high temperatures combined with high humidity, cooling with evaporative systems is likely to be limited. In those conditions, methods that are beneficial to the animal's natural mechanism of heat loss are preferable and recommended for adoption. Any cooling mechanism must take into consideration the intense solar radiation, high ambient temperature and the relative humidity, which increases to almost saturation during hot–humid (rainy) season. The important consideration in all the systems should be water economy and recycling. Sprinkler and fan cooling systems generate a large volume of wastewater which must be processed or recycled. The fan/sprinkler system uses about tenfold more water than the fan/mist system. Mechanical cooling, or air conditioning, is generally considered expensive and uneconomic in animal housing. During hot–humid season, use of evaporative cooling pads reduces air temperature of the barn without any effect on milk yield. Therefore, improved systems capable of either cooling the animal directly or cooling the surrounding environment are necessary to better control the animal's body temperature and maintain production in hot and hot–humid climates. With cooling devices, the temperature in the animal sheds may be kept low to cool animal, and THI can be kept around 72. The high-producing cows may be kept under low temperature, and an investment in mechanical cooling may be feasible.

References

Aggarwal A (2004) Effect of environment on hormones, blood metabolites, milk production and composition under two sets of management in cows and buffaloes. Ph.D. thesis submitted to National Dairy Research Institute, Karnal

Aggarwal A, Singh M (2005) Physiological response, milk production and composition in crossbred cows with and without mister system during hot-humid season. Egypt J Dairy Sci 32:175–186

Aggarwal A, Singh M (2007) Economics of using mist and fan system during summer and houses during winter for alleviating environmental stress in dairy animals. Indian J Agric Econ 62:272–279

Aggarwal A, Singh M (2008) Changes in skin and rectal temperature in lactating buffaloes provided with showers and wallowing during hot-dry season. Trop Anim Health Prod 40:223–228

Anderson N, Strader R, Davidson C (2003) Airborne reduced nitrogen: ammonia emissions from agriculture and other sources. Environ Int 29:277–286

Armstrong DV (1994) Symposium: heat stress interaction with shade and cooling. J Dairy Sci 77:2044–2050

Auvermann BW, Rogers WJ (2000) Documented human health effects of airborne emissions from intensive livestock operations: literature review. Special report submitted to Alberta Pork and Intensive Livestock Working Group, Alberta, p 40

Bond TE, Kelly CF, Morrison SR, Pereira N (1967) Solar, atmospheric, and terrestrial radiation received by shaded and unshaded animals. Trans ASAE 10:622–627

Bouwman AF, Lee DS, Asman WAH, Dentener FJ, Van Der Hoek KW, Olivier JGJ (1997) A global high-resolution emission inventory for ammonia. Global Biogeochemical Cycles 11:561–587

Bray DR, Bucklin RA (1997) Recommendations for cooling systems for dairy cattle, Fact sheet DS-29. Institute of Food and Agricultural Sciences, University of Florida, Gainesville, p 5

Bray DR, Bucklin RA, Carlos L, Cavalho V (2003) Environmental temperatures in a tunnel ventilated barn and in an air conditioned barn in Florida. Fifth international dairy housing conference. Texas, pp 235–242

Brockett BL, Albright LD (1987) Natural ventilation in single airspace buildings. J Agric Eng Res 37:141–154

Brouk MJ, Smith JF, Harner JP III (2001) Freestall barn design and cooling system. Kansas State University, Agricultural Experiment Station and Cooperative Extension Service, Kansas, p 1

Bucklin RA, Bray DR, Beede DK (1988) Methods to relive heat stress for Florida dairies. Florida cooperative extension service circular 782

Bucklin RA, Turner LW, Beede DK, Bray DR, Hemken RW (1991) Methods to relive heat stress for dairy cows in hot, humid environments. Appl Eng Agric 7:241–247

Buffington DE, Collier RJ, Canton GH (1983) Shade management systems to reduce heat stress for dairy cows. Trans ASAE 26:1798–1802

Chastain JP, Turner LW (1994) Practical result of a model of direct evaporative cooling on dairy cows. In: Dairy system for the 21st century, proceedings of the 3rd international dairy housing conference, ASAE, St. Joeph, pp 337–352

Chauhan TR, Dahiya SS, Gupta R, Hooda DK, Lall D, Punia BS (1998) Effect of climatic stress on nutrient utilization and milk production in lactating buffaloes. Buffalo J 16:45–52

Chiappini U, Christiaens JPA (1992) Cooling in animal houses. 2nd report of working group on climatization of animal houses. CIGR, State University of Gent, Belgium, pp 82–97

CHPS (2002) Building enclosure and insulation. CHPS Best practices manual, pp 258. http://www.chps.net/manual/documents/2002_updates/BldgEncl.pdf. 24 June 2004

CIGR – Commission Internationale de Génie Rural (1994) Aerial environment in animal housing. Concentration in and emissions from farm buildings. Working Group Report Series No 94.1

Collier RJ, Eley RM, Sharma AK, Pereira RM, Buffington DE (1981) Shade management in subtropical environment for milk yield and composition in Holstein and Jersey cows. J Dairy Sci 64:844–849

De Belie N, Lenehan JJ, Braam CR, Svennerstedt B, Richardson M, Sonck B (2000a) Durability of building materials and components in the agricultural environment, part III: concrete structures. J Agric Eng Res 76:3–16

De Belie N, Richardson M, Braam CR, Svennerstedt lenehan JJ, Sonck B (2000b) Durability of building materials and components in the agricultural environment: part I, the agricultural environment and timber structures. J Agric Eng Res 75:225–241

De Belie N, Sonck B, Braam CR, Svennerstedt B, Richardson M (2000c) Durability of building materials and components in the agricultural environment, part II: metal structures. J Agric Eng Res 75:333–347

Demmers TGM, Burgess LR, Short JL, Phillips VR, Clark JA, Wathes CM (1998) First experiences with methods to measure ammonia emissions from naturally ventilated cattle buildings in the UK. Atmos Environ 32:285–293

Duffie JA, Beckman WA (1974) Solar energy thermal processes. Wiley, New York, pp 10, 77-91, 849

Erisman JW, Grennfelt P, Sutton M (2003) The European perspective on nitrogen emission and deposition. Environ Int 29:311–325

Esmay ML (1978) Principles of animal environment, Textbook editionth edn. AVI Publishing Company, INC, Westport, pp 33–34, 150–157, 197–211, 250–253

Fisher AD, Roberts N, Bluett SJ, Verkerk GA, Matthews LR (2008) Effects of shade provision on the behaviour, body temperature and milk production of grazing dairy cows during a New Zealand summer. N Z J Agric Res 51:99–105

Frazzi E, Calamari L, Calegari F, Stefanini L (2000) Behavior of dairy cows in response to different barn cooling systems. Trans ASAE 43:387–394

Gralla Architects (2000–2001) Equestrian facility planning and design considerations. Stable wise, horse farm planning. http://www.stablewise.com/gralla/design.html

Green Building Encyclopedia. http://www.whygreenbuildings.com/passive_solar.php

Gustafsson G (1988) Luft och värmebalanser i djurstallar. Air and heat balances in animal houses. Dissertation, Swedish University of Agricultural Sciences, Department of Farm Buildings, report 59, Lund, p 70

Hafez ESE (1969) Adaptation of domestic animals. Lea and Febiger, Philadelphia

Haley DB, Rushen J, Passille AMD (2000) Behavioural indicators of cow comfort: activity and resting behaviour of dairy cows in two types of housing. Can J Anim Sci 80:257–263

Hamrin G (1996) Byggteknik, del B – Byggnadsfysik. AMG Hamrin, Göteborg, pp 9–38

Her E, Wolfenson D, Flamenbaum I, Folman Y, Kaim M, Berman A (1988) Thermal, productive, and reproductive responses of high yielding cows exposed to short-term cooling in summer. J Dairy Sci 71:1085–1092

Hillman PE, Lee CN, Willard ST (2005) Thermoregulatory responses associated with lying and standing in heat-stressed cows. Trans ASAE 48:795–801

Huber JT, Higginbotham G, Gomez-Alarcon RA, Taylor RB, Chen KH, Chan SC, Wu Z (1994) Heat stress interactions with protein, supplemental fat, and fungal cultures. J Dairy Sci 77:2080–2090

Igono MO, Steevens BJ, Shanklin MD, Johnson HD (1985) Spray cooling effects on milk production, milk, and rectal temperatures of cows during a moderate temperate summer season. J Dairy Sci 68:979–985

Igono MO, Johnson HD, Steevens BJ, Shanklin MD (1987) Physiological, productive, and economic benefits of shade, spray and fan systems versus shade for Holstein cows during summer heat. J Dairy Sci 70:1069–1079

Jones GM, Stallings CC (1999) Reducing heat stress for dairy cattle. Department of Dairy Science, Virginia Tech. Publication number 404-200. 1-4

Jungbluth T, Hartung E, Brose G (2001) Greenhouse gas emissions from animal houses and manure stores. Nutr Cycl Agroecosyst 60:133–145

Kammel DV, Raabe ME, Kappelman JJ (2003) Proceeding: design of high volume low speed fan supplemental cooling system in dairy free stall barns. Fifth international dairy housing conference. ASAE, St Joseph, pp 243–254

Kelly CF, Bond TE (1958) Effectiveness of artificial shade materials. Agric Eng 39(758–759):764

Kendall PE, Nielsen PP, Webster JR, Verkert GA, Littlejohn RP, Matthews LR (2006) The effects of providing shade to lactating dairy cows in a temperate climate. Livest Sci 103:148–157

Kinsman R, Sauer FD, Jackson HA, Wolynetz MS (1995) Methane and carbon dioxide emissions from dairy cows in full lactation monitored over a six-month period. J Dairy Sci 78:2760–2766

Langbein J, Nichelmann M (1993) Differences in behaviour of free-ranging cattle in the tropical climate. Appl Anim Behav Sci 37:197–209

Marcillac-Embertson NM, Robinson PH, Fadel JG, Mitlöhner FM (2008) Effects of shade and sprinklers on performance, behaviour, physiology and the environment of heifers. Am Dairy Sci Assoc 92:506–517

Moss AR, Jouany JP, Newbold J (2000) Methane production by ruminants: its contribution to global warming. Ann Zootech 49:231–254

Nevander LE, Elmarsson B (1994) Fukt handbok praktik och teori. Stockholm, pp 375–376

Nilsson L, Kangro A (1998) Field study of an underground counterflow heat exchanger for ventilation air. Swed J Agric Res 28:207–213

Overton MW, Sischo WM, Temple GD, Moore DA (2002) Using time-lapse video photography to assess dairy cattle lying behavior in a free-stall barn. J Dairy Sci 85:2407

Palmer R (2004) Selecting the right side. Dairy business communications. http://dairybusiness.com/midwest/Jan01/DMPG11.htm. 13 Aug 2004

Parker D, Sherwin J (1998) Comparative summer attic thermal performance of six roof constructions. The 1998 ASHRAE annual meeting, Toronto, 20–24 June

Phillips VR, Holden MR, Sneath RW, Short JL, White RP, Hartung J, Seedorf J, Schroder M, Linkert KH, Pedersen S, Takai H, Johnsen JO, Groot Koerkamp PWG, Uenk GH, Scholtens R, Metz JHM, Wathes CM (1998) The development of robust methods for measuring concentrations and emission rates of gaseous and particulate air pollutants in livestock buildings. J Agric Eng Res 70:11–24

Radon K, Danuser B, Iversen M, Monso E, Weber C, Hartung J (2002) Air contaminants in different European farming environments. Ann Agric Environ Med 9:41–48

Roman-Ponce H, Thatcher WW, Buffington DE, Wilcox CJ, Van Horn HH (1977) Physiological and production responses of dairy cattle to a shade structure in a subtropic environment. J Dairy Sci 60:424–430

Sakamotoa N, Tanib M, Umetsub K (2006) Effect of novel covering digested dairy slurry store on ammonia and methane emissions during subsequent storage. Int Congr Ser 1293:319–322

Schütz KE, Rogers AR, Cox NR, Tucker C (2009) Dairy cows prefer shade that offers greater protection against solar radiation in summer: shade use, behaviour, and body temperature. Appl Anim Behav Sci 116:28–34

Shearer JK, Bray DR, Bucklin RA, Beede DK (1991) Environmental modifications to reduce heat stress in dairy cattle. Agri-Pract 12:7–18

Shearer JK, Bray RA, Bucklin RA (1999) The management of heat stress in dairy cattle: what we have learned in Florida. Proceedings of the feed and nutritional management cow college, Virginia Tech, pp 1–13

Shinde S, Matsushige T, Mastumura H (1994) Kakibunnbenngyu deno syonetsusutoresu hannou kaishiyosoku to bousyotaisaku no kouka. Bull Hiroshima Livest Technol Res Cent 10:5–15

Shultz TA (1984) Weather and shade effects on corral cow activities. J Dairy Sci 67:868–873

Silanikove N (2000) Effects of heat stress on the welfare of extensively managed domestic ruminants. Livest Prod Sci 67:1–18

Smith JF, Armstrong DV, Gamroth MJ, Martin JG (1997) Symposium: dairy farms in transition. Planning the milking center in expanding dairies. J Dairy Sci 80:1866–1871

Sneath RW, Phillips VR, Demmers TGM, Burgess LR, Short JL, Welch SK (1997) Long term measurements of greenhouse gas emissions from UK livestock buildings. Livestock environment V, Proceedings of the fifth international symposium, Bloomington, pp. 146–153

Sommer SG, Zhang GQ, Bannink A, Chadwick D, Misselbrook T, Harrison R, Hutchings NJ, Menzi H, Monteny GJ, Ni JQ, Oenema O, Webb J (2006) Algorithms determining ammonia emission from buildings housing cattle and pigs and from manure stores. In: Donald LS (ed) Advances in agronomy. Academic, London, pp 261–335

Stowell RR, Gooch CA, Inglis S (2001) Performance of tunnel ventilation for freestall dairy facilities as compared to natural ventilation with supplemental cooling fans. Livestock environment VI, ASAE, Louisville, pp 29–40

Strickland JT, Bucklin RA, Nordstedt RA, Beede DK, Bray DR (1989) Sprinkler and fan cooling system for dairy cows in hot, humid climates. Appl Eng Agric 5:231–236

Tucker CB, Rogers AR, Schütz KE (2008) Effect of solar radiation on dairy cattle behaviour, use of shade and body temperature in a pasture-based system. Appl Anim Behav Sci 109:141–154

Turner LW, Chastain JP, Hemkin RW, Gates RS, Crist WL (1992) Reducing heat stress in dairy cows through sprinkler and fan cooling. Appl Eng Agric 8:251–256

Tyson JT, Graves RE, McFarland DF (1998) Tunnel ventilation for dairy tie stall barns, a companion guideline to NRAES/DPC 37, planning dairy stall barns. The dairy practices council and Northeast Regional Agricultural Engineering Service, p 2

Van Lieu P (2003) Building freestall barns and milking centers, methods and materials. Natural Resource, Agriculture, and Engineering Service (NRAES), Ithaca, pp 356–363

Verbeck R, Smith JF, Armstrong DV (1995) Heat stress in dairy cattle. New Mexico State University, Mexico

Verbeck RT, Ross T, Smith JF (1996) Effects of a spray and fan cooling system on milk yield and components, body condition, and respiration rates of early lactation cows in a hot dry climate. J Anim Sci 74(suppl 1):32

Worely JW (1999) Cooling systems for Georgia dairy cattle. The University of Georgia College of Agricultural and Environment Sciences, the U.S. Department of Agriculture Cooperating, p 3

Zähner M, Schrader L, Hauser R, Keck M, Langhans W, Wechsler B (2004) The influence of climatic conditions on physiological and behavioural parameters in dairy cows kept in open stables. Anim Sci 78:139–147

Zhang G, Strøm JS, Li B, Rom HB, Morsing S, Dahl P, Wang C (2005) Emission of ammonia and other contaminant gases from naturally ventilated dairy cattle buildings. Biosyst Eng 92:355–364

Zhu J, Jacobson L, Schmidt D, Nicolai R (2000) Daily variations in odor and gas emissions from animal facilities. Appl Eng Agric 16:153–158

About the Authors

Dr. Anjali Aggarwal is working as a Principal Scientist at National Dairy Research Institute, Karnal (India). She holds a Ph.D. degree in Animal Physiology and is involved in research and teaching at post-graduate level. Her area of research work is stress and environmental physiology. She has more than 50 publications, two technical bulletins, four manuals and many book chapters to her credit. She has successfully guided many post-graduate and Ph.D. students.

Her major research accomplishments are on microclimatic modification for alleviation of heat and cold stress, mist and fan cooling systems for cows and buffaloes, and use of wallowing tank in buffaloes. Her work involves the use of technology of supplementing micronutrients during dry period and early lactation to crossbred and indigenous cows for alleviating metabolic and oxidative stress and improved health and productivity. Studies are also done in her lab on partitioning of heat loss from skin and pulmonary system of cattle and buffaloes as a result of exercise or exposure to heat stress.

Dr. Ramesh Upadhyay is working as Head, Dairy Cattle Physiology Division at National Dairy Research Institute, Karnal (India). He graduated in Veterinary Sciences and obtained his Ph.D. degree in Animal Physiology. His area of recent research is climate change, stress, and environmental physiology. His major research accomplishment is on climate change impact assessment of milk production and growth in livestock. His work also involves studying methane conversion and emission factors for Indian livestock and use of IPCC methodology of methane inventory of Indian livestock. Heat shock protein-70 expression studies in cattle and buffaloes are also done in his lab. Draught animal power evaluation, fatigue assessment, work-rest cycle and work limiting factors form the highlights of his work. Studies on partitioning of heat loss from skin and pulmonary system of cattle and buffaloes and electrocardiographic studies in cattle, buffalo, sheep and goat are also undertaken in his lab.

He has more than 75 research papers, four books and several book chapters to his credit. Technologies developed and research done by him include methodology of methane measurement: open and closed circuit for cattle and buffaloes; inventory of methane emission from livestock using IPCC methodology; livestock stress index, thermal stress measurement based on physiological functions; and draught power evaluation system and large animal treadmill system. He received training in radio-nuclides in medicine at Australian School of Nuclear Technology, Lucas heights, NSW, Australia in 1985 and use of radioisotopes in cardiovascular investigations at CSIRO, Prospect, NSW, Australia, during 1985–1986. He has guided several post-graduate and Ph.D. students. He is recipient of Hari Om Ashram Award-1990 (ICAR) for outstanding research in animal sciences.

A. Aggarwal and R. Upadhyay, *Heat Stress and Animal Productivity*, DOI 10.1007/978-81-322-0879-2, © Springer India 2013

Index

A. Aggarwal and R. Upadhyay, *Heat Stress and Animal Productivity*,
DOI 10.1007/978-81-322-0879-2, © Springer India 2013

Lightning Source UK Ltd.
Milton Keynes UK
UKOW07n1033210116

266772UK00003B/25/P